International Association of Geodesy Symposia

Jeffrey T. Freymueller, Series Editor
Laura Sánchez, Series Assistant Editor

Series Editor
Jeffrey T. Freymueller
Endowed Chair for Geology of the Solid Earth
Department of Earth and Environmental Sciences
Michigan State University
East Lansing, MI, USA

Assistant Editor
Laura Sánchez
Deutsches Geodätisches Forschungsinstitut
Technische Universität München
Munich, Germany

International Association of Geodesy Symposia

Jeffrey T. Freymueller, Series Editor
Laura Sánchez, Series Assistant Editor

Beyond 100: The Next Century in Geodesy

Proceedings of the IAG General Assembly, Montreal, Canada, July 8–18, 2019

Edited by

Jeffrey T. Freymueller, Laura Sánchez

Volume Editors
Jeffrey T. Freymueller
Endowed Chair for Geology of the Solid Earth
Department of Earth and Environmental Sciences
Michigan State University
East Lansing, MI, USA

Laura Sánchez
Deutsches Geodätisches Forschungsinstitut
Technische Universität München
Munich, Germany

Series Editor
Jeffrey T. Freymueller
Endowed Chair for Geology of the Solid Earth
Department of Earth and Environmental Sciences
Michigan State University
East Lansing, MI, USA

Assistant Editor
Laura Sánchez
Deutsches Geodätisches Forschungsinstitut
Technische Universität München
Munich, Germany

ISSN 0939-9585
International Association of Geodesy Symposia
ISBN 978-3-031-09859-8
https://doi.org/10.1007/978-3-031-09857-4

ISSN 2197-9359   (electronic)

ISBN 978-3-031-09857-4   (eBook)

This Springer imprint is published by the registered company Springer Nature Switzerland AG
The registered company address is: Gewerbestrasse 11, 6330 Cham, Switzerland

Preface

The International Union of Geodesy and Geophysics (IUGG) held its 27th General Assembly from July 8 to 18, 2019 in Montreal, Canada. The general assembly theme was the celebration of the centennial of the establishment of the IUGG in Brussels, Belgium, in 1919. During the General Assembly, symposia were organized by all IUGG Associations, as well as joint and union symposia, offering a wide spectrum of oral and poster presentations. The participating Associations are: International Association of Cryospheric Sciences (IACS), International Association of Geodesy (IAG), International Association of Geomagnetism and Aeronomy (IAGA), International Association of Hydrological Sciences (IAHS), International Association of Meteorology and Atmospheric Sciences (IAMAS), International Association for the Physical Sciences of the Oceans (IAPSO), International Association of Seismology and Physics of the Earth's Interior (IASPEI), International Association of Volcanology and Chemistry of the Earth's Interior (IAVCEI). In total, 3952 participants registered, 437 of them with IAG priority. In total, there were 234 symposia and 18 Workshops with 4580 presentations, of which 469 were in IAG-associated symposia.

IAG organized one Union Symposium, 6 IAG Symposia, 7 Joint Symposia with other associations, and 20 business meetings. In addition, IAG co-sponsored 8 Union Symposia and 15 Joint Symposia. The IAG specific symposia were:

- G01 – Reference Systems and Frames;
- G02 – Static Gravity Field and Height Systems;
- G03 – Time-Variable Gravity Field;
- G04 – Earth Rotation and Geodynamics;
- G05 – Multi-Signal Positioning, Remote Sensing and Applications;
- G06 – Monitoring and Understanding the Dynamic Earth with Geodetic Observations.

The Union Symposium organized by IAG and IUGG jointly was dedicated to the Global Geodetic Observing System (GGOS) and was called

- U8 – Earth and Space Observations (IAG, GGOS),

The joint symposia are always led by an association with co-conveners from other associations. IAG lead the following Inter-Association Symposia:

- JG01 – Interactions of Solid Earth, Ice Sheets and Oceans;
- JG02 – Theory and Methods of Potential Fields;
- JG03 – Near-Real Time Monitoring of Regional to Global Scale Water Mass Changes;
- JG04 – Geodesy for Atmospheric and Hydrospheric Climate Research;
- JG05 – Remote Sensing and Modelling of the Atmosphere;
- JG06 – Monitoring Sea Level Changes by Satellite and In-Situ Measurements;
- JG07 – Monitoring, Imaging and Mapping of Volcanic Belts;
- JG08 – Earth Systems Literacy: Geophysics in K-16 Class Rooms, Outreach Projects, and Citizen Science Research Projects;

and IAG sponsored (with IAG co-conveners) the following Inter-Association Symposia led by other associations:

- JA01 – Geophysical Constraints on the Earth's Core and Its Relation to the Mantle;
- JA02 – Geophysical Data Assimilation;
- JA03 – Geophysical Records of Tectonic and Geodynamic Processes;
- JA06 – Space Weather Throughout the Solar System: Bringing Data and Models Together;
- JA07 – Geoscience Data Licensing, Production, Publication, and Citation (IAGA);
- JA08 – Probing the Earth's Lithosphere and Its Dynamics Using Geophysical Modeling;
- JH02 – Climate and Hydrological Services: Bridging from Science to Practice and Adaptation;
- JP01 – Tides of the Oceans, Atmosphere, Solid Earth, Lakes and Planets;
- JS01 – Cryoseismology;
- JS02 – Early Warning Systems for Geohazards;
- JS03 – Subduction Zone Deformation and Structure: Tracking the Sea Floor in Motion;
- JS04 – Seismo – Geodesy;
- JS05 – Probabilistic & Statistical Approaches in Geosciences;
- JS06 – Old Data for New Knowledge: Preservation and Utilization of Historical Data in the Geosciences;
- JS07 – Integrated Geophysical Programs for Earth Systems Monitoring;
- JV03 – Strain Localization and Seismic Hazards.

This volume contains 30 selected papers from the all of the symposia related to IAG. All published papers were peer-reviewed, and we warmly recognize the contributions and support of the Associated Editors and reviewers (see the list in later pages).

With this volume, the IAG Symposia Series begins a new phase with an open access publication policy. Now and into the future, all papers accepted to the series will be fully open access from the time that they are accepted. While the complete electronic book will still be available as a distinct volume, every paper is also available individually, free and open for all to read.

East Lansing, MI, USA   Jeffrey T. Freymueller
Munich, Germany   Laura Sanchez

Contents

Part I

Reference Systems and Frames

Towards an International Height Reference Frame Using Clock Networks

Hu Wu and Jürgen Müller

Abstract

Establishing an International Height Reference Frame (IHRF) has been a major goal of the International Association of Geodesy (IAG) for a long time. One challenge is to obtain the vertical coordinates, i.e., geopotential numbers, of the reference stations with high precision and global consistency. A promising approach is using clock networks, which are powerful in precisely obtaining geopotential or height differences between distant sites through measuring the gravitational redshift effect by comparing clocks' frequencies. We propose a hybrid clock network following a specific hierarchy. It includes stationary clocks as the backbone of the frame and transportable clocks for regional densifications. The vertical coordinates of the clock stations can be straightforwardly referenced to the unique benchmark by various long-distance frequency transfer techniques, like using optical fibers or free-space microwave and laser links via relay satellites. Another practical way towards an IHRF is to unify all local height systems around the world. Clock networks are considered as an alternative to classical geodetic methods. The idea was verified through closed-loop simulations. We found that the measurements acquired by a few 10^{-18} clocks, three or four in triangular or quadrangular distributions for each local system, are sufficient to adjust the discrepancies between local datums and the systematic slopes within local height networks.

Keywords

Chronometric levelling · Clock networks · Height system unification · IHRS/IHRF · Relativistic geodesy

1 Introduction

A consistent global height reference system is essential for monitoring change processes on Earth in the vertical direction, such as sea level rise, crustal motion, continental water storage variation and so on. Today, over 100 regional/national height systems are practically used. These height systems have their own datums relative to various local mean sea levels which exhibit discrepancies from dm

H. Wu (✉) · J. Müller
Institut für Erdmessung (IfE), Leibniz Universität Hannover, Hannover, Germany
e-mail: wuhu@ife.uni-hannover.de

to m (Sánchez and Sideris 2017). Other systematic errors like slopes appear along the levelling lines of a local height system (Gruber et al. 2014). They are inevitably introduced by spirit leveling where the error accumulates over long distances. In addition, different conventions lead to different types of local height systems, like the normal, orthometric, or dynamic height system (Jekeli 2000). It is therefore a challenging task to realize a height system with high precision and homogeneous consistency worldwide.

During the IUGG General Assembly in 2015, IAG released a resolution on the definition of an International Height Reference System (IHRS). The IHRS is defined with conventions on W_0 (the reference gravity potential value of an equipotential surface of the Earth's gravity field), the tide system for data reduction, the terrestrial coordinates of the

© The Author(s) 2020
J. T. Freymueller, L. Sanchez (eds.), *Beyond 100: The Next Century in Geodesy*,
International Association of Geodesy Symposia 152, https://doi.org/10.1007/1345_2020_97

reference stations and so on. More details are given in Ihde et al. (2017). We would like to underline the convention that the geopotential number (gravity potential difference w.r.t. to W_0) rather than the height has been adopted as the vertical coordinate. This will certainly avoid some inconsistencies between different types of height systems. The resolution also includes conventions on the realization of IHRS or the establishment of IHRF. The reference network of the IHRF shall follow the same hierarchy as the ITRF (International Terrestrial Reference Frame), i.e., a global network with regional/national densifications. The challenge is the precise determination of the geopotential numbers and their changes over time for all reference stations.

Classical geodetic methods have been widely used in past decades for the determination of geopotential numbers or gravity potential differences between distant points. One combines spirit levelling and terrestrial gravimetry, which is however time consuming and labour demanding and poses challenges in areas with a complex terrain, like mountain areas (Rummel and Teunissen 1988). Another method is geoid modelling. It estimates the potential value by solving a geodetic boundary value problem (Pavlis 1991). This method can achieve a precise and high-spatial-resolution geoid model in regions with high-quality data available. But the performance of the geoid model might be severely degraded in areas where no or sparse observations are available or data is in a poor quality. One can also use a global gravity field model to calculate the potential value through the procedure of spherical harmonic synthesis. Although there might be some smoothing effects due to a limited spatial resolution or omission errors due to a truncated degree (Gruber et al. 2014). Moreover, oceanographic modelling can potentially be used to estimate the potential differences between points separated by the sea (Woodworth et al. 2012).

Meanwhile, new measurement tools like clock networks have emerged and rapidly developed. The latest generation of optical clocks reaches the level of 10^{-18} (fractional frequency) and even beyond, which is expected to enable relativistic geodesy with cm-level accuracy. Under the theoretical framework of relativistic geodesy, gravity potential differences (resp. geopotential numbers or height differences) between distant sites can be obtained through measuring the gravitational redshift effect by comparing the clocks' frequencies (Bjerhammar 1985). The clock-based method shows advantages in several aspects: (1) delivering the potential values directly, well-fitting the definition of the IHRS; (2) precise measurements over long distances without an accumulated error as in spirit levelling; (3) point-wise measurements, without a smoothing effect due to the spatial resolution of gravity field models; (4) continuous observation, while a gravity field model represents an average signal within a time period. These points make clock networks a perfect candidate to realize a consistent and accurate global height system.

In this context, we propose to use a hybrid clock network for the establishment of the IHRF following some specified hierarchy. This clock network is composed of different types of clocks, where stationary clocks are used to build the global reference network of IHRF and transportable clocks are responsible for the regional densification. The vertical coordinates of reference clock stations can be obtained with respect to the unique datum of IHRF through the precise comparison of distant clocks by various frequency transfer techniques. Another way to achieve an IHRF is to unify all existing regional/local height systems. The challenges are the estimation of discrepancies between different height datums and the adjustment of systematic errors along levelling lines in each local height system. We propose to use an inter-connected clock network where a few clocks were assumed in each local region for the unification. The idea is verified through closed-loop simulations.

The paper is structured as follows. In Sect. 2, we review the theory of relativistic geodesy with clocks. Next, we focus on the realization of the IHRS using a hybrid clock network. Then, we run dedicated simulations to verify the idea by using clock networks for height system unification. In the last section, we address our conclusions and some future perspectives.

2 Relativistic Geodesy with Clocks

2.1 Relativistic Geodesy

Einstein's theory of general relativity predicts that clocks tick at different rates if they are transported with different speed or they are under the influence of different gravitational potential. Considering the case that two motionless clocks are operated at different points on the Earth's surface, the change of the clocks' frequencies is proportional to the difference of the gravity potential (sum of the gravitational potential and the centrifugal potential) at both points (Bjerhammar 1985). It reads

$$\frac{\Delta f_{21}}{f_1} = \frac{f_2 - f_1}{f_1} = \frac{W_2 - W_1}{c^2} + O(c^{-4}), \qquad (1)$$

where f_1 and f_2 are the proper frequencies at points 1 and 2 on the Earth's surface, while W_1 and W_2 are the corresponding gravity potentials, c is the speed of light, and higher order terms are neglected. Supposing that point 1 is on the equipotential surface of Earth's gravity field with W_0, i.e., the geoid, we obtain

$$\frac{\Delta f_{21}}{f_1} = \frac{W_2 - W_0}{c^2} = -\frac{C_2}{c^2}, \qquad (2)$$

where C_2 is the geopotential number (the vertical coordinate of IHRS) of point 2. For any two arbitrary points, we can further write

$$\frac{\Delta f_{21}}{f_1} = \frac{(W_2 - W_0) - (W_1 - W_0)}{c^2} = -\frac{C_2 - C_1}{c^2}. \quad (3)$$

The difference of the clocks' frequencies is proportional to the difference of the geopotential numbers for two different points. When the geopotential number (difference) is obtained, it can be converted to the physical height (difference). For example, the difference of the orthometric height ΔH_{21} can be written as (Müller et al. 2018)

$$\Delta H_{21} = \frac{C_2 - C_1}{\overline{g}} = -\frac{\Delta f_{21}}{f_1} \times \frac{c^2}{\overline{g}}. \quad (4)$$

Or the normal height difference ΔH_{21}^N is written as

$$\Delta H_{21}^N = \frac{C_2 - C_1}{\overline{\gamma}} = -\frac{\Delta f_{21}}{f_1} \times \frac{c^2}{\overline{\gamma}}, \quad (5)$$

where $\overline{g}$ is the mean gravity value along the plumb line between the Earth's surface and the equipotential surface of W_0, while $\overline{\gamma}$ is the mean normal gravity value.

The method to observe height or geopotential differences between two points through the comparison of proper frequencies is called relativistic geodesy or chronometric levelling (Vermeer 1983). The measurement scheme is illustrated in Fig. 1. For an approximate estimation, a fractional frequency inaccuracy of 1.0×10^{-18} corresponds to an uncertainty of about $1.0\,\mathrm{cm}$ in height or $0.1\,\mathrm{m^2/s^2}$ in geopotential. This method is ideal for the establishment of the IHRF, since it can directly obtain the vertical coordinates as geopotential numbers and shows advantages in connecting distant points without being affected by accumulated errors as in spirit levelling. It delivers measurements pointwise so that it can get rid of the smoothing effect of a global gravity field model with limited spatial resolution. Besides, this method is powerful to determine height differences between points that are separated by the sea or a big mountain. Therefore, high-performance clocks that can provide the proper frequency at a high accuracy level,

e.g., 1.0×10^{-18}, will be a novel tool for height measurements.

2.2 Clock Networks

Clock networks are expected to realize relativistic geodesy in practice. When the theory of relativistic geodesy was presented by Bjerhammar in the 1980s, the most accurate clock was probably the atomic hydrogen maser, with a 10^{-13} level of frequency uncertainty. Atomic clocks were rapidly developed in the past decades and their performance has significantly improved. An atomic clock that uses electron transition in the microwave range as the frequency standard reaches the level of 10^{-16} (Heavner et al. 2014). One best representative for this kind of clock is the cesium clock, which is used for the definition of the second. Another type of atomic clocks that received much attention in recent years is the optical clock, using electron transition in the optical range. The latest generation of optical clocks that were built in laboratories achieved the level of 10^{-18} and even beyond (Brewer et al. 2019; McGrew et al. 2018; Huntemann et al. 2016). The evolvement of the frequency uncertainty for both types of atomic clocks is displayed in Fig. 2. Moreover, efforts were directed to make clocks more compact so that they can be operated in transportable vehicles like a car trailer and used for field measurement campaigns. Such a transportable optical clock was reported with an uncertainty of 7.0×10^{-17} (Koller et al. 2017). Optical clocks were also developed for the application in space (Origlia et al. 2018).

For the comparison of distant clocks, frequency transfer techniques that use ultra-high frequency (UHF) microwaves, free-space laser links and optical fibers have been greatly developed. The TWSTFT (Two-Way Satellite Time and Frequency Transfer) method exchanges the signals of two distant clocks through a UHF microwave via a relay satellite like a navigation satellite or a geostationary satellite. This method can cancel or reduce some common error sources since the signals are transmitted along the same path. The frequency transfer instability can reach the level of 10^{-17} at several days averaging (Petit et al. 2014). By using free-space laser beams to replace the microwaves, the two-way link can

Fig. 1 The measurement scheme of chronometric levelling with connected clocks

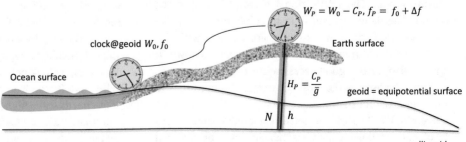

Fig. 2 The evolvement of the frequency uncertainty for optical clocks and Cs clocks

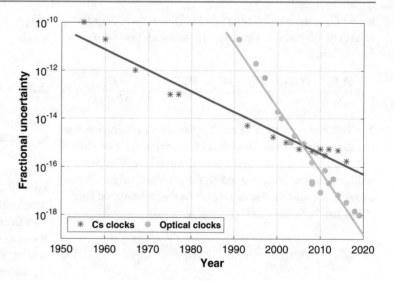

principally reduce the instability of frequency transfer much faster. An experiment with two ground clocks linked by a 4-km laser beam shows that the instability for frequency transfer reaches 1.2×10^{-17} at 1 s and drops to 6×10^{-20} at 850 s (Sinclair et al. 2018). The frequency transfer based on optical fibers is also extensively experimented. The link between clocks in Paris (France) and in Braunschweig (Germany) by 1,415 km of optical fibers reached the level of 10^{-19} at 1 day (Lisdat et al. 2016). With these various frequency transfer techniques, high-performance clocks can be combined in networks and applied for geodetic applications (Müller et al. 2018).

3 A Hybrid Clock Network for the Establishment of IHRF

According to the IAG convention, the reference network of IHRF should follow the same hierarchy as ITRF, i.e., a global reference network with regional/national densifications. Here, we propose to apply a hybrid clock network for the realization of the global network of IHRF. Different types of clocks are included in this network, and they are linked via different means of frequency transfer. The assumed clock network is illustrated in Fig. 3. We classify the used clocks into different groups based on their functionality, which are

– **Datum clock**: The IHRF global network has a unique datum. The datum station is in principle located on the equipotential surface of the global gravity field with W_0, or its potential difference w.r.t. W_0 is precisely known. It should also have a good accessibility to other reference stations. We assume the most accurate and stable clock at the unique datum station.
– **Core clocks**: The global network includes some high-level core stations which are benchmarks for regional

and national networks. We assume ultra-high performance clocks at these core stations. These clocks should be referenced to the datum clock so that their vertical coordinates, i.e., geopotential numbers, can be determined through the precise comparison of frequencies. The link between core clocks might be across continents, where the free-space frequency transfer via relay satellites is necessary.
– **Regional/national clocks**: For a local area, high-performance clocks are assumed on some permanent, continuous observatories to realize a regional or national network. These regional and national clocks are linked to the core clocks or directly referenced to the datum clock. The free-space frequency transfer and the optical fibers can be used for their link. Since clock-based levelling shows advantages in long-distance height measuring, a few permanent clocks might be sufficient for a regional network. This is not like the classic geodetic levelling network, where more nodal points are required. Also, considering the size, cost and operationality, clock-based levelling might not be superior to the classic geodetic levelling in a relatively small area or in some practical engineer construction. Nevertheless, the existing local levelling networks can be connected to those regional and national clock stations.
– **Transportable clocks**: The above mentioned clocks are more likely stationary and permanent. We still need transportable clocks for field measurement campaigns. Transportable clocks might also be useful for the connection of existing local levelling networks to regional and national clock networks.
– **Relay satellites**: The free-space link between distant clocks needs some high-orbit relay satellites, like the navigation satellites or geostationary satellites. These satellites can enable the comparison of distant clocks at any time.

Fig. 3 Different types of clocks that linked with various frequency transfer techniques are proposed to realize the global reference network of IHRF

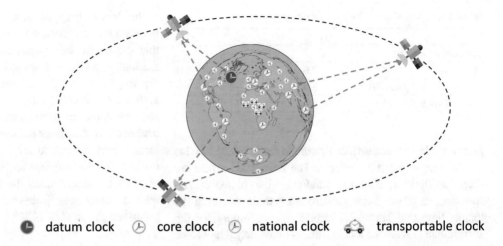

🕐 datum clock   ⏁ core clock   ⏁ national clock   🚗 transportable clock

4 Clock Networks for Height System Unification

Another way to achieve a global height system is to unify all practically-used local height systems around the world. Generally, a local height system has its own datum related to a local mean sea level which might however be biased to the global reference equipotential surface, i.e., the geoid. The discrepancies can result in offsets from dm to m between different local height datums (Sánchez and Sideris 2017). Within a local height system, reference networks are also suspectable to systematic errors that are introduced by spirit levelling. Levelling errors accumulate over long distances and can reach 1.0–3.0 cm every 100 km. These system-

atic errors might cause slopes in a local height system, see examples in Spain, France and Germany where linear slopes in the latitudinal and longitudinal directions were found (Gruber et al. 2014). More complex systematic errors, like distortions within a local height system, might also exist.

We propose to use clock networks for local height system unification and verify the idea through closed-loop simulations (Wu et al. 2019). We take the a-priori unified height system EUVN/2000 (European United Vertical Network) (Ihde et al. 2000), which contains 202 reference stations covering most countries in Europe. These stations are manually divided into four groups, with each own datum related to a historic tide gauge station, see Fig. 4. For each local

Fig. 4 The reference stations of EUVN/2000 are divided into four groups G1, G2, G3 and G4, which are marked in different colors

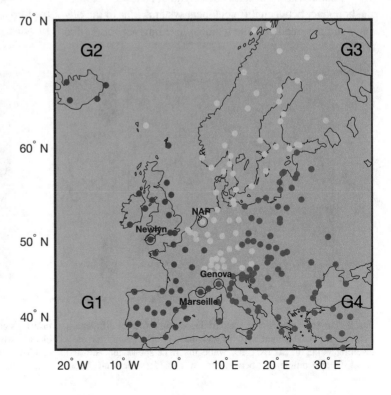

Table 1 Introduced errors for each local height system

	G1	G2	G3	G4
Random error [cm]		1.0		
Bias [cm]	−18.0	25.0	0.0	8.0
Slope along lat. [cm/100 km]	3.0	−2.0	1.5	−3.0
Slope along lon. [cm/100 km]	2.0	3.0	−1.5	−2.0

group, we introduced artificial errors to the levelling heights to generate a realistic scenario. The introduced errors are listed in Table 1, including random noise, biases (corresponding to offsets between datums) and slopes along the longitudinal and latitudinal directions (accounting for the accumulated errors for local levelling networks). The values of these errors are assumed based on the a-priori knowledge of realistic cases. Afterwards, a few clocks are assumed for every local system and all clocks are inter-connected to obtain the height differences between the corresponding levelling stations. The clock measurements and the levelling heights in the local systems are then jointly processed by a least-squares adjustment to re-unify all levelling heights to a unique datum. Here, the datum of the local height system G3 was chosen as the unique one for the re-unified system. Finally, the differences between the levelling heights in the re-unified system and the a-priori system are taken for the result evaluation.

Four clocks in quadrangular distributions are employed in each local system. They are assimilated to levelling stations and marked with different types of symbols, cf. Fig. 5b. The clock measurement accuracy is assumed at the level of 1.0×10^{-18}, about 1.0 cm in terms of height. As shown in Fig. 5, the boundary between different local systems (due to the offsets) and the trends for the regional levelling networks (due

to the slopes) disappeared after unification. The comparison between the re-unified system and the a-priori system shows that the height differences for all levelling points behave randomly and deviate around zero. The RMSs (Root Mean Square) of the height differences for each group are 0.84, 1.19, 1.29, 1.58 cm. That manifests that the systematic offsets and slopes are precisely estimated and adjusted. The result demonstrates that clock networks are a powerful tool to unify local height systems to achieve the IHRF. Furthermore, we ran simulations to optimize the number and the distribution of clocks, and evaluated the performance by using clocks with different magnitudes of accuracy. These results are published in Wu et al. (2019).

5 Conclusions and Future Perspectives

We investigated the potential of using clock networks for the realization of the International Height Reference System under its latest definition. Clock networks, with high-performance clocks (at the level of 1.0×10^{-18} or even beyond) linked by relevant frequency transfer techniques, are a powerful tool to directly observe the gravity potential differences or height differences between distant points. This fact makes clock networks a perfect candidate for the establishment of the IHRF. We proposed a hybrid clock reference network following the hierarchy of the ITRF, which contains high-performance stationary clocks as the backbone of the frame and transportable clocks for regional densifications. The vertical coordinates of the clock stations can be precisely referenced to a unique datum through frequency comparison by free-space links via relay satellites or by optical fibers. Because of the high performance in transmitting heights

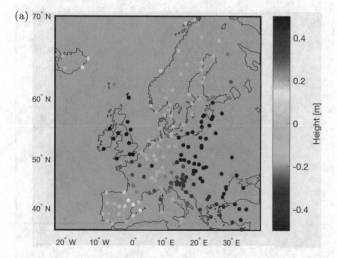

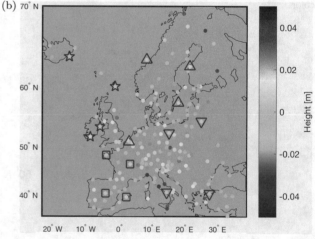

Fig. 5 The left figure shows the introduced errors (height differences between the local height systems and the a-priori system) for each levelling point of EUVN/2000, while the right shows the adjusted errors (height differences between the re-unified system and the a-priori system). The magenta symbols shown in the right figure represent clocks, where each of the four types of symbols indicates clocks employed for one local height system. (**a**) Before unification. (**b**) After unification

over long distances, clock networks are also suitable for the unification of local height systems. They can be used to estimate the offsets between different height datums and the systematic slopes within a local height system. We verified this idea through closed-loop simulations and found that measurements acquired by a few clocks, e.g., three or four in triangular or quadrangular distributions for each local system, can precisely remove those systematic discrepancies to achieve a unified system.

Today, extensive efforts are still ongoing to improve the performance of optical clocks and the frequency transfer techniques. Clocks developed in various laboratories worldwide will achieve the level of 10^{-19} in the next decade, and transportable clocks are promising to achieve the 10^{-18} level. More fibre links will facilitate the realization of an operational clock network in Europe. In addition, free-space links via satellites will be demonstrated soon by the upcoming ACES (Atomic Clock Ensemble in Space) mission. This makes us believe that clock networks will become a common geodetic measurement tool in the near future.

Acknowledgements This work is funded by the Deutsche Forschungsgemeinschaft (DFG, German Research Foundation) under Germany's Excellence Strategy – EXC-2123 QuantumFrontiers – 390837967, and the Sonderforschungsbereich (SFB) 1128 "geo-Q". We thank the two reviewers for their useful comments.

References

Bjerhammar A (1985) On a relativistic geodesy. Bull Géodés 59(3):207–220. https://doi.org/10.1007/BF02520327

Brewer SM, Chen JS, Hankin AM, Clements ER, Chou CW, Wineland DJ, Hume DB, Leibrandt DR (2019) $^{27}Al^+$ quantum-logic clock with a systematic uncertainty below 10^{-18}. Phys Rev Lett 123:033201. https://doi.org/10.1103/PhysRevLett.123.033201

Gruber T, Rummel R, Ihde J, Liebsch G, Rülke A, Schäfer U, Sideris M, Rangelova L, Woodworth P (2014) Height system unification with GOCE summary and final report. Technical Report GO-HSU-PL-0021

Heavner TP, Donley EA, Levi F, Costanzo G, Parker TE, Shirley JH, Ashby N, Barlow S, Jefferts S (2014) First accuracy evaluation of NIST-F2. Metrologia 51(3):174

Huntemann N, Sanner C, Lipphardt B, Tamm C, Peik E (2016) Single-ion atomic clock with 3×10^{-18} systematic uncertainty. Phys Rev Lett 116:063001. https://doi.org/10.1103/PhysRevLett.116.063001

Ihde J, Adam J, Gurtner W, Harsson B, Sacher M, Schlüter W, Wöppelmann G (2000) The height solution of the European Vertical Reference Network (EUVN). Veröff Bayer Komm Int Erdmessung Astronom Geod Arb 61:132–145

Ihde J, Sánchez L, Barzaghi R, Drewes H, Foerste C, Gruber T, Liebsch G, Marti U, Pail R, Sideris M (2017) Definition and proposed realization of the International Height Reference System (IHRS). Surv Geophys 38(3):549–570. https://doi.org/10.1007/s10712-017-9409-3

Jekeli C (2000) Heights, the geopotential, and vertical datums. Technical Report, Department of Civil and Environmental Engineering and Geodetic Science, Ohio State University

Koller SB, Grotti J, Vogt S, Al-Masoudi A, Dörscher S, Häfner S, Sterr U, Lisdat C (2017) Transportable optical lattice clock with 7×10^{-17} uncertainty. Phys Rev Lett 118:073601. https://doi.org/10.1103/PhysRevLett.118.073601

Lisdat C, Grosche G, Quintin N, Shi C, Raupach S, Grebing C, Nicolodi D, Stefani F, Al-Masoudi A, Dörscher S, et al (2016) A clock network for geodesy and fundamental science. Nat Commun 7:1–7. https://doi.org/10.1038/ncomms12443

McGrew WF, Zhang X, Fasano RJ, Schäffer SA, Beloy K, Nicolodi D, Brown RC, Hinkley N, Milani G, Schioppo M, Yoon TH, Ludlow AD (2018) Atomic clock performance enabling geodesy below the centimetre level. Nature 564(7734):87–90. https://doi.org/10.1038/s41586-018-0738-2

Müller J, Dirkx D, Kopeikin S, Lion G, Panet I, Petit G, Visser P (2018) High performance clocks and gravity field determination. Space Sci Rev 214(1):5. https://doi.org/10.1007/s11214-017-0431-z

Origlia S, Pramod MS, Schiller S, Singh Y, Bongs K, Schwarz R, Al-Masoudi A, Dörscher S, Herbers S, Häfner S, Sterr U, Lisdat C (2018) Towards an optical clock for space: compact, high-performance optical lattice clock based on bosonic atoms. Phys Rev A 98:053443. https://doi.org/10.1103/PhysRevA.98.053443

Pavlis NK (1991) Estimation of geopotential differences over intercontinental locations using satellite and terrestrial measurements. Technical Report 409, Department of Geodetic Science and Surveying, The Ohio State University, Colombus, Ohio

Petit G, Wolf P, Delva P (2014) Atomic time, clocks, and clock comparisons in relativistic spacetime: a review. In: Frontiers in relativistic celestial mechanics, vol 2, chap 7. De Gruyter, Berlin, pp 249–279

Rummel R, Teunissen P (1988) Height datum definition, height datum connection and the role of the geodetic boundary value problem. Bull Géodés 62(4):477–498. https://doi.org/10.1007/BF02520239

Sánchez L, Sideris MG (2017) Vertical datum unification for the International Height Reference System (IHRS). Geophys J Int 209(2):570. https://doi.org/10.1093/gji/ggx025

Sinclair LC, Bergeron H, Swann WC, Baumann E, Deschênes JD, Newbury NR (2018) Comparing optical oscillators across the air to milliradians in phase and 10^{-17} in frequency. Phys Rev Lett 120:050801. https://doi.org/10.1103/PhysRevLett.120.050801

Vermeer M (1983) Chronometric levelling. Reports of the Finnish Geodetic Institute, Geodeettinen Laitos, Geodetiska Institutet

Woodworth P, Hughes C, Bingham R, Gruber T (2012) Towards worldwide height system unification using ocean information. J Geodetic Sci 2(4):302–318. https://doi.org/10.2478/v10156-012-0004-8

Wu H, Müller J, Lämmerzahl C (2019) Clock networks for height system unification: a simulation study. Geophys J Int 216(3):1594–1607. https://doi.org/10.1093/gji/ggy508

Towards the Realization of the International Height Reference Frame (IHRF) in Argentina

Claudia Noemi Tocho, Ezequiel Dario Antokoletz, and Diego Alejandro Piñón

Abstract

This paper describes a practical implementation of the International Height Reference System (IHRS) in Argentina. The contribution deals with the determination of potential values $W(P)$ at five Argentinean stations proposed to be included in the reference network of the International Height Reference Frame (IHRF). All sites are materialized with GNSS stations of the Argentine continuous satellite monitoring network and most of them are included in the SIRGAS Continuously Operating Network. Not all the stations are connected to the National Vertical Reference System 2016 and most of them are near to an absolute gravity station measured with an A10 gravimeter.

This paper also discusses the approach for the computation of $W(P)$ at the IHRF stations using the Argentinean geoid model GEOIDE-Ar 16 developed by the Instituto Geográfico Nacional, Argentina together with the Royal Melbourne Institute of Technology (RMIT) University, Australia using the remove-compute-restore technique and the GOCO05s satellite-only Global Gravity Model. Then, geoid undulations (N) were transformed to height anomalies (ζ) in order to infer $W(P)$ at the stations located on the Earth's surface. The transformation from N to ζ must be consistent with the hypothesis used for the geoid determination. Special emphasis is made on the standards, conventions and constants applied.

Keywords

Argentinean stations to the IHRF reference network · GEOIDE-Ar 16 · International Height Reference System and Frame

1 Introduction

During the 26th General Assembly of the International Union of Geodesy and Geophysics (IUGG) held from June 22 to July 2, 2015 in Prague, Czech Republic, the International Association of Geodesy (IAG) released the Resolution No. 1 that outlines five conventions for the definition and real-ization of an International Height Reference System (IHRS; Drewes et al. 2016). The definition is given, cf. Ihde et al. (2017, Section 5) and cf. Sánchez and Sideris (2017, Section 1), by:

1. The vertical coordinate is given in terms of geopotential quantities: potential values $W(P)$ or geopotential numbers $C(P)$ referred to an equipotential surface of the Earth's gravity field realized by the conventional value $W_0 = 62636853.40 \, \mathrm{m^2 \, s^{-2}}$ (Sánchez et al. 2016):

$$C(P) = W_0 - W(P) = -\Delta W(P). \quad (1)$$

2. The spatial reference of the position P for the gravity potential $W(P) = W(X)$ is given by the coordinate

C. N. Tocho (✉) · E. D. Antokoletz
Facultad de Ciencias Astronómicas y Geofísicas, Universidad Nacional de La Plata, La Plata, Argentina
e-mail: ctocho@fcaglp.unlp.edu.ar

D. A. Piñón
Instituto Geográfico Nacional (IGN), Buenos Aires, Argentina

© The Author(s) 2020
J. T. Freymueller, L. Sanchez (eds.), *Beyond 100: The Next Century in Geodesy*,
International Association of Geodesy Symposia 152, https://doi.org/10.1007/1345_2020_93

vector X of the International Terrestrial Reference Frame (ITRF; Altamimi et al. 2016).

3. The estimation of $X(P)$, $W(P)$ or $C(P)$ includes their variation with time; i.e., $\dot{X}(P)$, $\dot{W}(P)$ or $\dot{C}(P)$.
4. This resolution also states that parameters, observations and data shall be related to the mean tidal system/mean crust.
5. The unit of length is the meter and the unit of time is the second (SI).

The realization of the IHRS is called International Height Reference Frame (IHRF) and it corresponds to a set of physical points (continuously operated stations) with precise potential values $W(P)$, or geopotential numbers $C(P)$ and geometrical coordinates $X(P)$, see Ihde et al. (2017).

Five sites have been selected along Argentina to be included in the IHRF, from north to south: Salta (UNSA), San Juan (OAFA), La Plata (AGGO), Rio Gallegos (UNPA)

and Rio Grande (RIO2). Figure 1 illustrates the location of each station and the space geodetic and gravimetric techniques that are operated in each station. Figure 1 also shows the topography.

The Argentinean-German Geodetic Observatory (AGGO) is a fundamental geodetic observatory located in the east-central part of Argentina, close to the city of La Plata. The observatory was moved in 2015 from Concepcion, Chile, to La Plata and is currently operated jointly by the German Federal Agency for Cartography and Geodesy (BKG) and the National Scientific and Technical Research Council of Argentina (CONICET). Very Long Baseline Interferometry (VLBI) and Satellite Laser Ranging (SLR) techniques are co-located with Global Navigation Satellite System (GNSS). A gravity laboratory is established at AGGO where the superconducting gravimeter (SG) SG038 has been continuously recording gravity changes since December 16th, 2015

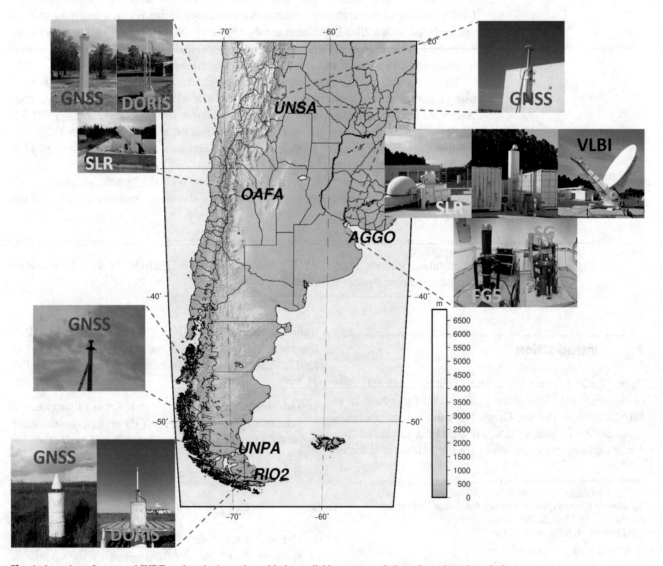

Fig. 1 Location of proposed IHRF stations in Argentina with the available space geodetic and gravimetric techniques

(Antokoletz et al. 2017). Moreover, on January 2018, an absolute gravimeter FG5 was installed to set an absolute gravity reference for the station. AGGO is a reference station of the new International Gravity Reference Frame (IGRF). As precise time keeping is essential, different atomic clocks are also installed at AGGO.

All GNSS stations with the exception of OAFA contribute to the continuously operating reference network of the Geocentric Reference System for the Americas (SIRGAS-CON; Sánchez and Brunini 2009) but all the stations are included in the Argentine Continuous Satellite Monitoring Network (RAMSAC; Piñón et al. 2018). UNSA, AGGO and RIO2 belong to the global network of the International GNSS Service (IGS; Johnston et al. 2017).

The current terrestrial reference frame of Argentina is Posiciones Geodesicas Argentina 2007 (POSGAR07), based on ITRF2005 with epoch 2006.632 (Cimbaro et al. 2009).

At present, only AGGO is connected to the National Vertical Reference System 2016 (SRVN16) (Piñón et al. 2016) for vertical datum unification.

2 Computation of Potential Values W(P)

This contribution presents the calculation of potential values recovered from the existing Argentinean geoid model GEOIDE-Ar 16 (Piñón 2016), described in the next section and also shows the computation of potential values based on Global Gravity Models (GGMs) of high-degree.

2.1 Potential Values W(P) Recovered from an Existing Geoid Model

The potential value $W(P)$ can be understood as the sum of the disturbing potential $T(P)$ determined as a solution of a geodetic boundary value problem (GBVP) at the known position $P(X)$ on the Earth's surface plus the normal gravity potential $U(P)$ at the same point using the formula (2-224) from Hofmann-Wellenhof and Moritz (2005):

$$W(P) = U(P) + T(P), \tag{2}$$

with:

$$U(P) = U_0 + \frac{\partial U_0}{\partial h} h_P = U_0 - \gamma_P h_P, \tag{3}$$

where U_0 is the normal potential at the reference ellipsoid, h_P is the ellipsoidal height and the gradient of the normal potential $\frac{\partial U_0}{\partial h}$ is the normal gravity value (γ_P) at P.

Equation (2) can be written as:

$$W(P) = T(P) + U_0 - \gamma_P h_P. \tag{4}$$

In Argentina, the disturbing potential and geoid were solved applying the classical Stokes approach (Piñón 2016). The disturbing potential was determined at a point P_0 on the geoid. Spherical harmonics of degrees zero and one were not considered in the geoid heights derived from the GBVP solution (N_{GBVP}). The zero-degree term (Eq. (5)) was added to N_{GBVP}. Then, geoid heights ($N = N_{GBVP} + N_0$) were converted to height anomalies (ζ). The zero-degree term takes into account the difference between the Earth's and reference ellipsoid's geocentric gravitational constant (GM) and also the difference between the reference potential W_0 value adopted by the IHRS and the normal potential U_0 on the reference ellipsoid.

The zero-degree term can be derived with:

$$N_0 = \frac{(GM_{GGM} - GM_{GRS80})}{\gamma_{Q_0} r_{P_0}} - \frac{W_0 - U_0}{\gamma_{Q_0}}, \tag{5}$$

where the GM_{GGM} is the geocentric gravitational constant of the GGM, GM_{GRS80} is the geocentric gravitational constant of the Geodetic Reference System 1980 (GRS80; Moritz 2000), γ_{Q_0} is the normal gravity on the reference ellipsoid and r_{P_0} is the geocentric radial distance of the point P_0. See Fig. 2 for the position of P_0, Q_0, P and Q.

The basic relation for the geoid–quasigeoid separation is obtained using the formula (8-113) of Hofmann-Wellenhof and Moritz (2005):

$$\zeta = N - \frac{\overline{g} - \overline{\gamma}}{\overline{\gamma}} H \approx N - \frac{\Delta g_B}{\overline{\gamma}} H, \tag{6}$$

where $\overline{\gamma}$ is the mean normal gravity between a point Q_0 on the ellipsoid and the corresponding point Q on the telluroid; $\overline{g}$ is the mean gravity along the real plumbline between P_0 on the geoid and P on the Earth's surface, H the orthometric height and Δg_B is the Bouguer gravity anomaly.

The transformation from N to ζ must be consistent with the hypothesis of masses applied for the geoid computation (Sánchez et al. 2018), that in the case of Argentina, it was the Helmert's second method of condensation: the topographic masses are shifted and condensed to a surface layer on the geoid (Hofmann-Wellenhof and Moritz 2005).

Once the geoid is transformed to quasigeoid, the potential values $W(P)$ can be inferred using Eq. (4) as:

$$W(P) = \gamma_Q \zeta + U_0 - \gamma_P h_P. \tag{7}$$

Since the IAG resolution No. 1 has also stated that parameters, observations, and data should be related to the mean tidal system/mean crust, the ITRF ellipsoidal heights

Fig. 2 Heights and reference surfaces (modified from Sánchez et al. 2018)

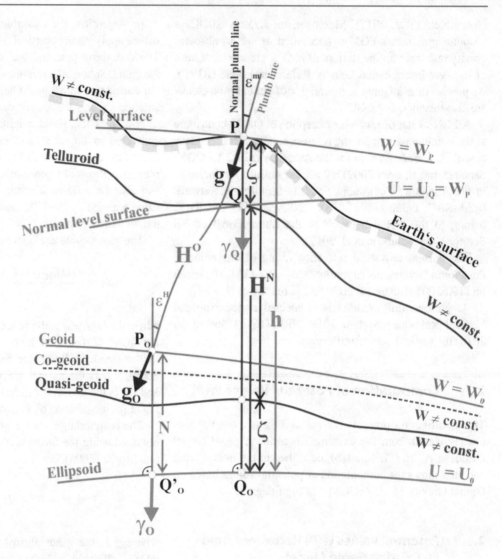

were transformed from tide-free system (TF) to mean-tide system (MT) following Petit and Luzum (2010) and Sánchez and Sideris (2017):

$$h_{MT} = h_{TF} + 0.0602 - 0.1790 \sin^2 \varphi - 0.0018 \sin^4 \varphi, \quad (8)$$

where φ is the ellipsoidal latitude.

Regarding geoid undulations, they were transformed from the tide-free system to the mean-tide system following (Ekman 1989):

$$N_{MT} = N_{TF} + (1 + k)(9.9 - 29.6 \sin^2 \varphi) \times 10^{-2}, \quad (9)$$

where $k = 0.30190$, being consistent with the Love numbers proposed in Petit and Luzum (2010).

2.2 Estimation of Potential Values W(P) with Global Gravity Models

Potential values $W(P)$ can be estimated by the combination of ITRF positions with global gravity models in terms of Stokes spherical harmonics coefficients (C_{nm}, S_{nm}) of high degree, where n is the degree, n_{max} the maximum degree and m the order (Eq. (10); Hofmann-Wellenhof and Moritz

2005):

$$W(P) = \frac{GM}{r} \left[\sum_{n=0}^{n_{max}} \sum_{m=0}^{n} \left(\frac{a}{r}\right)^n \times \right.$$

$$\left. P_{nm}(\sin\phi)\,(C_{nm}\cos m\lambda + S_{nm}\sin m\lambda) \right] \quad (10)$$

$$+ \Phi(r,\phi,\lambda),$$

where $P_{nm}(\sin\phi)$ represents the first class Legendre associated functions evaluated in $\sin\phi$, a is the semi-major axis of the Earth and Φ is the centrifugal potential at point P. (r, λ, ϕ) are the spherical geocentric coordinates of the computation point (P) transformed from ellipsoidal coordinates (h, λ, φ) using the transformation formulas found in Hofmann-Wellenhof and Moritz (2005).

3   Data Used

3.1   Geoid Model GEOIDE-Ar 16

GEOIDE-Ar 16 is the current official geoid model for Argentina (Piñón 2016). It is the result of a gravimetric geoid model with a spatial resolution of $1' \times 1'$, fitted to the Argentinean vertical datum though the determination of a corrective trend surface, which was computed using the co-located GPS/Levelling benchmarks. GEOIDE-Ar 16 was developed by the Instituto Geográfico Nacional (IGN), Argentina together with the Royal Melbourne Institute of Technology (RMIT) University, Australia using the remove-compute-restore technique (Schwarz et al. 1990) and the GOCO05s GGM (Mayer-Gürr et al. 2015) up to degree 280, together with 671,547 gravity measurements referred to the International Gravity Standardization Net 1971 (IGSN71; Morelli et al. 1972), on the Argentine continental territory, its neighboring countries, Islas Malvinas and the coastal (marine) areas. For the regions with lack of gravity observations, the DTU13 world gravity model (Andersen et al. 2013) was applied to fill in the gravity voids.

For the determination of the potential values in this contribution, the "pure gravimetric geoid" before fitting it to the Argentinean vertical datum is used. The zero-degree term previously computed with Eq. (5) has been added to the "pure gravimetric geoid" since it was not taken into account for the geoid computation (Piñón 2016).

3.2   Gravity Data Around the Proposed IHRF Stations

The distribution of relative terrestrial and shipborne gravity data used for the determination of the gravimetric geoid around each IHRF station can be seen in Fig. 3 (blue points). Figure 3 also shows the nearest absolute gravity stations that belong to the Red Argentina de Gravedad Absoluta (RAGA, Lauría et al. 2017), which were measured with a Micro-g LaCoste A10 (red points). The absolute gravity measurement made at AGGO with a Micro-g LaCoste FG5 is also included (yellow point).

Different buffer radius of 10, 50, 110 and 200 km are depicted in Fig. 3. Following Sánchez et al. (2017), the minimum amount of gravity points required is 5 inside the radius of 10 km, 15 inside the 50 km, 30 inside the 110 km and 45 inside the 210 km. Each circle is divided into 1, 4, 7 and 11 compartments, respectively. From Fig. 3, we can observe that the gravity data do not fulfill the requirements of density and distribution around each IHRF station. Approximately, one hundred stations homogeneously distributed around the IHRF stations up to a distance of 200 km are required (Sánchez et al. 2017). Since UNSA and OAFA are located in a rough area in the Andes, more stations homogeneously distributed are needed.

3.3   Standards

Some standards used for the computation of the "pure gravimetric geoid" were examined, following some agreements for the computation of the station potential values as IHRF coordinates, geoid undulations and height anomalies within the Colorado 1 cm geoid experiment (Sánchez et al. 2018). The numerical values for the constant of gravitation (G), the geocentric gravitational constant (GM), the mean angular velocity of the Earth (ω), the average density of the topographic masses (ρ) used to apply gravity reductions for geoid computation were the same as those proposed in Sánchez et al. (2018).

The GRS80 that provides the numerical value for the parameters of the geodetic Earth model was used. GOCO05s was the GGM taken into account for the remove-compute-restore technique in the geoid computation of Argentina. First-degree Stokes coefficients were assumed to be zero (Earth's center of masses aligned with the ITRF coordinates).

In Sect. 2.1, the parameters involved in the equations are listed below:

- $W_0 = 62636853.40 \, \mathrm{m^2\,s^{-2}}$, is the reference potential value adopted by the IHRS.
- $U_0 = 62636860.85 \, \mathrm{m^2\,s^{-2}}$, is the normal potential on the GRS80 reference ellipsoid.
- $GM_{GGM} = 3.986004415 \times 10^{14} \, \mathrm{m^3\,s^{-2}}$.
- $GM_{GRS80} = 3.986005 \times 10^{14} \, \mathrm{m^3\,s^{-2}}$.

Atmospheric correction was applied to the terrestrial gravity data (Piñón 2016).

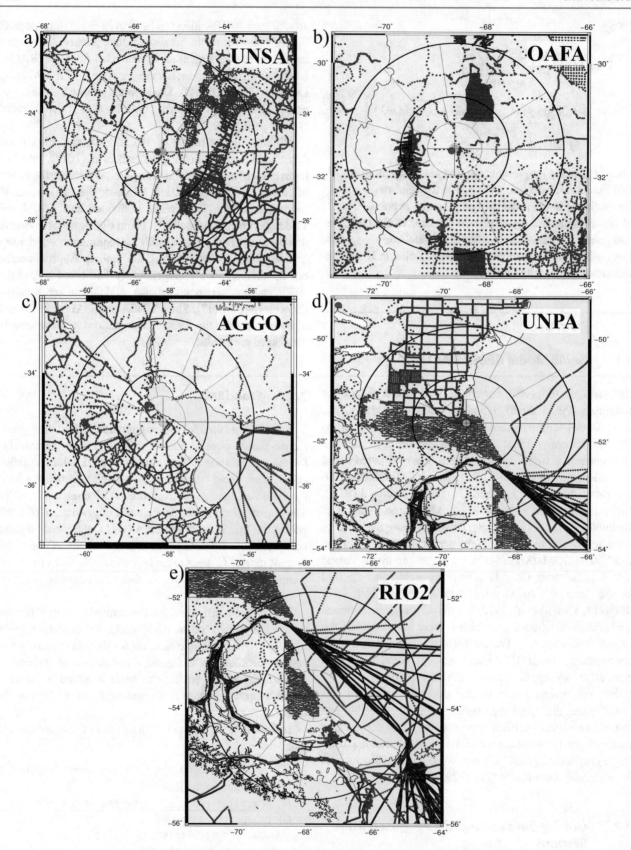

Fig. 3 Distribution of relative gravity data around the proposed IHRF stations for Argentina (blue), RAGA stations (red) and the absolute and superconducting gravity station at AGGO (yellow) over topography. (**a**) UNSA, (**b**) OAFA, (**c**) AGGO, (**d**) UNPA, (**e**) RIO2

4 Results and Discussion

To determine the potential values $W(P)$, the "pure gravimetric" geoid (see Sect. 3.1) was used and the zero-degree term of the geoid (Eq. (5)) was added. The resulted geoid was then transformed from tide-free system to mean-tide system using Eq. (9).

Geoid undulations were converted into height anomalies taking into account the N-ζ transformation according to Eq. (6), using the refined Bouguer gravity anomalies computed with SRTMv4.1 and SRTM30_plus_v10 (Jarvis et al. 2008; Becker et al. 2009).

Table 1 shows N and ζ for each IHRF station in Argentina and the potential value $W(P)$ computed applying Eq. (7). It can be seen that for those stations located near the coast, N and ζ are practically identical while for those stations located near the Andes the differences reach 10 cm for OAFA and 20 cm for UNSA.

For AGGO, IGN has provided a geopotential number $C(P) = 230.284\,\mathrm{m^2\,s^{-2}}$ from gravity and levelling survey (Piñón et al. 2016). Then, the potential value can be estimated with $W(P) = W_0 - C(P) = 62636623.12\,\mathrm{m^2\,s^{-2}}$. The difference of $0.67\,\mathrm{m^2\,s^{-2}}$ (~6 cm) with the potential value estimated in this contribution can be attributed to the fact that $C(P)$ was obtained from levelling survey referred to a local vertical datum (long-term mean sea level measured at a selected tide-gauge).

Potential values from several high-degree (up to $n_{max} = 2190$) GGMs were obtained in order to compare them with those computed in this contribution. EGM2008 model (Pavlis et al. 2012), EIGEN-6C4 model (Förste et al. 2014) and the experimental gravity field model XGM2019e_2159 (Zingerle et al. 2019), were evaluated. The computations were done using the International Centre for Global Earth Models computation service (ICGEM; http://icgem.gfz-potsdam.de/; Ince et al. 2019). The tidal system and reference ellipsoid were selected being consistent with what is previously discussed. Table 2 and Fig. 4 shows the differences between the obtained potential values and those derived from the GGMs. Differences are consistent between models: those stations located near the coast (AGGO, UNPA and RIO2) present differences between 1 to $2\,\mathrm{m^2\,s^{-2}}$, while for stations located near the Andes (UNSA and OAFA), differences are larger. These results become more clear analyzing the differences between potential values computed from the selected GGMs (see also Table 2). Regarding the expected accuracy of the potential values derived from GGMs, Rummel et al. (2014) and Sánchez and Sideris (2017) proposed that the mean accuracy applying one GGM is ± 0.4 to $\pm 0.6\,\mathrm{m^2\,s^{-2}}$ in well-surveyed areas, and about ± 2 to $\pm 4\,\mathrm{m^2\,s^{-2}}$ with extreme cases of $\pm 10\,\mathrm{m^2\,s^{-2}}$ in sparsely surveyed regions. In this sense, the results shown in Table 2 allow to conclude that: more terrestrial gravity data should be included to improve the accuracy of the potential values, especially in regions with rough heights; and, at present, GGMs of high-degree are not accurate enough to derive potential values of the IHRF stations in Argentina.

Finally, the certainty of the potential values presented in this paper is mainly limited by three aspects:

1. The accuracy of the geoid model taken into account. In order to improve it, more terrestrial gravity data should be included in the geoid computation, especially in the vicinity of the selected IHRF stations;
2. the approximation used in Eq. (6) to transform from N to ζ, which could cause errors of several cm in mountain

Table 1 Unfitted geoid undulation (N), height anomaly (ζ) and potential values ($W(P)$) for each proposed station applying Eq. (7)

Station	N [m]	ζ [m]	$W(P)$ [$\mathrm{m^2\,s^{-2}}$]
AGGO	15.353	15.352	62636622.44
UNSA	33.189	33.413	62624879.30
OAFA	24.430	24.535	62629978.70
UNPA	9.022	9.022	62636607.04
RIO2	11.751	11.751	62636662.28

Table 2 Comparison between computed $W(P)$ and those obtained from GGMs

Station	Computed $W(P)$ vs.			Differences between GGMs		
	EGM2008	EIGEN-6C4	XGM2019e_2159	EIGEN-6C4 vs. EGM2008	XGM2019e_2159 vs. EGM2008	EIGEN-6C4 vs. XGM2019e_2159
AGGO	−1.92	−1.92	−1.33	0.01	−0.59	0.59
UNSA	4.82	3.12	3.07	1.71	1.75	−0.04
OAFA	5.59	0.17	2.37	5.42	3.22	2.20
UNPA	−1.55	−2.07	−1.03	0.52	−0.51	1.03
RIO2	−1.57	−0.96	−1.34	−0.61	−0.23	−0.38
Max.	5.59	3.12	3.07	5.42	3.22	2.20
Min.	−1.92	−2.07	−1.34	−0.61	−0.59	−0.38
Mean	1.08	−0.33	0.35	1.41	0.73	0.68
Std. Dev.	3.78	2.13	2.18	2.40	1.69	1.01

Units in [$\mathrm{m^2\,s^{-2}}$]

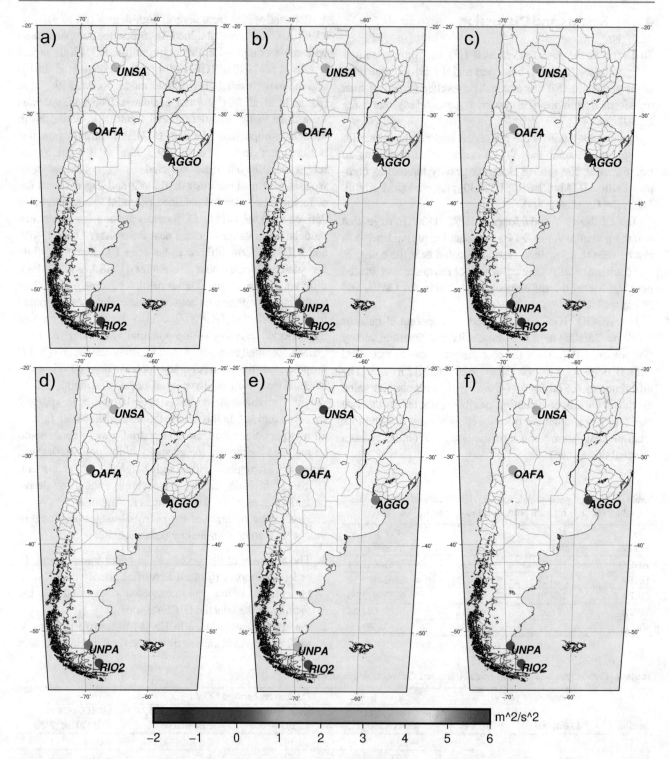

Fig. 4 Differences between computed $W(P)$ and those obtained from GGMs. (**a**) Computed $W(P)$ vs. EGM2008. (**b**) Computed $W(P)$ vs. EIGEN-6C4. (**c**) Computed $W(P)$ vs. XGM2019e_2159. (**d**) EIGEN-6C4 vs. EGM2008. (**e**) EIGEN-6C4 vs. XGM2019e_2159. (**f**) XGM2019e_2159 vs. EGM2008

areas (Flury and Rummel 2009). An extensive approach (e.g. Flury and Rummel 2009; Sjöberg 2010) for the transformation should be evaluated in the future; and,

3. the transformation from N to ζ itself. More reliable $W(P)$ could be obtain by computing a local quasigeoid model for each station.

5 Conclusions and Future Work

This contribution presents the five Argentinean stations that were selected to belong to the global reference network of the IHRF. These stations are named UNSA, OAFA, AGGO, UNPA and RIO2.

All these stations are continuously monitored to detect deformations of the reference frame and they are referred to the ITRS/ITRF to know with high-precision the geometric coordinates. It is desirable that OAFA would be included in the SIRGAS-CON network (Sánchez and Brunini 2009). Currently, UNSA, UNPA, OAFA and RIO2 are not connected to the local vertical datum (SRVN16; Piñón et al. 2016). The connection will be done in the future.

AGGO is a fundamental geodetic observatory where several geodetic techniques are co-located with absolute and superconducting gravity meters, enabling the connection between X, W and gravity.

Preliminary potential values were obtained for the stations selected. They were recovered from the existing geoid model for Argentina GEOIDE-Ar 16 without fitting it to GPS/Levelling benchmarks. Potential values were also derived from high-degree GGMs. Differences between models show that present GGMs are not accurate enough for the estimation of potential values of the selected stations in Argentina to integrate the IHRF.

For a precise transformation from geoid values to height anomalies, orthometric heights and gravity observations should be available for all stations. Moreover, two aspects should be evaluated in the future: (a) the transformation applied from geoid undulations to height anomalies, which could be not accurate enough; and (b) the gravity data around the stations. In this sense, homogeneously gravity data should be distributed around the IHRF reference stations up to $210\,\mathrm{km}$ ($\sim2°$) with a minimum accuracy of the gravity values of $\pm20\,\mu\mathrm{Gal}$, especially for those stations located in the Andes region (OAFA and UNSA).

As a consistent comparison of the obtained potential values, geopotential numbers should also be available from gravity and levelling surveys connected to the national vertical reference system (SRVN16).

Acknowledgements The authors would like to thank the anonymous reviewers for their careful reading of our manuscript and their many useful and constructive comments.

References

Altamimi Z, Rebischung P, Métivier L, Collilieux X (2016) ITRF2014: a new release of the international terrestrial reference frame modeling nonlinear station motions. J Geophys Res Solid Earth 121:6109–6131. https://doi.org/10.1002/2016JB013098

Andersen OB, Knudsen P, Kenyon SC, Factor JK, Holmes S (2013) The DTU13 global marine gravity field. Ocean Surface Topography Science Team meeting 2013. Boulder, USA

Antokoletz ED, Wziontek H, Tocho C (2017) First six months of superconducting gravimetry in Argentina. In: Vergos G, Pail R, Barzaghi R (eds) International symposium on gravity, geoid and height systems 2016. International association of geodesy symposia, vol 148. Springer, Cham. https://doi.org/10.1007/1345_2017_13

Becker JJ, Sandwell DT, Smith WHF, Braud J, Binder B, Depner J, Fabre D, Factor J, Ingalls S, Kim S-H, Ladner R, Marks K, Nelson S, Pharaoh A, Trimmer R, Von Rosenberg J, Wallace G, Weatherall P (2009) Global bathymetry and elevation data at 30 arc seconds resolution: SRTM30_PLUS. Mar Geod 32(4):355–371. https://doi.org/10.1080/01490410903297766

Cimbaro SR, Lauría EA, Piñón DA (2009) Adopción del Nuevo Marco de Referencia Geodésico Nacional. Paper presented to Instituto Geográfico Militar, Buenos Aires, Argentina

Drewes H, Kuglitsch F, Adám J, Rózsa S (2016) The geodesist's handbook 2016. J Geod 90:907. https://doi.org/10.1007/s00190-016-0948-z

Ekman M (1989) Impacts of geodynamic phenomena on systems for height and gravity. Bull Géod 63(3):281–296. https://doi.org/10.1007/BF02520477

Flury J, Rummel R (2009) On the geoid–quasigeoid separation in mountain areas. J Geod 83(9):829–847. https://doi.org/10.1007/s00190-009-0302-9

Förste C, Bruinsma SL, Abrikosov O, Lemoine J-M, Marty JC, Flechtner F, Balmino G, Barthelmes F, Biancale R (2014) EIGEN-6C4 the latest combined global gravity field model including GOCE data up to degree and order 2190 of GFZ Potsdam and GRGS Toulouse. GFZ Data Services. http://doi.org/10.5880/icgem.2015.1

Hofmann-Wellenhof B, Moritz H (2005) Physical geodesy. Springer, New York

Ihde J, Sánchez L, Barzaghi R, Drewes H, Foerste C, Gruber T, Liebsch G, Marti U, Pail R, Sideris MG (2017) Definition and proposed realization of the International Height Reference System (IHRS). Surv Geophys 38(3):549–570. https://doi.org/10.1007/s10712-017-9409-3

Ince ES, Barthelmes F, Reißland S, Elger K, Förste C, Flechtner F, Schuh H (2019) ICGEM – 15 years of successful collection and distribution of global gravitational models, associated services and future plans. Earth Syst Sci. Data 11:647–674. http://doi.org/10.5194/essd-11-647-2019

Jarvis A, Reuter HI, Nelson A, Guevara E (2008) Hole-filled SRTM for the globe Version 4. http://www.cgiar-csi.org/data/srtm-90m-digital-elevation-database-v4-1

Johnston G, Riddell A, Hausler G (2017) The international GNSS service. In: Teunissen PJG, Montenbruck O (eds) Springer handbook of global navigation satellite systems, 1st edn. Springer, Cham, pp 967–982. https://doi.org/10.1007/978-3-319-42928-1

Lauría E, Pacino MC, Blitzkow D, Cimbaro S, Piñón DA, Miranda S, Bonvalot S, Gabalda G, Tocho CN (2017) Red Argentina de Gravedad Absoluta (RAGA). In I Simposio Internacional de Geomática Aplicada y Soluciones Geoespaciales (GEODATA 2017). Rosario, 3 al 7 de abril de 2017. http://sedici.unlp.edu.ar/handle/10915/75806

Mayer-Gürr T, Pail R, Gruber T, Fecher T, Rexer M, Schuh W-D, Kusche J, Brockmann J-M, Rieser D, Zehentner N, Kvas A, Klinger B, Baur O, Höck E, Krauss S, Jäggi A (2015) The combined satellite gravity field model GOCO05s. Presentation at EGU 2015, Vienna, April 2015

Morelli C, Gantar C, McConnell RK, Szabo B, Uotila U (1972) The international gravity standardization net 1971 (IGSN 71). Osservatorio Geofisico sperimentale Trieste (Italy). Available at https://apps.dtic.mil/dtic/tr/fulltext/u2/a006203.pdf. Accessed 20 Dec 2019

Moritz H (2000) Geodetic reference system 1980. J Geod 74(1):128–133. https://doi.org/10.1007/s001900050278

Pavlis NK, Holmes SA, Kenyon SC, Factor JK (2012) The development of the Earth gravitational model 2008 (EGM2008). J Geophys Res 117:B04406. https://doi.org/10.1029/2011JB008916

Petit G, Luzum B (eds) (2010) IERS conventions 2010. IERS Technical Note 36. Verlag des Bundesamtes für Kartographie und Geodäsie, Frankfurt a.M.

Piñón D (2016) Development of a precise gravimetric geoid model for Argentina. Masters by Research, Mathematical and Geospatial Sciences, RMIT University. http://researchbank.rmit.edu.au/view/rmit:161742

Piñón DA, Guagni H, Cimbaro SR (2016) Nuevo Sistema Vertical de la República Argentina. Simposio SIRGAS 2016. https://www.ign.gob.ar/descargas/geodesia/2016_Nuevo_Sistema_Vertical_de_Referencia.pdf

Piñón DA, Gómez DD, Smalley JrR, Cimbaro SR, Lauría EA, Bevis MG (2018) The history, state, and future of the Argentine continuous satellite monitoring network and its contributions to geodesy in Latin America. Seismol Res Lett 89(2A):475–482. https://doi.org/10.1785/0220170162

Rummel R, Gruber TH, Ihde J, Liebsch G, Rülke A, Schäfer U, Sideris M, Rangelova E, Woodworth PH, Hughes CH (2014) STSE-GOCE+, height system unification with GOCE, Doc. No. GOHSU-PL-002, issue 1, 24-02-2014

Sánchez L, Brunini C (2009) Achievements and challenges of SIRGAS. In: Drewes H (ed) Geodetic reference frames. International association of geodesy symposia, vol 134. Springer, Berlin/Heidelberg. https://doi.org/10.1007/978-3-642-00860-3_25

Sánchez L, Sideris MG (2017) Vertical datum unification for the International Height Reference System (IHRS). Geophys J Int 209(2):570–586. https://doi.org/10.1093/gji/ggx025

Sánchez L, Cunderlík R, Dayoub N, Mikula K, Minarechová Z, Sima Z, Vatrt V, Vojtíšková M (2016) A conventional value for the geoid reference potential W0. J Geod 90(9):815–835. https://doi.org/10.1007/s00190-016-0913-x

Sánchez L, Ihde J, Pail R, Gruber T, Barzaghi R, Marti U, Ågren J, Sideris M, Novák P (2017) Towards a first realization of the International Height Reference System (IHRS). Geophysical Research Abstracts, vol 19, EGU2017-17104. European Geosciences Union General Assembly 2017. Available at https://ihrs.dgfi.tum.de/fileadmin/JWG_2015/Sanchez_et_al_Towards_a_IHRS_Realization_EGU2017.pdf. Accessed 19 Dec 2019

Sánchez L, Ågren J, Huang J, Wang YM, Forsberg R (2018) Basic agreements for the computation of station potential values as IHRS coordinates, geoid undulations and height anomalies within the Colorado 1 cm geoid experiment. Work document. Version 0.5, October 30, 2018. Available at https://ihrs.dgfi.tum.de/fileadmin/JWG_2015/Colorado_Experiment_Basic_req_V0.5_Oct30_2018.pdf. Accessed 20 Dec 2019

Schwarz KP, Sideris MG, Forsberg R (1990) The use of FFT techniques in physical geodesy. Geophys J Int 100(3):485–514. https://doi.org/10.1111/j.1365-246X.1990.tb00701.x

Sjöberg LE (2010) A strict formula for geoid-to-quasigeoid separation. J Geod 84(11):699–702. https://doi.org/10.1007/s00190-010-0407-1

Zingerle P, Pail R, Gruber T, Oikonomidou X (2019) The experimental gravity field model XGM2019e. GFZ Data Services. http://doi.org/10.5880/ICGEM.2019.007

Comparing Vienna CRF solutions to Gaia-CRF2

David Mayer and Johannes Böhm

Abstract

We are using various models and analysis strategies, such as galactic aberration, ray-tracing etc., to create different Vienna celestial reference frame (CRF) solutions. These solutions are then compared against the Gaia reference frame (Gaia-CRF2). This is done using a degree 2 vector spherical harmonics approach. The estimated parameters are used to investigate the impact of the various analysis methods on the differences between Gaia and the Very Long Baseline Interferometry (VLBI) CRF. We find that correcting for galactic aberration reduces the difference between the Gaia-CRF2 and the VLBI CRF significantly (30 µas in D_2 and 13 µas in D_3). Furthermore, we find that using a priori ray-traced tropospheric delays in addition with low absolute constraints on tropospheric gradients reduces the a^e_{20} parameter by 20 µas. Using these analysis strategies we can explain almost all significant differences between the Gaia-CRF2 and the VLBI CRF. However, the vector spherical harmonic (VSH) parameter a^e_{20} is still highly significant and can not be explained by modeling and analysis choices from the VLBI technique.

Keywords

Gaia · Galactic aberration · ICRF · Reference systems

1   Introduction

The rotation about the Galactic center causes an acceleration of the Solar System Barycenter (SSB) towards the center of the Galaxy. The galactic aberration (GA) is the aberration of positions of distant objects resulting from the revolution of the SSB about the Galactic center with a period of 250 million years. Over decades of observing, it imprints an apparent source proper motion of a few µas per year.

Source positions estimated from geodetic Very Long Baseline Interferometry (VLBI) are on an accuracy level where an effect of this magnitude can be calculated and consequently has to be corrected in the analysis. Several groups estimated the GA from VLBI data (see Titov et al. 2011; Xu et al. 2012; Titov and Lambert 2013; Titov and Krásná 2018). The reported values range from 5.2 ± 0.2 to 6.4 ± 1.1 µas per year with the center of the Galaxy at 17 h 45 min 40 s in right ascension and −29°00′28″ in declination.

The International VLBI Service for Geodesy and Astrometry (IVS) established a working group with the general purpose of investigating the issues concerning the incorporation of the GA in IVS analysis, see MacMillan et al. (2019). This working group agreed upon a value of 5.8 µas per year which was also applied in the creation of the ICRF3, see Charlot et al. (2020).

One of the largest error sources in geodetic VLBI is the troposphere. It was demonstrated by Mayer et al. (2017) that using a priori ray-traced tropospheric delays in a global

D. Mayer (✉)
Department Control Survey, Federal Office of Metrology and Surveying, Vienna, Austria
e-mail: david.mayer@bev.gv.at

J. Böhm
Department of Geodesy and Geoinformation, Technische Universität Wien, Vienna, Austria

© The Author(s) 2020
J. T. Freymueller, L. Sanchez (eds.), *Beyond 100: The Next Century in Geodesy*, International Association of Geodesy Symposia 152, https://doi.org/10.1007/1345_2020_99

CRF solution significantly influences the source coordinates. Further, they showed that constraining tropospheric gradients to their a priori values influences the source coordinates as well.

The second data release from the Gaia satellite, Gaia Data Release 2 (Gaia DR2), includes a celestial reference frame (Gaia-CRF2) of comparable accuracy to VLBI, see Lindegren et al. (2018) and Mignard et al. (2018) for more information. Since Gaia was launched at the end of 2013 the effect of GA on its positions is negligible. Further, Gaia is a satellite in space and, therefore, unaffected by tropospheric disturbance. Additionally, the satellite's rotation and precession (the so called scanning law) are designed to maximize the uniformity of the sky coverage. The scanning law introduces some systematic effects, presented in Lindegren et al. (2018), however, they can most likely be ignored when large scale global systematic effects are concerned.

VLBI suffers from an uneven network distribution, which could result in a global deformation of the frame when the troposphere is not modeled correctly. Further, the unmodeled effects of GA introduce a global systematic deformation of the frame over the years. Since the Gaia-CRF2 is not affected by these effects it provides a perfect independent source for external validation of the effects of GA and the troposphere.

2   Data and Analysis

We generated five global geodetic VLBI solutions with different analysis options and compared them to the Gaia-CRF2. The basic solution setup is described in the beginning of this section. Other solutions are based on the basic solution with some changes in modeling and analysis described at the end of this section.

All VLBI solutions presented here utilize all geodetic VLBI sessions that were used for the creation of the ICRF3. In total this data set includes about 13 million observations of about 4,500 sources. The VLBI CRF solutions were generated using the Vienna VLBI and Satellite Software (Böhm et al. 2018), which is developed by TU Wien. These solutions follow the IERS Conventions 2010 by Petit and Luzum (2010) for reducing the observations and geophysical modeling. Also, antenna thermal deformation (Nothnagel 2009) and atmospheric pressure loading (Wijaya et al. 2013) were taken into account.

A priori positions for stations (including velocities) and sources are taken from the ITRF2014 (Altamimi et al. 2016) and ICRF2 (see Ma et al. 2009; Fey et al. 2015) respectively. A standard geodetic analysis is performed. This results in an updated ITRF and ICRF as well as EOP time series. Station

Table 1 Comparison of five investigated solutions

Solution 1	Reference solution
Solution 2	Solution 1 + a priori ray-traced delays
Solution 3	Solution 2 + removing of absolute constraints on tropospheric gradients
Solution 4	Solution 3 + GA model
Solution 5	Solution 4 + error scaling

coordinate adjustments were estimated as global parameters with No-Net-Rotation and No-Net-Translation conditions applied to the positions and velocities of a group of 21 stations. Most of the radio source adjustments were estimated as global parameters after a No-Net-Rotation constraint is applied to the positions of the 295 ICRF2 defining sources. The coordinates of the 39 special handling sources (Fey et al. 2015) were, however, estimated once per session.

A priori hydrostatic zenith delays were determined from local pressure values and then mapped to the elevation using the Vienna Mapping Functions 1, see Böhm et al. (2006). Additionally, tropospheric gradients from the NASA/GSFC Data Assimilation Office (DAO) model (see MacMillan 1995; MacMillan and Ma 1997) are utilized for all solutions. The wet zenith delays, north and east troposphere gradients, and clock values were estimated every 30 min, 6 h and 1 h, respectively. Tropospheric gradients are constrained to their a priori values. This was realized with absolute and relative constraints of 0.5 mm (after 6 h), which removes unrealistic gradient estimates but affects the declination of estimated sources. Other solutions are based on the same parameterization with slight amendments, see Table 1 for an overview. In the second solution a priori ray-traced tropospheric delays were included. Since the absolute constraints on a priori tropospheric gradients tend to influence source declination we also created the same solution where we removed these constraints. In the fourth solution GA was corrected a priori with the recommended value of 5.8 µas/year with the center of the Galaxy at 17 h 45 min 40 s in right ascension and $-29°00'28''$ in declination. As a reference epoch we chose 2015.0 since the Gaia positions epoch is close to this epoch. In VieVS the correction was realized by modification of the conventional group delay equation as proposed by Titov et al. (2011). Additionally, we generated a solution where we scaled the errors to more realistic values. This was done using a scaling factor of 1.5 and a noise floor of 30 µas, see Charlot et al. (2020) for more information on these values.

As a reference a subset of sources from the Gaia-CRF2 solution described in Mignard et al. (2018) was used. This subset consists of the positions of 2,820 sources which have an ICRF3 counterpart. A detailed analysis of the differences of the catalogue and an ICRF3 prototype solution can be found in Mignard et al. (2018) and Petrov et al. (2018).

3 Methodology

The resulting catalogs are compared to the Gaia-CRF2 using vector spherical harmonics (VSH) as described in Mignard and Klioner (2012). Global features of the differences such as a rotation of the catalogs and the so called glide parameters are reflected in degree 1. Degree 2 describes the quadrupole deformations between the catalogs. The whole transformation reads

$$\Delta\alpha\cos\delta = R_1\cos\alpha\sin\delta + R_2\sin\alpha\sin\delta - R_3\cos\delta \quad (1)$$
$$- D_1\sin\alpha + D_2\cos\alpha$$
$$+ a_{20}^m\sin 2\delta$$
$$+ \left(a_{21}^{e,\mathrm{Re}}\sin\alpha + a_{21}^{e,\mathrm{Im}}\cos\alpha\right)\sin\delta$$
$$- \left(a_{21}^{m,\mathrm{Re}}\cos\alpha - a_{21}^{m,\mathrm{Im}}\sin\alpha\right)\cos 2\delta$$
$$- 2\left(a_{22}^{e,\mathrm{Re}}\sin 2\alpha + a_{22}^{e,\mathrm{Im}}\cos 2\alpha\right)\cos\delta$$
$$- \left(a_{22}^{m,\mathrm{Re}}\cos 2\alpha - a_{22}^{m,\mathrm{Im}}\sin 2\alpha\right)\sin 2\delta,$$

$$\Delta\delta = -R_1\sin\alpha + R_2\cos\alpha \quad (2)$$
$$- D_1\cos\alpha\sin\delta - D_2\sin\alpha\sin\delta + D_3\cos\delta$$
$$+ a_{20}^e\sin 2\delta$$
$$- \left(a_{21}^{e,\mathrm{Re}}\cos\alpha - a_{21}^{e,\mathrm{Im}}\sin\alpha\right)\cos 2\delta$$
$$- \left(a_{21}^{m,\mathrm{Re}}\sin\alpha + a_{21}^{m,\mathrm{Im}}\cos\alpha\right)\sin\delta$$
$$- \left(a_{22}^{e,\mathrm{Re}}\cos 2\alpha - a_{22}^{e,\mathrm{Im}}\sin 2\alpha\right)\sin 2\delta$$
$$+ 2\left(a_{22}^{m,\mathrm{Re}}\sin 2\alpha + a_{22}^{m,\mathrm{Im}}\cos 2\alpha\right)\cos\delta$$

where R_i are the three rotation parameters, D_i are the three glide parameters and $a_{lm}^{m,e}$ are the quadrupole parameters of electric (e) and magnetic (m) type.

The data set used here is rather small (about 2,800 sources) and we are only interested in global effects. Therefore, we decided to stop the VSH expansion at degree 2. Outliers were eliminated, see Sect. 3.1, and transformation parameters were estimated using the classical least squares method. Variances and correlations are used to weigh the differences.

3.1 Outlier Detection

There are many options to choose from when eliminating outliers. Using the normalized separation was proposed by Mignard et al. (2016). This approach takes the correlation between right ascension and declination into account and

introduces an arbitrary cut off of 10 mas (angular separation). However, we expect to have differences in positions due to galactic aberration which are on the level of the formal errors. This is especially critical for sources, which have a long observing history with VLBI. These sources tend to be most affected by GA. Further, their long observing history implies that they have been observed many times and, therefore, have small error bars.

The outlier test proposed by Mignard et al. (2016) is based on the assumption that the differences follow a Rayleigh distribution. However, it was found by Petrov and Kovalev (2017) that the differences between Gaia-CRF2 and VLBI CRF deviate from a Rayleigh distribution. For the reasons mentioned above we decided to not use the outlier test proposed by Mignard et al. (2016) but rather use our own method of outlier detection.

Since the number of intersecting sources is rather small the outlier detection can be realized in the parameter space. This was accomplished by removing each source once from the standard solution and calculating the VSH parameters. At the end we have a set of about 2,800 VSH parameters which can be used to calculate the standard deviation of each VSH parameter. Outliers can then be found with a simple three sigma cut off for each parameter. The source is excluded when one of the VSH parameters is outside of this cut off. This means that a source that significantly changes one of the VSH parameter by itself is considered an outlier. With this approach we find that about 7% of sources are flagged as outliers. This list of outliers is then used for every solution presented here. In order to compare the different transformation parameters we decided to stick to one list of good sources. This was realized by doing the outlier elimination on the standard solution and then using this list of outliers for each other solution.

When plotting the detected outliers one can see that the outlier test proposed by Mignard et al. (2016) clearly removes some kind of systematic effect, see right part of Fig. 1. When removing the outliers with our technique the systematic is not that clear but might still be there, see left part of Fig. 1. It is important to note here that other outlier elimination techniques that were tested showed similar results.

4 Results and Discussion

During our research we created five global VLBI solutions. For each solution one analysis option was changed, see Table 1. However, the previous changes were not revoked but rather applied alongside. The VSH parameters of these five solutions with respect to Gaia-CRF2 are depicted in Fig. 2. The center of the bar represents the VSH estimate while

Fig. 1 The outliers found in the parameter space and with the normalized separation technique are depicted in (**a**) and (**b**) respectively. The galactic plane is illustrated by a red dashed line with the center of the Galaxy depicted as a black circle. The ecliptic is illustrated as a black dashed line

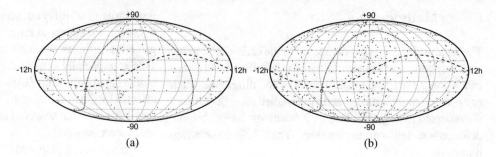

(a)   (b)

the length of the bar reflects the formal uncertainties of the estimate. Different solutions have different color codes and different names. Solution 1 reflect the standard Vienna CRF solution, as described in Sect. 2. This solution experiences significant deviations from Gaia-CRF2. Solution 2 is similar to solution 1 with the difference that a priori ray-traced tropospheric delays are used in the analysis. Using this model succeeds in reducing the a^e_{20} parameter. However, at the same time the D_3 parameter is increased. The D_3 parameter can be decreased again by loosening the constraints on the tropospheric gradients, which is reflected by solution 3. We further succeed in reducing the D_2 and D_3 parameter when adding the GA model, see solution 4. Scaling the formal uncertainties (solution 5), which is a mandatory task when creating a VLBI CRF, does not change the parameters significantly. Note that correlations between the VSH parameters are relatively low with a maximum of -0.37 between D_2 and R_1.

When looking at Fig. 2 the deformations of degree 1 and degree 2 are particularly interesting, since they describe real differences between the frames. The rotation between Gaia-CRF2 and the VLBI CRF is statistically significant, but less interesting because they do not reflect real effects. During the creation of Gaia-CRF2 the frame was rotated onto an ICRF3 prototype solution, therefore, no rotations should be present between the two frames. However, the number of sources used for the rotation differs. The Gaia-CRF2 was rotated onto 2,844 matching ICRF3 sources while we use 2,588 sources after outlier detection. Further, the sources selected as outliers are the ones affecting the parameters most, therefore, a large rotation can be expected. However, the meaning of this rotation is questionable and will not be further discussed. Note that the ICRF3 prototype solution was rotated onto the same 295 ICRF2 defining sources that were used for the solutions discussed here.

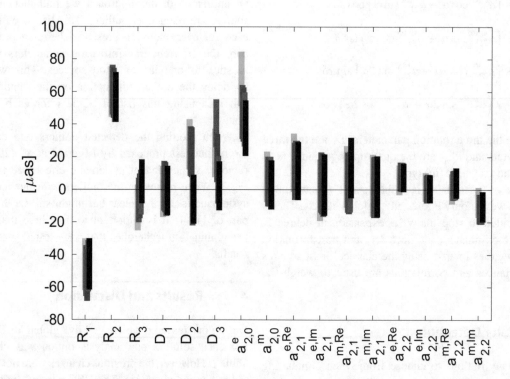

Fig. 2 VSH parameters of five VLBI CRF solutions w. r. t. Gaia-CRF2. The center of the bar represents the estimate and the length of the bar is twice the formal uncertainty of the parameter. In each consecutive solution an analysis step was added. The standard solution, see solution 1 in Table 1, is depicted in green. In red the standard solution with a priori ray-traced delays is depicted. This is solution 2. The blue solution (solution 3) illustrates the effect of removing absolute constraints on tropospheric gradient estimation. In the magenta solution the GA model was applied, see solution 4. The final solution (solution 5), where errors are scaled, is depicted in black

Fig. 3 Deformations of degree 1 (**a**) and deformations of degree 2 (**b**) of the standard Vienna CRF solution w.r.t. Gaia-CRF2. The largest arrow in (**a**) and (**b**) is 36 and 91 μas, respectively. The galactic plane is depicted by a red dashed line with the center of the Galaxy illustrated as a black circle. The ecliptic is depicted as a black dashed line

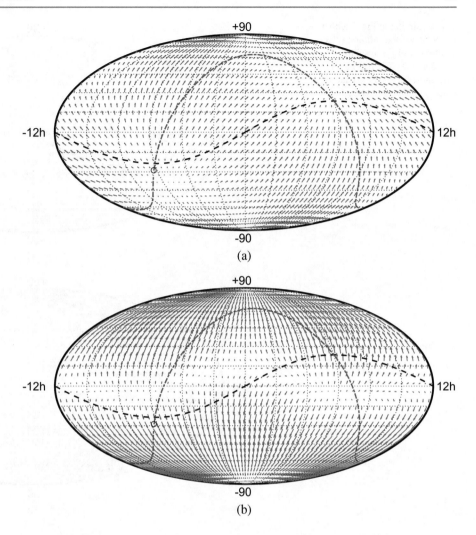

Using the GA model clearly removes systematic effects between the two frames (difference of 30 μas in D_2 and 13 μas in D_3). This can be explained by the fact that the two techniques have very different time spans. The first VLBI data used for the creation of the VLBI CRF dates back to 1979, which is a total of 40 years of data. This means the effects of GA had time to accumulate, which in turn effects the source positions calculated with VLBI. The Gaia satellite was only launched in 2013. Therefore, GA did not have enough time to affect source positions calculated from Gaia data. This means that a VLBI CRF corrected for GA (with a reference epoch close to Gaia) should agree better with the Gaia-CRF2 than a VLBI CRF not corrected for GA. This is exactly what we see in Fig. 2.

Using a priori ray-traced tropospheric delays succeeds in reducing the a_{20}^e parameter, which is the most significant deformation between the frames, by 20 μas. This is particularly important, since no other model or analysis choice affects this parameter. Unfortunately, the result is not that clear because the D_3 parameter is increased by about the same amount at the same time. However, we found that the D_3 parameter, which is directly connected to the source declination, is highly susceptible to models and analysis choices. Reducing the absolute constraints on tropospheric gradient estimation, for example, succeeds in reducing the D_3 parameter by 20 μas.

The results from Fig. 2 can also be plotted on the sphere for easier interpretation. Figure 3 depicts the deformations of the standard Vienna CRF solution, marked as green in Fig. 2. Plot (a) depicts the glide parameters (D_1, D_2 and D_3) with the addition of the galactic plane (red dashed line), the galactic center (black circle) and the ecliptic (black dashed line). One can clearly see the dominant effect of GA with arrows roughly from the galactic center to the anti center. Plot (b) depicts the deformations of degree 2, which are dominated by the a_{20}^e parameter. Figure 3 can now be compared to Fig. 4 which depicts the deformations of the final best fitting solution, which is marked in black in Fig. 2. One can see that deformations of degree 1 are insignificant. Further, the deformations of degree 2 are decreased but still significant.

For completeness the VSH are also calculated using the outlier detection method proposed by Mignard et al. (2016), see Fig. 5. The error bars and estimates are much smaller,

Fig. 4 Similar to Fig. 3 with a different Vienna solution (black solution in Fig. 2) used to calculate the VSH parameters. The largest arrow in (**a**) and (**b**) is 6 and 71 µas respectively

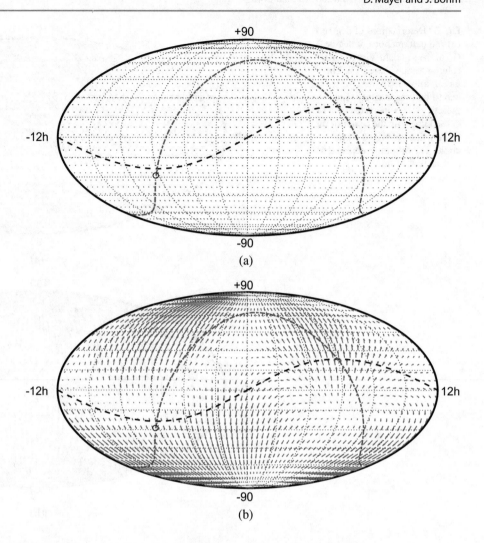

5   Conclusions and Outlook

when using this technique. This can be explained by the number of outliers found by the different techniques. With the technique proposed by Mignard et al. (2016) about 500 sources are flagged as outliers while only about 200 are flagged by the outlier elimination technique described in this paper. In this comparison the outliers were detected for each solution separately. Unfortunately, also the effect of the models is slightly different. Using a priori ray-traced tropospheric delays does also affect other deformations of degree 2. The correction of GA does move the parameters by roughly the same amount. However, since the parameters are small to begin with the correction moves the parameter into the negative. Scaling the formal uncertainties seems to affect the parameters when using this outlier elimination technique. This demonstrates that the VSH parameters between the VLBI CRF and Gaia-CRF2 are very susceptible to the outlier elimination technique used.

We produced several VLBI CRF solutions with different models and analysis choices. These solutions were then compared to the Gaia-CRF2 using VSH of degree 2.

We find three (D_2, D_3 and a_{20}^e) significant parameters between the standard Vienna CRF solution and the Gaia-CRF2. The a_{20}^e can be reduced by 20 µas by using a priori ray-traced tropospheric delays in the analysis. However, this also increases the D_3 parameter by the same amount. We find that the D_3 parameter, which is directly connected to the source declination, is very susceptible to models and analysis choices. The parameter can be significantly reduced (in this case about 20 µas) when lowering the absolute constraints on tropospheric gradients. Further, we find that using GA succeeds in reducing the D_2 parameter by 30 µas and D_3

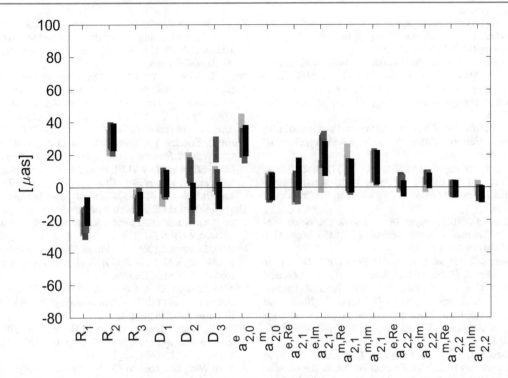

Fig. 5 The figure is similar to Fig. 2 with the exception that the outlier elimination technique proposed by Mignard et al. (2016) is used to determine the VSH parameters

parameter by 13 µas. We can explain all the deformations of degree 1 between the VLBI CRF and the Gaia-CRF2 with analysis choices made by the VLBI analyst. However, the a_{20}^e can only be reduced but not fully explained by choices made by the VLBI analyst.

Furthermore, we find that using the normalized separation method proposed by Mignard et al. (2016) to detect outliers removes sources that are needed to see the systematic differences between the VLBI and Gaia frame.

A more in depth analysis with more VLBI CRF solutions can be found in Mayer (2019).

Acknowledgements We thank the Austrian Science Fund (FWF) for supporting projects I 2204 (SORTS) and T 697 (Galactic VLBI).

References

Altamimi Z, Rebischung P, Métivier L, Collilieux X (2016) ITRF2014: a new release of the International Terrestrial Reference Frame modeling nonlinear station motions. J Geophys Res Solid Earth 121(8):6109–6131. https://doi.org/10.1002/2016JB013098

Böhm J, Werl B, Schuh H (2006) Troposphere mapping functions for GPS and very long baseline interferometry from European Centre for Medium-Range Weather Forecasts operational analysis data. J Geophys Res Solid Earth 111(B2):n/a–n/a. https://doi.org/10.1029/2005JB003629

Böhm J, Böhm S, Boisits J, Girdiuk A, Gruber J, Hellerschmied A, Krásná H, Landskron D, Madzak M, Mayer D, McCallum J, McCallum L, Schartner M, Teke K (2018) Vienna VLBI and satellite software (VieVS) for geodesy and astrometry. Publ Astron Soc Pac 130(986):044503. https://doi.org/10.1088/1538-3873/aaa22b

Charlot P, Jacobs CS, Gordon D, Lambert S, de Witt A, Böhm J, Fey AL, Heinkelmann R, Skurikhina E, Titov O, Arias EF, Bolotin S, Bourda G, Ma C, Malkin Z, Nothnagel A, Mayer D, MacMillan DS, Nilsson T, Gaume R (2020) The third realization of the international celestial reference frame by very long baseline interferometry (in preparation)

Fey AL, Gordon D, Jacobs CS, Ma C, Gaume RA, Arias EF, Bianco G, Boboltz DA, Böckmann S, Bolotin S, Charlot P, Collioud A, Engelhardt G, Gipson J, Gontier AM, Heinkelmann R, Kurdubov S, Lambert S, Lytvyn S, MacMillan DS, Malkin Z, Nothnagel A, Ojha R, Skurikhina E, Sokolova J, Souchay J, Sovers OJ, Tesmer V, Titov O, Wang G, Zharov V (2015) The second realization of the international celestial reference frame by very long baseline interferometry. Astron J 150(2):58. http://stacks.iop.org/1538-3881/150/i=2/a=58

Lindegren L, Hernández J, Bombrun A, Klioner S, Bastian U, Ramos-Lerate M (2018) Gaia Data Release 2. The astrometric solution. Astron Astrophys. https://doi.org/10.1051/0004-6361/201832727

Ma C, Arias EF, Bianco G, Boboltz DA, Bolotin SL, Charlot P, Engelhardt G, Fey AL, Gaume RA, Gontier AM, Heinkelmann R, Jacobs CS, Kurdubov S, Lambert SB, Malkin ZM, Nothnagel A, Petrov L, Skurikhina E, Sokolova JR, Souchay J, Sovers OJ, Tesmer V, Titov OA, Wang G, Zharov VE, Barache C, Boeckmann S, Collioud A, Gipson JM, Gordon D, Lytvyn SO, MacMillan DS, Ojha R (2009) The second realization of the international celestial reference frame by very long baseline interferometry. IERS Technical Note 35. https://www.iers.org/IERS/EN/Publications/TechnicalNotes/tn35.html

MacMillan DS (1995) Atmospheric gradients from very long baseline interferometry observations. Geophys Res Lett 22(9):1041–1044. https://doi.org/10.1029/95GL00887

MacMillan DS, Ma C (1997) Atmospheric gradients and the VLBI terrestrial and celestial reference frames. Geophys Res Lett 24(4):453–456. https://doi.org/10.1029/97GL00143

MacMillan DS, Fey A, Gipson JM, Gordon D, Jacobs CS, Krásná H, Lambert SB, Malkin Z, Titov O, Wang G, Xu MH (2019) Galactocentric acceleration in VLBI analysis - Findings of IVS WG8. Astron Astrophys 630:A93. https://doi.org/10.1051/0004-6361/201935379

Mayer D (2019) VLBI celestial reference frames and assessment with Gaia. Technische Universität Wien, Vienna. https://resolver.obvsg.at/urn:nbn:at:at-ubtuw:1-121147

Mayer D, Böhm J, Krásná H, Landskron D (2017) Tropospheric delay modelling and the celestial reference frame at radio wavelengths. Astron Astrophys 606:A143. https://doi.org/10.1051/0004-6361/201731681

Mignard F, Klioner S (2012) Analysis of astrometric catalogues with vector spherical harmonics. Astron Astrophys 547:A59. https://doi.org/10.1051/0004-6361/201219927

Mignard F, Klioner S, Lindegren L, Bastian U, Bombrun A, Hernández J, Hobbs D, Lammers U, Michalik D, Ramos-Lerate M, Biermann M, Butkevich A, Comoretto G, Joliet E, Holl B, Hutton A, Parsons P, Steidelmüller H, Andrei A, Bourda G, Charlot P (2016) Gaia Data Release 1. Reference frame and optical properties of ICRF sources. Astron Astrophys 595:A5. https://doi.org/10.1051/0004-6361/201629534

Mignard F, Klioner S, Lindegren L, Hernández J, Bastian U, Bombrun A, Hobbs D, Lammers U (2018) Gaia Data Release 2. The celestial reference frame (Gaia-CRF2). Astron Astrophys. https://doi.org/10.1051/0004-6361/201832916

Nothnagel A (2009) Conventions on thermal expansion modelling of radio telescopes for geodetic and astrometric VLBI. J Geod 83(8):787–792. https://doi.org/10.1007/s00190-008-0284-z

Petit G, Luzum B (eds) (2010) IERS conventions 2010, Frankfurt am Main: Verlag des Bundesamts für Kartographie und Geodäsie. IERS Technical Note No. 36. http://iers-conventions.obspm.fr/updates/2010updatesinfo.php

Petrov L, Kovalev YY (2017) Observational consequences of optical band milliarcsec-scale structure in active galactic nuclei discovered by Gaia. Mon Not R Astron Soc 471(4):3775–3787. https://doi.org/10.1093/mnras/stx1747. http://oup.prod.sis.lan/mnras/article-pdf/471/4/3775/19536705/stx1747.pdf

Petrov L, Kovalev YY, Plavin AV (2018) A quantitative analysis of systematic differences in the positions and proper motions of Gaia DR2 with respect to VLBI. Mon Not R Astron Soc 482(3):3023–3031. https://doi.org/10.1093/mnras/sty2807. http://oup.prod.sis.lan/mnras/article-pdf/482/3/3023/26618176/sty2807.pdf

Titov O, Krásná H (2018) Measurement of the solar system acceleration using the Earth scale factor. Astron Astrophys. https://doi.org/10.1051/0004-6361/201731901

Titov O, Lambert S (2013) Improved VLBI measurement of the solar system acceleration. Astron Astrophys 559:A95. https://doi.org/10.1051/0004-6361/201321806

Titov O, Lambert SB, Gontier A-M (2011) VLBI measurement of the secular aberration drift. Astron Astrophys 529:A91. https://doi.org/10.1051/0004-6361/201015718

Wijaya D, Böhm J, Karbon M, Krásná H, Schuh H (2013) Atmospheric pressure loading. Springer, Berlin, pp 137–157. https://doi.org/10.1007/978-3-642-36932-2_4

Xu MH, Wang GL, Zhao M (2012) The solar acceleration obtained by VLBI observations. Astron Astrophys 544:A135. https://doi.org/10.1051/0004-6361/201219593

Co-location of Space Geodetic Techniques: Studies on Intra-Technique Short Baselines

Iván Herrera Pinzón and Markus Rothacher

Abstract

The goal of the project "Co-location of Space Geodetic Techniques on Ground and in Space", in the DFG funded research unit on reference systems and founded by the Swiss National Foundation (SNF), is the improvement of existing and the establishment of new ties between the space geodetic techniques, together with the assessment and reduction of technique-specific biases. To achieve this, the wealth of co-located instruments at the Geodetic Observatory Wettzell (Germany) are used, where systematic errors in the space geodetic techniques can be detected, assessed and removed on very short, well-known baselines. Within this paper we summarise results for three *intra-technique* co-location experiments in Wettzell. Firstly, an assessment of the GNSS to GNSS baselines in relation to the surveyed local ties shows discrepancies of up to 9 mm, for solutions based on the ionosphere-free linear combination. Secondly, an analysis of the short VLBI baseline shows that the use of a clock tie achieves a sub-mm agreement with respect to the local tie. And finally, initial results on the usage of differencing approaches on the short SLR baseline show that double-difference residuals are within ±4 mm. The results of this work show the potential of intra-technique studies on short baselines for the understanding of technique-specific biases and errors and for the monitoring of local ties.

Keywords

Co-location · Geodetic reference systems · Local ties · Space geodetic techniques

1 Introduction

The combination of the space geodetic techniques constituting the ITRF is performed using the local ties at fundamental sites (Ray and Altamimi 2005). However, multiple sites show discrepancies beyond the requirements of the Global Geodetic Observing System (GGOS): positions ≤1 mm and velocities ≤0.1 mm/yr (Rothacher et al. 2009). For instance, based on the tie discrepancies of the ITRF2014, Fig. 1 shows that the differences in east, north and up components for a GNSS-to-GNSS baseline surpasses largely the 1 mm

requirement at several sites (Altamimi et al. 2016). The same can be observed for baselines with GNSS to SLR, and GNSS to VLBI, with discrepancies in the cm-level. Some sites, such as the Geodetic Observatory Wettzell in Germany (Fig. 2), are equipped with more than one instrument of the same technique. Thus, very short baselines of the same technique can be formed. These short baselines provide the perfect opportunity to study technique-specific systematic and time-dependent biases, as the baselines are known precisely from terrestrial measurements (local ties), the relative atmospheric delays can be modelled and a common clock can be used. In particular for Wettzell, a VLBI short baseline, a SLR short baseline, and multiple GNSS short baselines are available. Within the scope of this project, several experiments with short baselines have been performed to continuously monitor

I. Herrera Pinzón (✉) · M. Rothacher
Swiss Federal Institute of Technology in Zurich (ETHZ), Zurich, Switzerland
e-mail: Ivan.Herrera@geod.baug.ethz.ch; Markus.Rothacher@ethz.ch

J. T. Freymueller, L. Sanchez (eds.), *Beyond 100: The Next Century in Geodesy*,
International Association of Geodesy Symposia 152, https://doi.org/10.1007/1345_2020_95

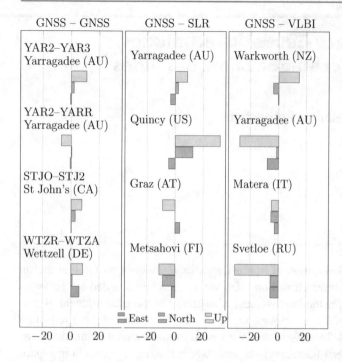

Fig. 1 ITRF2014 tie discrepancies [mm] at selected co-location sites according to Altamimi et al. (2016)

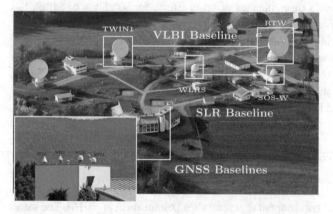

Fig. 2 Co-located instruments at the Geodetic Observatory Wettzell. Credits: IAPG TU-München

the local ties, and detect technique-specific systematic and time-dependent biases which are affecting the performance of the different geodetic techniques. The study of intra-technique experiments is expected to lead to a better understanding of system-specific error sources, biases and delays and constitutes an essential step for the realisation of a highly precise terrestrial reference frame that fulfils the demanding requirements of today and the future.

2   Multi-Year Analysis of GNSS Short Baselines at Co-location Sites

The first of these intra-technique co-location experiments deals with GNSS short baselines. For the network of GNSS stations in Wettzell, 15 years of GNSS data were reprocessed, using a tailored parameterisation, based on double differences with ambiguity fixing, with six different solutions: Single frequency L1 and L2, ionosphere-free linear combination (L3), with (TR) and without (NT) the estimation of relative tropospheric delays. The assumption is that for such a small distance and small height difference, tropospheric delays can be modelled (Beutler et al. 1987; Dilßner et al. 2008; Saastamoinen 1972), or cancelled out. This reprocessing yielded highly consistent time series, with repeatabilities for the east and north component below 1 mm, and 2 mm for the up component. Figure 3 shows the repeatability for the up component of the station WTZZ with respect to WTZR, for each investigated solution, where seasonal outliers associated with snow on the antennas, noise variations due receiver changes and, in general, site-specific events can be observed. This analysis shows that single-frequency solutions without estimation of the relative troposphere have better performance in terms of repeatabilities, than the linear combinations or the solutions with the estimation of relative troposphere. In particular, the solution L3-TR shows an

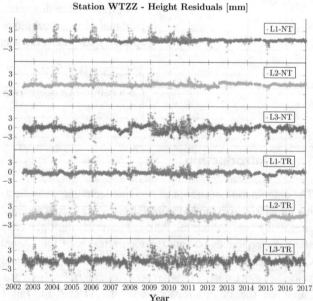

Fig. 3 Repeatabilities for the up component of station WTZZ (with respect to WTZR)

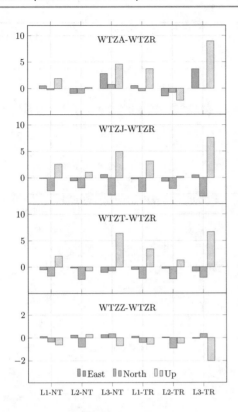

Fig. 4 Differences between GNSS-based baselines and local ties [mm] in Wettzell, at the epoch of the local tie

amplified noise and considerably larger outliers. It is worth mentioning that this solution is equivalent to that used for global solutions, hence the relevance of its characterisation.

In addition, we compare the GNSS-based solutions with the local tie at the epoch of the local tie. According to the log files of the stations, the Wettzell site has reported local ties with a precision better than 2 mm (IGS 2017). Figure 4 shows the differences for all the baselines at Wettzell, where the worst performance for the baseline WTZZ-WTZR is given by the L3 solution with troposphere estimates. The general performance at the site includes discrepancies up to 9 mm for the height component, when the estimation of the relative troposphere is involved. A detailed analysis for this and

several other co-location sites part of the ITRF2014 solution is shown in Herrera Pinzón and Rothacher (2018).

3 Assessment of the VLBI Short Baseline at Wettzell

The second type of intra-technique co-location experiment performed uses the short VLBI baseline at Wettzell, realised by the 20 m RTW telescope and the 13m TWIN1 telescope. We used 57 VLBI sessions of the IVS campaign, which contain the baselineRTW–TWIN1, between July 2015 and June 2016 (Behrend 2013). The local Wettzell baseline is not present after January 26th, 2016. Similarly to GNSS, four different approaches have been studied, where the modelling of the dry atmosphere, the solid Earth tides and ocean loading were common for all these solutions. The first approach (GLO) is a global solution, where all VLBI observations are used. Zenith wet delays are estimated as a piece-wise linear function with 2 h intervals using the wet VMF model for mapping. Receiver clock offsets are parametrised as a linear polynomial during the session, for each station except for RTW. The second processing approach (BAS) is a short baseline solution, where only the RTW-TWIN1 baseline observations are used. Receiver clock offsets are calculated each 24 h for TWIN1, for each session. The main feature of this approach is that the troposphere delays between the two stations are not estimated, based on the assumption that for such a small distance and small height difference, differences in tropospheric delays can be modelled, e.g. with the dry part of the Saastamoinen (1972) model. Similarly, the third approach (BA2) is also a baseline solution. However, in this solution zenith wet tropospheric delays are estimated piece-wise linearly with a time resolution of 2 h and mapped with the wet VMF model for RTW. Finally, for the outage of data in 2016, the (BA3) solution uses the station NYALES20, in Ny-Alesund (Svalbard, Norway) to connect the two Wettzell telescopes. Receiver clock offsets are defined as in the BA2 solution and zenith wet tropospheric delays are set up as for the GLO solution. Figure 5 shows the standard

Fig. 5 Time series of the standard deviation of the residuals of the VLBI processing for each investigated solution

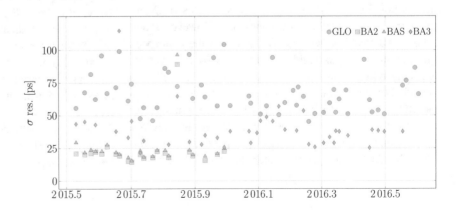

Fig. 6 Comparison of the
VLBI-based baseline and the
baseline derive from the local ties
between the telescopes at
Wettzell

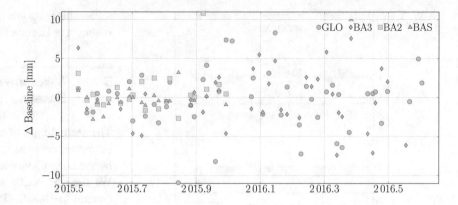

Fig. 6 Comparison of the VLBI-based baseline and the baseline derive from the local ties between the telescopes at Wettzell

deviation of the residuals of the estimation process, where the local solutions have evidently lower level of noise. A deeper explanation of the modelling and parameterisation used to obtain these results can be found in Herrera Pinzón et al. (2018). Based on these solutions, we performed the comparison of the VLBI-based solution and the local ties, regarding the baseline length (123.3070 m ± 0.7 mm) (Kodet et al. 2018). The differences obtained for these solutions have an overall satisfactory mean behaviour, with a mean over the time series below 1 mm, even for the global solution. However, the largest difference is the scatter of these time series: The global solutions have standard deviations of about 5 mm, while the local solutions display a standard deviation of around 1 mm.

Similarly to the processing of GNSS short baselines, the BAS solution (without estimation of relative troposphere) shows the best time series of results (Fig. 6). The mean of the differences over time for each solution are at the sub-mm level, namely GLO: −0.8 ± 4.9, BA3: −0.2 ± 4.6, BAS: −0.3 ± 0.8, BA2: −0.1 ± 1.3. A more comprehensive discussion of these results can be found in Herrera Pinzón et al. (2018).

4   Differencing Approaches for SLR Short Baselines

Forming differences is a standard approach in GNSS processing. But simultaneous SLR observations from one telescope to two satellites are impossible. However, Pavlis (1985) and Svehla et al. (2013) introduced the concept of quasi-simultaneity to build differences. Two observations are considered quasi-simultaneous if they lie within a specified time window. Figure 7 shows the concept of quasi-simultaneity for an SLR baseline, where time windows for the observation from telescope 2 with respect to telescope 1 are t_1 and t_2 for satellite 1 and 2, respectively. With this idea, the goal is to test the potential of the differencing

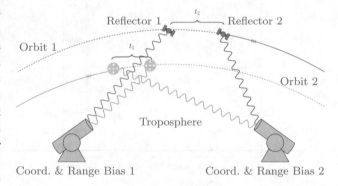

Fig. 7 Concept of quasi-simultaneity for the differencing SLR observations, together with the error sources targeted with these approaches

approaches, namely single- and double-differences, for the estimation of geodetic parameters. It is expected that single-difference observations from two stations to one satellite will remove biases related to the satellite orbit and the retro-reflectors. Similarly, quasi-simultaneous single-differences to two satellites can remove station-dependent range errors. These differences, together with the original ranges (zero-differences), are used to get estimates of both satellite- and station-specific error sources, so that systematic effects common to both stations can be identified at mm-level. Moreover, this approach can be potentially used to improve the processing of classical SLR observations of GNSS and LEO satellites, and to estimate accurate local ties. Initially, the residuals of the zero-difference processing are used to build the single- and double-differences. These observables are then used in a so-called zero test, where no geodetic parameters are estimated. Instead, coordinates of the stations are fixed to the local tie values and the atmospheric parameters are calculated with the standard model of Marini and Murray (1973). The zero-difference residuals for the short SLR baseline at Wettzell, realised by the telescopes WLRS and SOS-W, for July 3rd, 2018, are displayed in Fig. 8. Only GLONASS and Galileo satellites were used for this initial assessment.

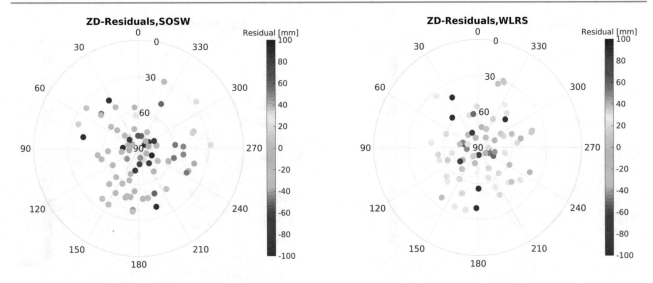

Fig. 8 Skyplot of the residuals of the zero-test, as seen at each SLR station, using the original observations (zero-differences)

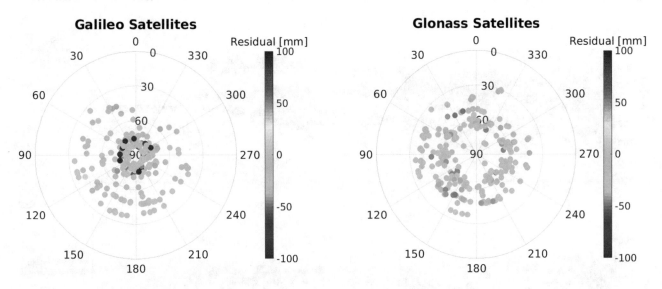

Fig. 9 Single-differences from 2 telescopes to 1 satellite. Left: Galileo satellites. Right: GLONASS satellites

4.1 Single-Difference Residuals [2 Telescopes to 1 Satellite]

Figure 9 shows the single-difference residuals grouped by constellation, allowing for mis-synchronisation of 3 h (quasi-simultaneity). This analysis reveals that the residuals are evidently biased, with a mean value of −26.3 mm, a value related to the range biases of WLRS and SOS-W. An extensive analysis of these biases is discussed in details in Riepl et al. (2019). Not only range biases can be observed within these differences, but, after removing the mean bias, errors associated to the orbits are also noticeable (Fig. 10), especially for Galileo satellites. The identification of orbital errors is therefore an advantage of this approach.

4.2 Single-Difference Residuals [1 Telescope to 2 Satellites]

Single-differences of residuals from the same telescope to two satellites can also be built. Allowing a quasi-simultaneity of 24 h, the time series of differenced residuals per station is depicted in Fig. 11. The blue coloured residuals indicate two Galileo satellites, the red coloured two GLONASS satellites, and the green colour is used when the difference is built using one satellite from each system. One feature stands out in the time series: the poor performance of some Galileo satellites, due to their orbit errors, produces the largest residuals throughout the time series. In particular Galileo satellites E01 and E05 show the largest residuals. This is observed for both

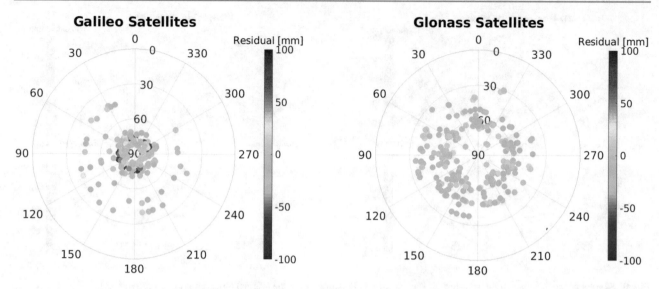

Fig. 10 Single-differences from 2 telescopes to 1 satellite, after removing the mean bias. Left: Galileo satellites. Right: GLONASS satellites

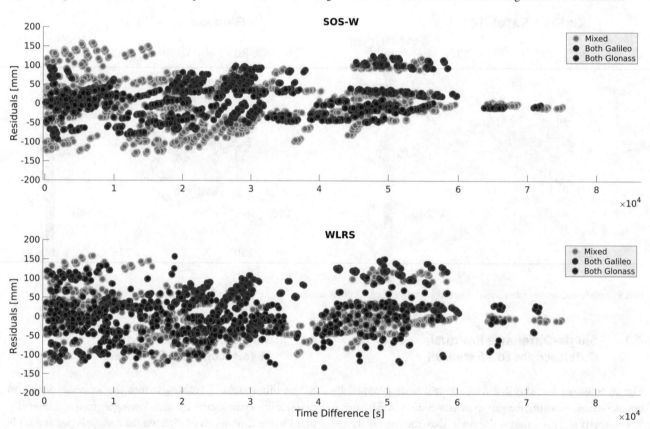

Fig. 11 Time series of single-differences from 1 telescope to 2 satellites. Top: Telescope SOS-W. Bottom: WLRS. The x-axis indicates the time difference between the two observations used to build the differences, namely the quasi-simultaneity

telescopes. This identification, and also the removal of orbital issues is a great advantage of the differencing approaches.

4.3 Double-Difference Residuals

Finally, in the same fashion, the double-difference residuals are built. Figure 12 shows all the possible differences that

can be built when allowing a quasi-simultaneity of 24 h. Based on the single differences from one telescope to two satellites, for SOS-W in the x-axis and WLRS in the y-axis, 63,452 differences were available. These residuals range from −10 to 10 cm, with a mean value of 1.2 mm and a scatter of 24 mm. This behaviour is heavily influenced by the bad performances of the aforementioned Galileo satellites. Considering only those double differences of GLONASS

Fig. 12 Double-differences of SLR residuals. The x-axis indicates the time difference allowed to build the single differences from the SOS-W telescope to two satellites. Similarly, the y-axis shows the time difference allowed to build the single-differences from the WLRS telescope to two satellites. Finally, the colour bar indicates the value of the residual

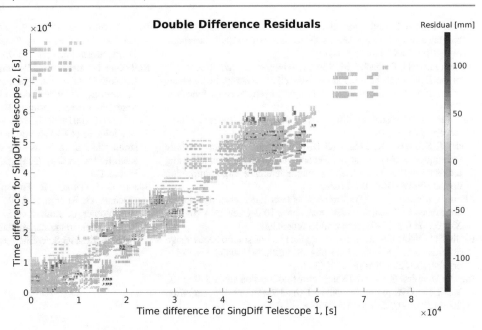

satellites during the first 30 min of the simultaneity ($t_1 <$ 30 & $t_2 <$ 30), the differenced residuals improve considerably in terms of scatter, with an standard deviation of 5.3 mm, with a mean value of -0.7 mm, for 150 differences. These results indicate that quasi-simultaneous SLR differences are feasible, and that it is possible to obtain double-difference residuals close to the sub-mm level. In turn, the use of SLR differenced observations constitutes a valuable observable for the estimation of geodetic parameters through SLR.

5 Conclusions and Outlook

The overarching goal of this work lies in determining the potential of intra-technique studies on short baselines for the understanding of technique-specific biases and errors and the monitoring of local ties. Experiments on GNSS to GNSS, SLR to SLR and VLBI to VLBI short baselines are assessed, where multiple local and environmental effects are investigated. In particular, the analysis of GNSS short baselines showed cm-level discrepancies with respect to local ties, for a processing strategy which is equivalent to that used in global solutions. On the other hand, the study of a short VLBI baseline showed mm to sub-mm agreement of the estimated baseline with the local tie. The benefits from the accurate and common timing estimation, for the determination of height and troposphere, were investigated. Finally, a concept for the differencing of SLR observables was studied, where mm-level double-difference residuals were found. Besides allowing the identification of station- and orbital biases, and based on the size of these residuals, this method is expected to be suitable for the estimation of geodetic parameters. These findings are expected to be extended in future experiments.

Replacing the estimation of clock corrections, by a Two Way Optical Time and Frequency System is expected to make the estimation of VLBI clock corrections unnecessary. The SLR differencing approaches are expected to be useful for the estimation of coordinates and the assessment of local ties. Finally inter-technique experiments on very short baselines, including GNSS and VLBI observations are foreseen, where the challenge will be the assessment of biases among the space geodetic techniques and the study of the benefits from a rigorous GNSS-VLBI combination of all common parameter types.

Acknowledgements This work has been developed within the project "Co-location of Space Geodetic Techniques on Ground and in Space" in the frame of the DFG funded research unit on reference systems, and founded by the Swiss National Foundation (SNF). Additionally, the authors would like to thank the team at the GO-Wettzell, in particular to Dr. Jan Kodet and Prof. Dr. Ulrich Schreiber, for their continuous support in the realisation of this work.

References

Altamimi Z, Rebischung P, Métivier L, Collilieux X (2016) ITRF2014: a new release of the international terrestrial reference frame modeling nonlinear station motions. J Geophys Res Solid Earth 121(8):6109–6131. http://dx.doi.org/10.1002/2016JB013098

Behrend D (2013) Data handling within the international VLBI service. http://doi.org/doi:10.2481/dsj.WDS-011

Beutler G, Bauersima I, Botton S, Gurtner W, Rothacher M, Schild-knecht T, Geiger A (1987) Accuracy and biases in the geodetic application of the global positioning system, vol 1. Springer, Berlin/Heidelberg/New York, pp 28–35

Dißner F, Seeber G, Wübbena G, Schmitz M (2008) Impact of near-field effects on the GNSS position solution. In: Proceedings of the international technical meeting, ION GNSS, Institute of Navigation

Herrera Pinzón I, Rothacher M (2018) Assessment of local GNSS baselines at co-location sites. J Geod 92(9):1079–1095. http://doi.org/10.1007/s00190-017-1108-9

Herrera Pinzón I, Rothacher M, Kodet J, Schreiber KU (2018) Analysis of the short VLBI baseline at the Wettzell observatory. Proceedings of the 10th IVS general meeting, Longyearbyen, Norway, June 3–8, 2018

IGS (2017) IGS stations list. http://www.igs.org/network/list.html. Accessed 01-05-2017

Kodet J, Schreiber K, Eckl J (2018) Co-location of space geodetic techniques carried out at the Geodetic Observatory Wettzell using a closure in time and a multi-technique reference target. http://doi.org/10.1007/s00190-017-1105-z

Marini J, Murray C (1973) Correction of laser range tracking data for atmospheric refraction at elevations above 10 degrees. (NASA-TM-X-70555):60 p. NASA technical memorandum

Pavlis EC (1985) On the geodetic applications of simultaneous range differences to LAGEOS. J Geophys Res (90):9431–9438. http://doi.org/doi:10.1029/JB090iB11p09431

Ray J, Altamimi Z (2005) Evaluation of co-location ties relating the VLBI and GPS reference frames. J Geod 79(4):189–195. http://dx.doi.org/10.1007/s00190-005-0456-z

Riepl S, Müller H, Mähler Sea (2019) Operating two SLR systems at the Geodetic Observatory Wettzell: from local survey to space ties. J Geod. https://doi.org/10.1007/s00190-019-01243-z

Rothacher M, Beutler G, Behrend D, Donnellan A, Hinderer J, Ma C, Noll C, Oberst J, Pearlman M, Plag HP, Richter B, Schöne T, Tavernier G, Woodworth PL (2009) The future global geodetic observing system. Springer, Berlin/Heidelberg, pp 237–272. https://doi.org/10.1007/978-3-642-02687-4_9

Saastamoinen J (1972) Atmospheric correction for the troposphere and stratosphere in radio ranging of satellites. In: The use of artificial satellites for geodesy. Geophysical Monograph, vol 15. AGU, Washington, D.C.

Svehla D, Haagmans R, Floberghagen R, Cacciapuoti L, Sierk B, Kirchner G, Rodriguez J, Wilkinson M, Appleby G, Ziebart M, Hugentobler U, Rothacher M (2013) Geometrical SLR approach for reference frame determination: the first SLR double-difference baseline. In: IAG Scientific Assembly 2013, Potsdam

Status of IGS Reprocessing Activities at GFZ

Benjamin Männel, Andre Brandt, Markus Bradke, Pierre Sakic, Andreas Brack, and Thomas Nischan

Abstract

Based on a large network of continuously operated GNSS tracking stations the International GNSS Service (IGS) has a valuable contribution for the realization of the International Terrestrial Reference System (ITRS). In order to contribute to its next realization, the IGS is preparing for a new reprocessing of the GNSS data from 1994 to 2020 including GPS, GLONASS, and – for the first time – Galileo. A first test campaign including single- and multi-system solutions for 2017 and 2018 was performed to derive consistent transmitter phase center corrections for all systems. Preliminary results of the test solutions derived at GFZ show well determined orbits with overlaps of 28 mm for GPS, 67 mm for GLONASS, and 40 mm for Galileo and an overall RMS of satellite laser ranging residuals for Galileo of 58 mm. Using multi-GNSS antenna calibrations (including also E5a and E5b calibrations) horizontal coordinate differences are almost zero between a GPS+GLONASS and a Galileo-only solutions. Due to the mixture of estimated (GPS, GLONASS) and measured (Galileo) transmitter phase center offsets a scale difference of 1.16 ± 0.27 ppb is found between both solutions which agrees nicely to results derived by other analysis centers.

Keywords

GNSS · Orbit determination · Reprocessing · Terrestrial Reference Frame

1 Introduction

To provide the best possible GNSS solution for the realization of the International Terrestrial Reference System, the Analysis Centers (ACs) of the International GNSS Service (IGS, Johnston et al. 2017) are preparing for a full reprocessing of GNSS data from 1994 to 2020. Like the previous efforts (`repro1` and `repro2`) the upcoming reprocessing will provide a fully consistent set of orbits, station coordinates and Earth rotation parameters derived with the best and most consistent models available. It is well known that in terms of reference frame parameters the most critical issues for GNSS are, firstly, the transmitter phase center offsets (which are highly correlated with the terrestrial scale, e.g., Zhu et al. 2003) and, secondly, the modeling of the solar radiation pressure on the orbits (main reason for draconitic period in geocenter results, e.g., Meindl et al. 2013). While trying to reduce or to solve both issues several additional topics have to be considered like the 13.63/13.66 day signal in GNSS time series (see for example Ray et al. 2013) or remaining modeling inconsistencies compared to other space geodetic techniques. Compared to the last reprocessing, new satellite systems like Galileo and BeiDou became almost fully operational. As their signals were tracked by an increasing number of IGS stations during the past years, the set of considered systems has to be redefined from GPS+GLONASS in `repro2` to an up-to-date multi-GNSS solution.

B. Männel (✉) · A. Brandt · M. Bradke · P. Sakic · A. Brack · T. Nischan
Helmholtz Centre Potsdam - German Research Centre for Geosciences, Potsdam, Germany
e-mail: benjamin.maennel@gfz-potsdam.de

© The Author(s) 2020
J. T. Freymueller, L. Sanchez (eds.), *Beyond 100: The Next Century in Geodesy*,
International Association of Geodesy Symposia 152, https://doi.org/10.1007/1345_2020_98

Fig. 1 Stability of selected IGS stations (according to IGS-ACS-1235 mail category c1 to c4 as well as GFZ stations in other categories): height repeatability derived by GPS-only PPP processing between 1999.0 and 2018.0

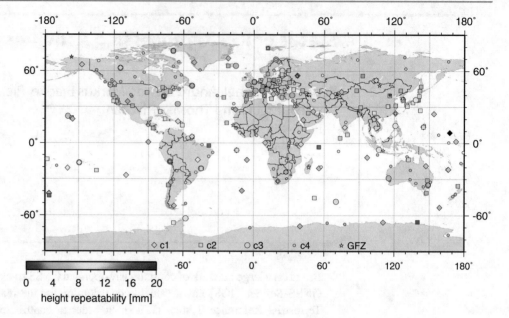

During the IGS Analysis Workshop 2019 held in Potsdam, Germany, the IGS ACs agreed to strive for an combined GPS (G), GLONASS (R), and Galileo (E) solution in the upcoming `repro3`. However, to avoid systematic distortions, so far missing, receiver antenna corrections for the Galileo signals E5a, E5b, and E6 and the GPS L5 frequency as well as consistent transmitter phase center offsets (PCOs) are required (see e.g. Schmid et al. 2016). Whereas the first issue was solved for many antenna types used in the IGS as Geo++ provided robot-based calibrations for these signals it was agreed to solve the second issue by setting up a test campaign. This campaign includes multi- and single-system solutions (if possible GRE, GR, G, R, and E) for 2017 and 2018 which will be used to derive phase center offsets for GPS and GLONASS based on the Galileo offsets which are known thanks to published chamber calibrations (GSA 2017). As these Galileo PCOs are measured – and not estimated from observations itself – they are independent of the terrestrial scale which has to be fixed to the ITRF scale otherwise (Schmid et al. 2007). Therefore, an independent GNSS scale will become available if the final `repro3` could be performed with this consistent set of re-estimated and calibrated PCOs. It was also agreed to run a second test campaign using the final `repro3` setup including the station selection as benchmark test before starting the processing tasks.

This paper summarizes the current reprocessing status at GFZ during the first test campaign and highlights preliminary outcomes. Section 2 describes investigations regarding station selection and testing some models. Initial results based on the different GFZ solutions in the first test campaign will be presented in Sect. 3. Section 4 provides an outlook to the upcoming tasks.

2 Data and Processing

This section discusses the processed data, the station network, and the selected models for the test campaign but also for the final reprocessing.

2.1 Data Selection

The station selection process is based on the pre-selection and station classification which was provided to the Analysis Centres by Paul Rebischung via the IGS AC mailing list (IGS-ACS-1235).[1] According to software and time capabilities we will process stations listed in Categories 1 (revised set of IGS14 core stations), 2 (stations with local ties to other techniques), 3 (redundant local tie stations), and 4 (remaining IGS14 stations) as well as IGS stations operated by GFZ placed in lower categories. In order to assess this selection we re-imported the whole data set into our archive with dedicated checks for formal correctness and consistency with the provided site logs. In addition, we processed the GPS observations from all selected IGS stations using the EPOS.P8 software in PPP mode to identify the stations temporal behavior.[2] The processing was done based on orbit and clock products derived within a GFZ internal reprocessing effort which we carried out in 2018 to derive consistent products in the IGS14 frame. Figure 1 shows

---

[1] Available also at http://acc.igs.org/repro3/repro3, accessed January 2020.

[2] Stations provided by other networks, like SONEL (3), OAFA (1), GREF (1), EPN (4), NGS (6), UNAVCO (2), are not considered for this initial assessment but will be processed in the final reprocessing.

the derived height coordinate repeatability for each station with the symbol identifying the station's category. Overall, 325 stations are contained while the average repeatability is 7.5 ± 1.9 mm (median is 7.2 mm). The GFZ contribution to the first test campaign was, however, processed using the station selection used in GFZ's operational processing (IGS final line, 220 stations) supported by additional 30 IGS stations selected to achieve a basic coverage for Galileo also in early 2017. However, the number of selected Galileo sites was still rather low for an independent Galileo-only solution. For January 1st, 2017 the number of stations was 222/137/68 for GPS, GLONASS, and Galileo while rising to 210/145/124 for Dec. 31st, 2018 due to ongoing station upgrades within the IGS tracking network. With respect to the number of Galileo satellites it might be interesting to know that the number of satellites increased from 13 to 22 within the same time span.[3]

2.2 Processing of the First Test Campaign

Table 1 provides the processing strategy applied for the test campaign. The same parametrization can be expected for the final reprocessing. However, the orbit parametrization might be modified for some satellite blocks as tests regarding the optimal setup are performed currently. In general, the settings follow the IERS Conventions 2010 (Petit and Luzum 2010) and the repro2 standards. Using the EPOS.P8 software GFZ provided four solutions (GRE, GR, G, and E).

2.3 Updating Models for Repro 3

According to the discussions between the IGS ACs and the IERS, several models have to be updated for the final reprocessing compared to Table 1. For computing the rotational deformation (pole tide) the linear mean pole will be used as adopted by the IERS in 2018. Regarding the gravity field, a static gravity field up to degree and order 12 was used whereas a time-variable gravity field should be used in the reprocessing. The ocean tides and ocean loading model will be updated to a more recent FES2014b model (Carrere et al. 2016). In order to consider high-frequency variations in Earth orientation parameters (EOP) it was agreed to use the model provided by Desai and Sibois (2016) instead of the model provided in the IERS Conventions.

---

[3]The four satellites launched in July 2018 are not included as they were not operational before January 2019.

Table 1 Summary of estimation and processing strategy (repro3 test campaign); time span 2017.0–2019.0; the used ANTEX was provided by A. Villiger and the IGS ANTEX working group (IGS-ACS-1233 mail)

Modeling and a-priori information	
Observations	Ionosphere-free linear combination formed by undifferenced GPS observations
Tropospheric correction	GPT2 meteo values mapped with VMF1 (Böhm et al. 2006)
Ionospheric correction	First order effect considered with ionosphere-free linear combination, second order effect corrected using the International Magnetic Reference Field (11th realization, Finlay et al. 2010)
GNSS phase center	Dedicated multi-GNSS ANTEX applied (igsR3_2057.atx)
Clock datum	Zero-mean condition for satellites and selected stations
Gravity potential	EGM2008 up to degree and order 12 (Pavlis et al. 2012)
Solid Earth tides	According to IERS 2010 Conventions (Petit and Luzum 2010)
Permanent tide	Conventional tide free
Ocean tide model	FES2004 (Lyard et al. 2006)
Ocean loading	FES2004 (Lyard et al. 2006)
Atmospheric loading	Tidal: S_1 and S_2 corrections (Ray and Ponte 2003)
High-frequent EOP model	Desai-Sibois model (Desai and Sibois 2016)
Mean pole tide	Linear mean pole as adopted by the IERS in 2018
Parametrization	
Station coordinates	No-net-rotation w.r.t. IGS14 (Rebischung and Schmid 2016)
Troposphere	Zenith wet delays for 0.5 h intervals; two gradient pairs per station and day
GPS orbit modeling	Six initial conditions + nine ECOM2 parameters, pulses at 12 h
Earth rotation	Rotation pole coordinates, pole-rates and LOD for 24 h intervals, UT1 tightly constrained to a priori Bulletin A
Receiver clock	Pre-eliminated every epoch, ISB per station for Galileo, per station and satellite for GLONASS
Satellite clocks	Epoch-wise estimated
GNSS ambiguities	Ambiguity fixing for GPS and Galileo
Antenna Phase Center	Estimated for GPS, GLONASS, and Galileo but tightly constrained to values given in ANTEX

3 Initial Results

Within this section initial results derived within the test campaign will be discussed with respect to the derived orbits and stations coordinates. Figure 2 shows the orbit misclosures (orbit overlap error) for all satellites processed in

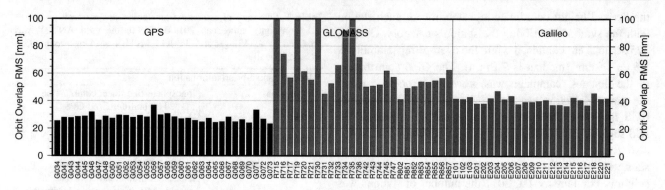

Fig. 2 Orbit overlaps errors; daily overlap RMS averaged for time span 2017.0–2019.0

Table 2 Orbit overlaps: average and standard deviation over all satellites and weeks; time span 2017.0–2019.0; unit: mm

Solution	GPS	GLONASS	Galileo
Full sol. (GRE)	27.8 ± 2.9	67.6 ± 26.3	40.5 ± 2.9
GPS+GLO (GR)	28.5 ± 2.8	65.4 ± 27.4	–
GPS (G)	28.4 ± 2.8	–	–
GAL (E)	–	–	40.5 ± 3.6

Table 3 Orbit overlaps: large GLONASS values; time span 2017.0–2019.0; as reference the overlaps provided in Table 1 of Dach et al. (2019) are given; unit: mm

SVN	PRN	GRE	GR	Dach et al. (2019)
R715	R14	157.2	157.9	–
R719	R20	114.3	114.0	–
R730	R01	103.5	97.2	103
R734	R05	91.2	89.7	112
R735	R24	110.4	111.0	118

the full solution (GRE). The overlaps are computed for 2 h each while estimating transformation parameters between the two orbit solutions. An averaged RMS of 28 mm is achieved for the GPS satellites. As shown in Table 2 the RMS is independent of the solution type (GRE, GR, or G) for GPS with differences of 1 mm. It is obvious, that the GPS orbits are not downgraded in terms of overlap errors by adding other systems to the solutions. For GLONASS a larger mean RMS of 67 mm is observed while five satellites exceed 90 mm overlap error (see Table 3). As shown in this table, Dach et al. (2019) reported similarly large overlaps for two of the three satellites (R715 and R719 are not contained in Table 1 of Dach et al. (2019)). Without these satellites the remaining average is 57.8 mm. Galileo orbit overlaps are in general larger compared to GPS mis-closures. Without large variations between the satellites an averaged RMS of 40 mm is determined for Galileo which is also achieved by the Galileo-only solution.

In order to further asses the satellite orbits a validation based on satellite laser ranging (SLR) was performed. While fixing the SLR telescope positions to their ITRF2014 coordi-nates (Altamimi et al. 2016) and estimating no other param-eters, the derived residuals (i.e., the differences between observed and computed ranges) provide insights into the absolute orbit accuracy. The number of SLR observations varies between 350 normal points (E217 and E218) and more than 17,000 (R802, R853). On average 5,230 normal points are collected per satellite within the 2 years. During the processing 3.6 % and 2.1 % of the observations where excluded as outliers for GLONASS and Galileo, respectively. Figure 3 shows the derived statistics. The determined biases, i.e., mean values over all residuals, reach up to 25 mm for GLONASS with some large variations between the satellites. For R856 with only 742 observations a larger bias of 44 mm is determined. In general, remaining biases indicate system-atics contained in the orbits (or applied sensor offsets). In the current solution positive biases are visible for most of the GLONASS satellites which needs further investigations. The biases for Galileo are small (few millimeters) but also mostly positive for the whole constellation. The larger bias of −13 mm for E218 might be related to the low number of 350 SLR observations available for this satellite launched in December 2017. The standard deviation of all SLR residuals reaches 91.5 ± 13.3 mm for GLONASS and 58.3 ± 8.7 mm for Galileo.

With respect to the different solutions a comparison of the station coordinates is very important. In theory, the estimated station coordinate should be independent of the processed GNSS. However, it was shown for example by Villiger et al. (2019) that one has to expect significant coordinate differ-ences between system-specific solutions. These differences are mostly related to the considered antenna corrections for transmitter and receiver. As stated earlier, transmitter PCOs are either estimated in a global adjustment or – as for Galileo – chamber calibrated. In addition, robot-calibrations of receiving antennas were not available for several GNSS signals so far. Figure 4 shows the mean differences (and standard deviation) in North, East, and height coordinates derived in a combined (GR) and a Galileo-only solution using the provided multi-GNSS antenna corrections. It has to

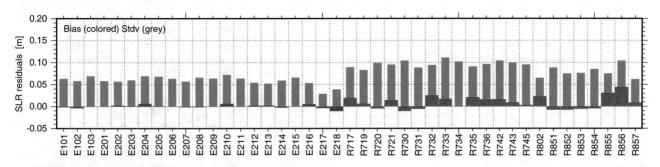

Fig. 3 SLR residuals: mean and standard deviation for time span 2017.0–2019.0

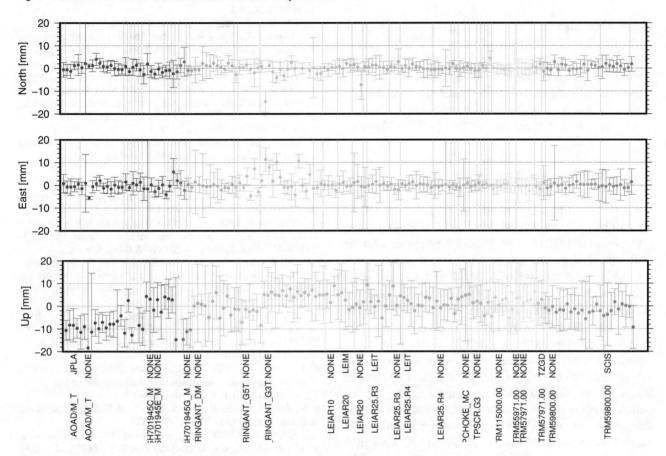

Fig. 4 Coordinate differences: GR-E mean and standard deviation for time span 2017.0–2019.0; from top to bottom: North, East, Up; sorted and color-coded by antenna types; antenna names are provided for types with at least three results, in addition a seven parameter Helmert transformation was applied

be mentioned that a seven parameter Helmert transformation was applied to determine also global systematics. The differences are sorted according to the antenna type of each station. However, some stations included in the processing are equipped with antennas having only L1/L2 calibrations (like the AOAD antennas). For these antennas larger height differences can be observed. In North and East direction overall no offset is visible with differences clearly below 5 mm. Larger differences are visible only for stations equipped with the JAVRINGANT_G3T and JAVRINGANT_G5T antennas where some stations show differences of 10 mm in East direction (two stations showed also differences larger than 20 mm in the North direction). Overall the coordinate differences agree well to differences computed from the test solutions provided for example by CODE and TU Graz. The height component shows, as expected, larger standard deviations but almost no significant offsets between the GR and the Galileo solution except for antennas with only L1/L2 calibrations which show differences of around −10 mm. Instead of comparing height differences, Fig. 5 shows the scale estimated as part of a Helmert transformation between the GR and the E solution (E-GR). Over the assessed 2 years, a scale difference of 1.16 ± 0.27 ppb was found. Due to the lower number of stations in the Galileo-only solution the

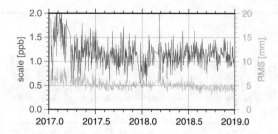

Fig. 5 Scale and RMS of the transformation between GR and E solution for time span 2017.0–2019.0

variation and also the RMS of the transformation is larger in the first half of 2017 compared to the following period. However, probably related to the lower number of Galileo satellites a somehow larger scale is derived for the first months in 2017. Again, the derived scale agrees well to the scale estimated between the solutions provided for example by CODE (1.10 ± 0.21 ppb) or by ESA (1.09 ± 0.18 ppb).

4 Summary and Outlook

The presented `repro3` test campaign was performed at GFZ as contribution to the re-determination of transmitter phase center offsets for GPS and GLONASS in preparation for the final reprocessing. The current setup for this reprocessing is a three-system solution (GPS, GLONASS, Galileo) while some models like the time-variable gravity field are still in discussion between the IGS ACs and the IERS. The preliminary results show acceptable overlap errors for the individual satellites with on average 28 mm, 67 mm and 40 mm for GPS, GLONASS, and Galileo. An SLR orbit validation revealed also good orbit quality without significant biases for the Galileo satellites. Comparing the derived station coordinates a good agreement between the GR- and the Galileo-only solution can be found in the horizontal components with a scale difference of around 1.1 ppb. According to the reprocessing schedule, a second test campaign will be performed after final decisions on the models. In addition, a few open questions have to be addressed like the reason for large orbit overlap errors for some GLONASS satellites or the coordinate difference for some stations equipped with `JAVRINGANT_G3T` and `JAVRINGANT_G5T` antennas.

References

Altamimi Z, Rebischung P, Métivier L, Collilieux X (2016) ITRF2014: a new release of the International Terrestrial Reference Frame modeling non-linear station motions. J Geophys Res 121:6109–6131. https://doi.org/10.1002/2016JB013098

Böhm J, Werl B, Schuh H (2006) Troposphere mapping functions for GPS and VLBI from European Centre for medium-range weather forecasts operational analysis data. J Geophy Res 111(B2):B02406. https://doi.org/10.1029/2005JB003629

Carrere L, Lyard F, Cancet M, Guillot A, Picot N (2016) FES 2014, a new tidal model – validation results and perspectives for improvements. In: Presentation to ESA living planet conference, Prague 2016

Dach R, Susnik A, Grahsl A, Villiger A, Schaer S, Arnold D, Prange L, Jäggi A (2019) Improving GLONASS orbit quality by re-estimating satellite antenna offsets. Adv Space Res 63(12):3835–3847. https://doi.org/10.1016/j.asr.2019.02.031

Desai SD, Sibois AE (2016) Evaluating predicted diurnal and semidiurnal tidal variations in polar motion with GPS-based observations. J Geophys Res Solid Earth 121(7):5237–5256. https://doi.org/10.1002/2016JB013125

Finlay CC, Maus S, Beggan CD, Bondar TN, Chambodut A, Chernova TA, Chulliat A, Golovkov VP, Hamilton B, Hamoudi M, Holme R, Hulot G, Kuang W, Langlais B, Lesur V, Lowes FJ, Lühr H, Macmillan S, Mandea M, McLean S, Manoj C, Menvielle M, Michaelis I, Olsen N, Rauberg J, Rother M, Sabaka TJ, Tangborn A, Toffner-Clausen L, Thebault E, Thomson AWP, Wardinski I, Wei Z, Zvereva TI (2010) International geomagnetic reference field: the eleventh generation. Geophys J Int 183(3):1216–1230. https://doi.org/10.1111/j.1365-246X.2010.04804.x

GSA (2017) Galileo satellite metadata. https://www.gsc-europa.eu/support-to-developers/galileo-satellite-metadata. Accessed 19 Aug 2019

Johnston G, Riddell A, Hausler G (2017) The international GNSS service. Springer International Publishing, Cham, pp 967–982. https://doi.org/10.1007/978-3-319-42928-1

Lyard F, Lefevre F, Letellier T, Francis O (2006) Modelling the global ocean tides: modern insights from FES2004. Ocean Dyn 56(5–6):394–415. https://doi.org/10.1007/s10236-006-0086-x

Meindl M, Beutler G, Thaller D, Dach R, Jäggi A (2013) Geocenter coordinates estimated from GNSS data as viewed by perturbation theory. Adv Space Res 51:1047–1064. https://doi.org/10.1016/j.asr.2012.10.026

Pavlis NK, Holmes SA, Kenyon SC, Factor JK (2012) The development and evaluation of the Earth Gravitational Model 2008 (EGM2008). J Geophys Res Solid Earth 117(B4). https://doi.org/10.1029/2011JB008916

Petit G, Luzum B (2010) IERS conventions (2010). IERS Technical Note 36. Verlag des Bundesamts für Kartographie und Geodäsie, Frankfurt am Main, iSBN 3-89888-989-6

Ray R, Ponte R (2003) Barometric tides from ECMWF operational analyses. Ann Geophys 21(8):1897–1910

Ray J, Griffiths J, Collilieux X, Rebischung P (2013) Subseasonal GNSS positioning errors. Geophys Res Lett 40(22):5854–5860. https://doi.org/10.1002/2013GL058160

Rebischung P, Schmid R (2016) IGS14/igs14.atx: a new framework for the IGS products. In: Presentation to AGU fall conference, San Francisco, 2016

Schmid R, Steigenberger P, Gendt G, Ge M, Rothacher M (2007) Generation of a consistent absolute phase center correction model of GPS receiver and satellite antennas. J Geod 81(12):781–798. https://doi.org/10.1007/s00190-007-0148-y

Schmid R, Dach R, Collilieux X, Jäggi A, Schmitz M, Dilssner F (2016) Absolute IGS antenna phase center model igs08.atx: status and potential improvements. J Geod 90(4):343–364. https://doi.org/10.1007/s00190-015-0876-3

Villiger A, Dach LPA, Zimmermann F, Kuhlmann H, Jäggi A (2019) Consistency of antenna products in the MGEX environment. In: IGS Workshop 2018, Wuhan

Zhu YS, Massmann FH, Yu Y, Reigber C (2003) Satellite antenna phase center offsets and scale errors in GPS solutions. J Geod 76(11):668–672. https://doi.org/10.1007/s00190-002-0294-1

A Wavelet-Based Outlier Detection and Noise Component Analysis for GNSS Position Time Series

Kunpu Ji and Yunzhong Shen

Abstract

Various signals of crustal deformation and mass loading deformation are contained in a GNSS position time series. However, a GNSS position time series is also polluted by outliers and various colored noise, which must be reasonably modelled before estimating deformation signals. Since temporal signals of the GNSS position time series are non-linear and complicated, we propose a wavelet-based approach for outlier detection, which first retrieves the temporal signals from the GNSS position time series by using wavelet analysis, and then detect outliers in the residual position time series by using the interquartile range. After the detected outliers are eliminated from the residual time series, the noise components, including white noise and flicker noise, are estimated by using MINQUE approach. Our proposed approach is used to process the real GNSS position time series of the Crustal Movement Observation Network of China (CMONOC) over the period spanning 1999–2018. The results demonstrate that our approach can detect the outliers more efficiently than the traditional approach, which retrieves the temporal signals by using a functional model with trend and periodic variations. As a result, the noise components estimated with our proposed approach are smaller than those with the traditional approach for the GNSS position time series of all CMONOC stations.

Keywords

GNSS position time series · MINQUE · Outlier detection · Wavelet analysis

1   Introduction

The position time series of various GNSS station networks are widely used to study the geophysical phenomena such as plate tectonics (Tobita 2016), post-glacial rebound (Peltier et al. 2015) and sea level change (Wöppelmann et al. 2007). Due to multipath effects, station-related error (such as electromagnetic interference), orbital anomaly and other unknown reasons, outliers inevitably exist in the GNSS position time series, which will lead to bias estimates in both functional and stochastic models (Koch 1999;

Khodabandeh et al. 2012). There are several approaches for detecting outliers in the GNSS position time series, such as three sigma method (3σ) (Mao et al. 1999), Bayesian method (Zhang and Gui 2013), as well as Detection Identification Adaptation (DIA) procedure (Amiri-Simkooei et al. 2015). Besides these methods, the window-opening test algorithm based on the Interquartile Range (IQR) statistic is another commonly used approach for outlier detection in the GNSS position time series (Nikolaidis 2002; Li and Shen 2018). This algorithm is fast and robust since the median and IQR values of a time series are less affected by outliers. Due to its superior performance, the outlier detection approach based on IQR criterion has been widely applied in the open source software or packages for GNSS position time series analysis, such as iGPS (Tian 2011), Hector (Bos et al. 2013) and TSAnalyzer (Wu et al. 2017).

K. Ji · Y. Shen (✉)
College of Surveying and Geo-Informatics, Tongji University, Shanghai, China
e-mail: yzshen@tongji.edu.cn

© The Author(s) 2020
J. T. Freymueller, L. Sanchez (eds.), *Beyond 100: The Next Century in Geodesy*,
International Association of Geodesy Symposia 152, https://doi.org/10.1007/1345_2020_106

Apart from outliers, the GNSS position time series are also polluted by temporally correlated noise, which is a combination of white noise plus flicker noise (Mao et al. 1999). The maximum likelihood estimation (MLE) is widely used for estimating the noise components of a GNSS time series. Besides, the existing methods of Variance Component Estimation (VCE), such as Helmert (1907), Minimum Norm Quadratic Unbiased Estimation (MINQUE) (Rao 1971), Best Invariant Quadratic Unbiased Estimation (BIQUE) (Koch 1999), as well as LS_VCE (Teunissen and Amiri-Simkooei 2008), are identical under the normal distribution (Teunissen and Amiri-Simkooei 2008). Therefore we use MINQUE method to estimate noise components in this paper.

The traditional least squares (LS) outlier detection based on IQR criterion (LS_IQR) and noise component estimation based on MINQUE method (LS_MINQUE) are all based on the harmonic functional model (Nikolaidis 2002) in which a position time series is described as a combination of linear trend, quasi-annual and semi-annual signals with constant amplitude and phase. However, the amplitudes and phases of seasonal variation signals in GNSS position time series also vary slightly over time due to the variation of surface-mass loading (Blewitt and Lavallée 2002), atmospheric and hydrological loadings (Bogusz and Figurski 2014). Consequently, a harmonic model isn't sufficient to reflect the nonlinear variation signals of GNSS position time series, especially the time-varying seasonal variation due to the irrationality of the model itself. Therefore, when a harmonic functional model is used to describe the GNSS position time series, the LS residuals still contains partial signal, which will affect the performance of outlier detection and lead to imprecise estimation of noise components. For this reason, we propose a wavelet-based algorithm for outlier detection and noise component analysis, which extracts the time variable signals by wavelet analysis and thereby named as WA_IQR and WA_MINQUE for the correspondent outlier detection and noise component algorithm. The remainder of the paper is organized as follows. Section 2 presents the main methodology, including dyadic wavelet analysis, outlier detection based on IQR criterion and noise component estimation using MINQUE method. Section 3 presents the results of real data analysis of CMONOC over the period from 1999 to 2018, and conclusions are summarized in Sect. 4.

2   Methodology

2.1   Dyadic Wavelet Analysis

When $\varphi(t)$ is denoted as a basic wavelet function, a set of wavelet functions can be derived by means of dilation a and

translation b of $\varphi(t)$ as (Daubechies 1992)

$$\varphi_{a,b}(t) = \frac{1}{\sqrt{a}} \varphi \left(\frac{t-b}{a} \right) \tag{1}$$

Taking $a = 2^j$, $b = 2^j k$, where j, k are integers, we can obtain the dyadic wavelet functions as

$$\varphi_{j,k}(t) = 2^{-j/2} \varphi \left(2^{-j} t - k \right) \tag{2}$$

For a discrete time series $\mathbf{x} = [x_0 \ x_1 \ \cdots \ x_{N-1}]^T$, its j-th dyadic wavelet transform is defined as (Walnut 2013)

$$w(j,k) = \sum_{i=0}^{N-1} x_i \int_{S_i} \varphi_{j,k}(t) dt \tag{3}$$

where $w(j,k)$ is the k-th value of j-th wavelet coefficient and S_i is the i-th sampling interval. Rewriting Eq. (3) with vector and matrix form as

$$\mathbf{w}_j = \mathbf{W}_j \mathbf{x} \tag{4}$$

where $\mathbf{w}_j$ is a vector of j-th wavelet coefficient with the size of $n_j = N/2^{j+1}$ and $\mathbf{W}_j = \left[\mathbf{W}_{j,0}^T \ \mathbf{W}_{j,1}^T \cdots \mathbf{W}_{j,n_{j-1}}^T \right]^T$ is j-th wavelet transform matrix with the size of $n_j \times N$, where, $\mathbf{W}_{j,k} = \left[\int_{S_0} \varphi_{j,k}(t) dt \ \int_{S_1} \varphi_{j,k}(t) dt \cdots \int_{S_{N-1}} \varphi_{j,k}(t) dt \right]$.

Stacking the wavelet coefficients from small to large scale and subjoining the scale coefficient $\mathbf{v}_{J-1}$, where J denotes the number of layers to be decomposed. For the GNSS position time series, the reconstructed seventh and eighth components of basic wavelet function represent time-varying signals with periods of about 182 and 365 days, which denote the semi-annual and annual signals (Klos et al. 2018), respectively. For this reason, we take J as 8. Then we obtain the wavelet transform of $\mathbf{x}$ in matrix form as

$$\mathbf{w} = \mathbf{W}\mathbf{x} \tag{5}$$

where $\mathbf{w} = \left[\mathbf{w}_0^T \cdots \mathbf{w}_{J-1}^T \mathbf{v}_{J-1}^T \right]^T$, $\mathbf{W} = \left[\mathbf{W}_0^T \cdots \mathbf{W}_{J-1}^T \mathbf{V}_{J-1}^T \right]^T$. $\mathbf{V}_{J-1}$ is the scale transform matrix, which is orthogonal to $\mathbf{W}_j$ and the wavelet transform matrix $\mathbf{W}$ is a standard orthogonal matrix. The original time series $\mathbf{x}$ can be reconstructed by the wavelet coefficients and transform matrix as follows:

$$\mathbf{x} = \mathbf{W}^T \mathbf{w} = \sum_{j=0}^{J-1} \mathbf{d}_j + \mathbf{a}_{J-1} \tag{6}$$

where $\mathbf{d}_j = \mathbf{W}_j^T \mathbf{w}_j$ represents the j-th detail component and $\mathbf{a}_{J-1} = \mathbf{V}_{J-1}^T \mathbf{v}_{J-1}$ represents the appropriate component of the time series.

2.2 Outlier Detection in Residual with IQR

The original time series $\mathbf{x}$ can be decomposed into components of different frequencies which represent either signal or noise after multi-resolution analysis (Mallat 1988). The signal and noise can be separated by the correlation coefficient method (Zhang et al. 2018), which calculates the correlation coefficient between the original time series and the reconstructed component of each layer, and the layer where the correlation coefficient firstly appears local minimum is considered to be the boundary layer. The correlation coefficient between $\mathbf{x}$ and i-th reconstructed component d_i can be calculated as

$$R(x, d_i) = \frac{\sum_{t=1}^{N}(x_t - \overline{x})\left(d_{t,i} - \overline{d}_i\right)}{\sqrt{\sum_{t=1}^{N}(x_t - \overline{x})^2}\sqrt{\sum_{t=1}^{N}\left(d_{t,i} - \overline{d}_i\right)^2}} \quad (7)$$

where x_t and $d_{t,i}$ represent t-th element of $\mathbf{x}$ and d_i, $\overline{x}$ and $\overline{d}_i$ represent the average value of x and d_i, respectively.

After the multi-resolution analysis of the original time series, we obtain the residual vector $\mathbf{v}$, in which outliers are mostly reflected. Sorting residual in ascending order, and then dividing it into several equal parts with the window length L, which was commonly taken as 182 (Nikolaidis 2002; Wu et al. 2017). Performing a window check on each part of the data set using the Z-ratio statistic (Nikolaidis 2002).

$$Z = \frac{v_i - med\left(v_{i-L/2}, v_{i+L/2}\right)}{IQR\left(v_{i-L/2}, v_{i+L/2}\right)} \quad (8)$$

where v_i represents the i-th residual, $med(*)$ and $IQR(*)$ denote the operators for computing the median and interquartile range of a series, respectively. According to IQR criterion (Nikolaidis 2002; Bos et al. 2013), when $Z > 3$, the i-th value of the original time series is detected as an outlier.

2.3 Noise Component Estimation Using MINQUE Approach

After the outliers are detected and then eliminated, the noise amplitudes of residual time series, including white noise and flicker noise are estimated by MINQUE method. The fundamental equation of variance component estimation (VCE) is (Li et al. 2010)

$$\mathbf{R}\boldsymbol{\Sigma}_\mathbf{y}\mathbf{R}^T = \mathbf{v}\mathbf{v}^T \quad (9)$$

where $\mathbf{v} = \mathbf{R}\mathbf{y}, \mathbf{R} = \mathbf{I} - \mathbf{A}\left(\mathbf{A}^T\boldsymbol{\Sigma}_y^{-1}\mathbf{A}\right)^{-1}\mathbf{A}^T\boldsymbol{\Sigma}_y^{-1}$, $\mathbf{A}$ is the coefficient matrix of the observational equation. The covariance matrix $\boldsymbol{\Sigma}_\mathbf{y}$ is a combination of two cofactor matrices for white noise and flicker noise as

$$\boldsymbol{\Sigma}_\mathbf{y} = \sigma_w^2\mathbf{I} + \sigma_f^2\mathbf{Q}_f \quad (10)$$

where σ_w^2, σ_f^2 are the white and flicker noise components to be estimated, $\mathbf{Q}_f$ is the cofactor matrix of flicker noise. For the calculation of $\mathbf{Q}_f$, one can refer to Mao et al. (1999).

According to the MINQUE estimation by Rao (1971), the equation to compute the white and flicker noise components is given as follows

$$\mathbf{N}\boldsymbol{\theta} = \mathbf{q} \quad (11)$$

where, $\boldsymbol{\theta} = \left(\sigma_w^2, \sigma_f^2\right)^T$. $\mathbf{N}$ is a 2×2 matrix and $\mathbf{q}$ is a 2 vector, the elements are given by

$$
\begin{aligned}
&n_{11} = tr\left(\mathbf{W}\mathbf{W}\right), \quad n_{12} = n_{21} = tr\left(\mathbf{W}\mathbf{W}\mathbf{Q}_f\right), \\
&n_{22} = tr\left(\mathbf{W}\mathbf{Q}_f\mathbf{W}\mathbf{Q}_f\right), \quad q_1 = \mathbf{v}^T\mathbf{W}\mathbf{W}\mathbf{v}, \\
&q_2 = \mathbf{v}^T\mathbf{W}\mathbf{Q}_f\mathbf{W}\mathbf{v}
\end{aligned} \quad (12)
$$

where $\mathbf{W} = \boldsymbol{\Sigma}_\mathbf{y}^{-1}\mathbf{R}$, $tr(*)$ is the operator for computing the trace of a matrix. Since $\mathbf{R}$ contains unknown noise components, Eq. (11) needs to be iteratively solved with given initial value of noise components.

3 Real GNSS Position Time Series Analysis

The real position time series of 27 permanent GNSS stations of CMONOC are processed with our proposed approach and their locations are shown in Fig. 1. All the GNSS position

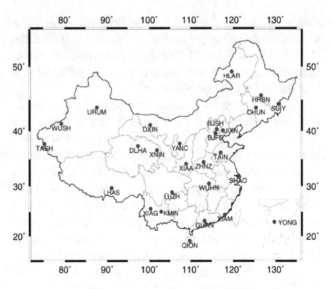

Fig. 1 Geographic locations of 27 stations in CMONOC (Shen et al. 2014)

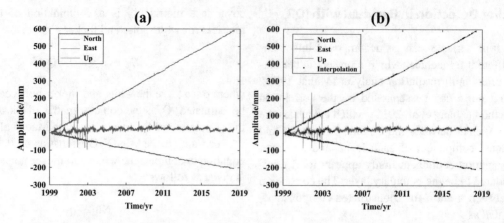

Fig. 2 Position time series of BJFS station. (**a**): Original one; (**b**): After missing values complemented and discontinuities corrected

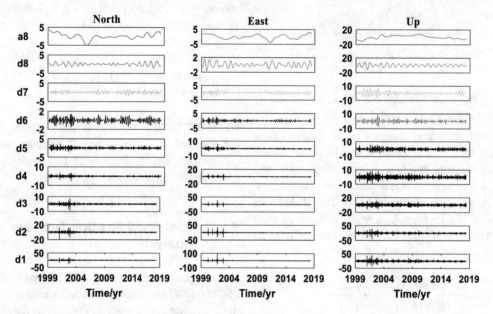

Fig. 3 Reconstructed 9 layer's components of BJFS station using wavelet analysis

time series are processed by a homogeneous state-of-the-art method using the processing package GAMIT/GLOBK (Ver.10.4) in the frame of ITRF 2000 (see processing details in ftp://ftp.cgps.ac.cn/doc/processing_manual.pdf and download data in http://www.cgps.ac.cn/).

Figure 2a presents position time series of Up, North, and East coordinates for BJFS station and it shows that position time series of three coordinates contain some outliers. Wavelet analysis requires that involved time series should be stable and equally spaced (Walnut 2013), however missing data inevitably occur in the position time series (Shen et al. 2014). We adopt the iterative interpolation scheme to handle data missing problem. Besides, some abrupt changes called discontinuities or offsets occur in the GNSS position time series due to various reasons such as brakes in station operation and change of antennas. Vitti (2012) provided a tool (*sigseg*) for the detection of position discontinuities in geodetic time series based on Blake-

Zisserman variational model. This tool is used to detect and repair the discontinuities in position time series. The new position time series after complementing the missing values and correcting the discontinuities are presented in Fig. 2b.

3.1 Signal and Noise Separation

The detrend BJFS time series in Fig. 2b is then decomposed with coif-5 wavelet, and the reconstructed components of each layer are presented in Fig. 3 and correlation coefficients between the original time series and the reconstructed component of each layer are presented in Table 1. Signals extracted by WA and LS estimation are presented in Fig. 4. Obviously, WA can well capture the nonlinear variation of position time series, while LS estimation based on harmonic model characterizes the nonlinear variation as a periodic

Table 1 Correlation coefficient between the original time series and reconstructed component of each wavelet layer

d_i	$R(x, d_i)$		
	North	East	Up
d_1	0.4375	0.6102	0.3741
d_2	0.3276	0.4430	0.3206
d_3	0.2400	0.3424	0.2280
d_4	0.1935	0.1900	0.2007
d_5	0.1825	0.1738	**0.1686**
d_6	**0.1738**	**0.1544**	0.1909
d_7	0.2119	0.1700	0.2377
d_8	0.2979	0.1809	0.4618

Bold values stand for the boundary layers of three components of BJFS station

signal with constant amplitude, which is clearly inconsistent with the actual change.

3.2 Outlier Detection

The IQR criterion is used to detect outliers in the residuals of three coordinates by WA and LS estimation, and results are presented in Fig. 5. Obviously, WA_IQR can detect much more outliers than LS_IQR. In Fig. 5, LS_IQR fails to detect a lot of outliers, especially in the epochs of the non-stationary part, which are caused by the poor fitting to the harmonic model. Figure 5 also presents the detected outliers by the 3σ method, it seems that the 3σ method can only detect a few outliers. The new time series after eliminating outliers from the original position time series are presented in Fig. 6, from which we can see that more outliers remain in the LS_IQR and the 3σ detected time series (i.e. between 1999 and 2003) than WA_IQR detected time series. However, none of them can recognize some outliers, of which the magnitude is quite small (i.e. outliers near epoch of 2015).

Figure 7 presents the proportion of detected outliers in position time series of 27 stations for three coordinates. For the BJFS station, the proportion of detected outliers for the whole data for three coordinates are 0.77%, 0.19% and 0.84% by 3σ, 1.78%, 1.47% and 2.11% by LS_IQR, and 4.50%, 5.55% and 3.65% by WA_IQR, respectively. From the remaining stations in Fig. 7, we can clearly see that WA_IQR can detect more outliers than LS_IQR and 3σ for all stations, the mean detected proportion of 27 stations are 0.16%, 0.50% and 0.39% by 3σ, 1.62%, 1.92% and 1.62% by LS_IQR, and 4.61%, 4.65% and 2.59% by WA_IQR, respectively.

3.3 Noise Components Estimation

After detected outliers are eliminated, WA_MINQUE and LS_MINQUE are employed to estimate noise components for the 27 stations, the results are presented in Figs. 8 and 9. The noise amplitude estimates of σ_w and σ_f in Figs. 8 and 9 clearly show that the noise component of Up coordinate is much larger than those of horizontal coordinates, and the flicker noise is larger than white noise. Therefore, as confirmed by Amiri-Simkooei et al. (2007), flicker noise is dominant in the GNSS position time series. Also, both the white and flicker noise derived by WA_MINQUE are all smaller than those by LS_MINQUE.

4 Conclusions and Remarks

The traditional LS_IQR for outlier detection and LS_MINQUE for noise component estimation are all based on the harmonic functional model, which cannot well describe the time-variable seasonal signals of GNSS position time series.

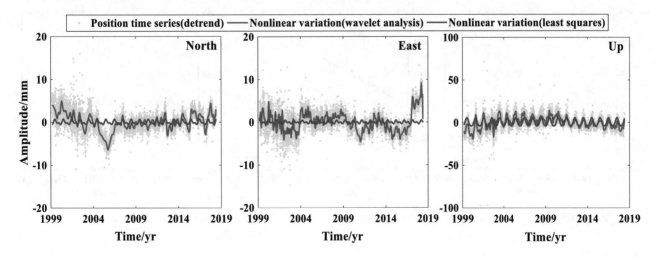

Fig. 4 Signals of BJFS station extracted by WA and LS method

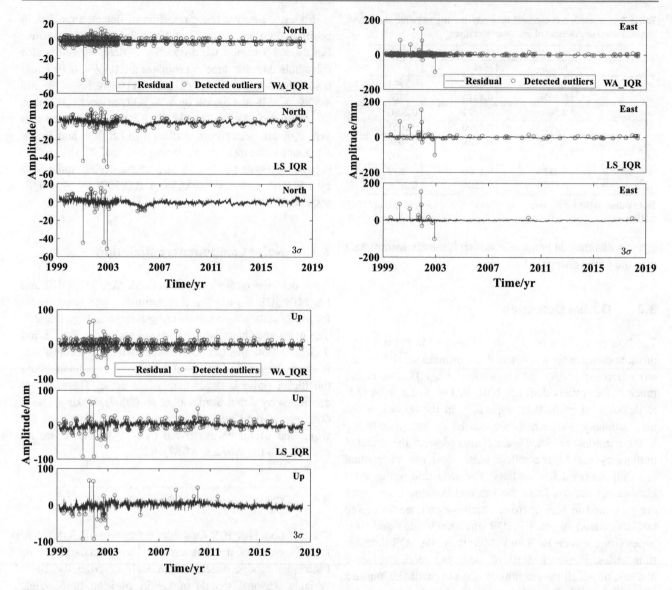

Fig. 5 Residuals derived by WA (Up) and LS (middle) for North, East and Up coordinates of BJFS station, and outliers detected by IQR and 3σ are marked with red dot

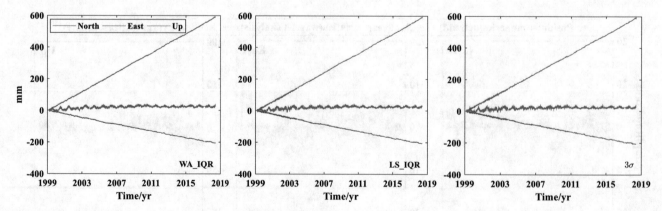

Fig. 6 The new time series of BJFS station after outliers removed based on WA_IQR(left), LS_IQR(middle) and 3σ (right)

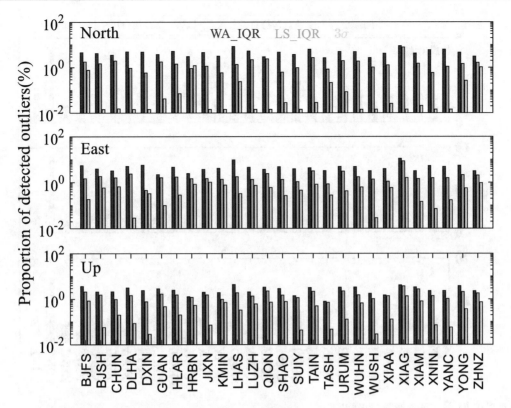

Fig. 7 Proportions of detected outliers in 27 stations using WA_IQR, LS_IQR and 3σ

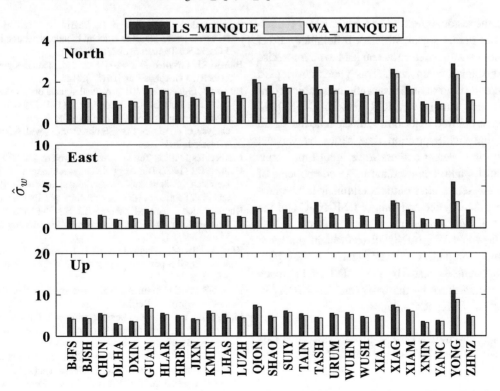

Fig. 8 Estimates of σ_w for 27 stations by two algorithms

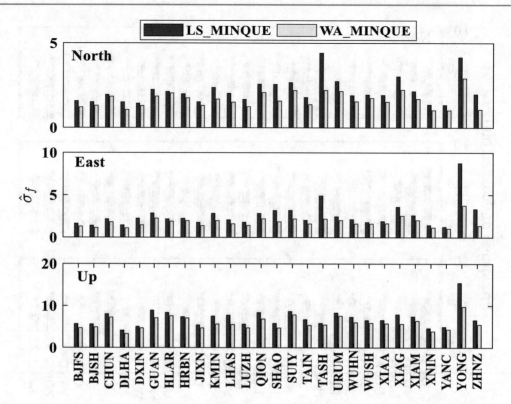

Fig. 9 Estimates of σ_f for 27 stations by two algorithms

Consequently, the residuals derived by traditional LS estimation still contain partial signal, which will definitely affect the performance of outlier detection and lead to an imprecise estimate of the noise component. This paper develops a wavelet-based algorithm of outlier detection and noise component estimation, namely WA_IQR and WA_MINQUE. The basic idea of our new algorithm is to separate the signal and noise of the GNSS position time series by wavelet analysis firstly, then detect outliers in residual time series using IQR statistic and then estimate noise components of the residual time series after outliers eliminated. The new algorithm is verified by the real data of CMONOC and the results show that WA_IQR is more effective than LS_IQR to detect outliers and WA_MINQUE can obtain the more reasonable noise component estimates than LS_MINQUE. The noise components estimated by WA_MINQUE approach are all smaller than those by the traditional LS_MINQUE approach for all 27 CMONOC stations.

Acknowledgments This work is sponsored by the National Natural Science Foundation of China (41731069 and 41974002).

References

Amiri-Simkooei AR, Tiberius CCJM, Teunissen PJG (2007) Assessment of noise in GPS coordinate time series: methodology and results. J Geophys Res 112:B07413

Amiri-Simkooei AR, Ansari H, Sharifi MA (2015) Application of recursive least squares to efficient blunder detection in linear models. J Geom Sci Technol 5(2):258–267

Blewitt G, Lavallée D (2002) Effect of annual signals on geodetic velocity. J Geophys Res 107(B7):2145

Bogusz J, Figurski M (2014) Annual signals observed in regional GPS networks. Acta Geodyn Geomater 11(174):125–131

Bos MS, Fernandes RMS, Williams SDP, Bastos L (2013) Fast error analysis of continuous GNSS observations with missing data. J Geod 87(4):351–360

Daubechies I (1992) Ten lectures on wavelets. SIAM, Philadelphia

Helmert FR (1907) Die Ausgleichungsrechnung nach der Methode der kleinsten Quadrate: mit Anwendungen auf die Geodäsie, die Physik und die Theorie der Messinstrumente. BG Teubner, Stuttgart

Khodabandeh A, Amiri-Simkooei AR, Sharifi MA (2012) GPS position time-series analysis based on asymptotic normality of M-estimation. J Geod 86(1):15–33

Klos A, Bos MS, Bogusz J (2018) Detecting time-varying seasonal signal in GPS position time series with different noise levels. GPS Solutions 22(1):21

Koch KR (1999) Parameter estimation and hypothesis testing in linear models. Springer, Berlin

Li W, Shen Y (2018) The consideration of formal errors in spatiotemporal filtering using principal component analysis for regional GNSS position time series. Remote Sens 10(4):534

Li B, Shen Y, Lou L (2010) Efficient estimation of variance and covariance components: a case study for GPS stochastic model evaluation. IEEE Trans Geosci Remote Sens 49(1):203–210

Mallat SG (1988) Multiresolution representations and wavelets. Ph.D. dissertation, University of Pennsylvania, Philadelphia

Mao A, Harrison CGA, Dixon TH (1999) Noise in GPS coordinate time series. J Geophys Res 104(B2):2797–2816

Nikolaidis RM (2002) Observation of global and seismic deformation with the Global Positioning System. Ph.D. thesis, University of California, San Diego

Peltier WR, Argus DF, Drummond R (2015) Space geodesy constrains ice age terminal deglaciation: the global ICE-6G_C (VM5a) model. J Geophys Res Solid Earth 120:450–487

Rao C (1971) Estimation of variance and covariance components – MINQUE theory. J Multivar Anal 1:257–275

Shen Y, Li W, Xu G (2014) Spatiotemporal filtering of regional GNSS network's position time series with missing data using principle component analysis. J Geod 88:1–12

Teunissen PJG, Amiri-Simkooei AR (2008) Least-squares variance component estimation. J Geod 82(2):65–82

Tian Y (2011) iGPS: IDL tool package for GPS position time series analysis. GPS Solutions 15(3):299–303

Tobita M (2016) Combined logarithmic and exponential function model for fitting postseismic GNSS time series after 2011 Tohoku-Oki earthquake. Earth Planets Space 68:41

Vitti A (2012) SIGSEG: a tool for the detection of position and velocity discontinuities in geodetic time-series. GPS Solutions 16:405–410

Walnut DF (2013) An introduction to wavelet analysis. Springer Science & Business Media, Berlin

Wöppelmann G, Aarup T, Schoene T (2007) An inventory of collocated and nearly-collocated CGPS stations and tide gauges, Progress report on the survey (25 July 2007). http://www.sonel.org/stations/cgps/surv_update.html

Wu D, Yan H, Shen Y (2017) TSAnalyzer, a GNSS time series analysis software. GPS Solutions 21:1389–1394

Zhang Q, Gui Q (2013) Bayesian methods for outliers detection in GNSS time series. J Geod 87(7):609–627

Zhang S, Li Z, He Y, Hou X, He Z, Wang Q (2018) Extracting of periodic component of GNSS vertical time series using EMD. Sci Surv Mapp 43(08):80–84. +96(in chinese)

Part II

Gravity Field Modelling

International Combination Service for Time-Variable Gravity Fields (COST-G)

Start of Operational Phase and Future Perspectives

Adrian Jäggi, Ulrich Meyer, Martin Lasser, Barbara Jenny, Teodolina Lopez, Frank Flechtner, Christoph Dahle, Christoph Förste, Torsten Mayer-Gürr, Andreas Kvas, Jean-Michel Lemoine, Stéphane Bourgogne, Matthias Weigelt, and Andreas Groh

Abstract

The International Combination Service for Time-variable Gravity Fields (COST-G) is a new Product Center of IAG's International Gravity Field Service (IGFS). COST-G provides consolidated monthly global gravity fields in terms of spherical harmonic coefficients and thereof derived grids of surface mass changes by combining existing solutions or normal equations from COST-G analysis centers (ACs) and partner analysis centers (PCs). The COST-G ACs adopt different analysis methods but apply agreed-upon consistent processing standards to deliver time-variable gravity field models, e.g. from GRACE/GRACE-FO low-low satellite-to-satellite tracking (ll-SST), GPS high-low satellite-to-satellite tracking (hl-SST) and Satellite Laser Ranging (SLR). The organizational structure of COST-G and results from the first release of combined monthly GRACE solutions covering the entire GRACE time period are discussed in this article. It is shown that by combining solutions and normal equations from different sources COST-G is taking advantage of the statistical properties of the various solutions, which results in a reduced noise level compared to the individual input solutions.

Keywords

Combined solutions · COST-G · IGFS Product Center · Time variable gravity

A. Jäggi (✉) · U. Meyer · M. Lasser · B. Jenny · T. Lopez
Astronomical Institute, University of Bern, Bern, Switzerland
e-mail: adrian.jaeggi@aiub.unibe.ch

F. Flechtner · C. Dahle · C. Förste
GFZ German Research Centre for Geosciences, Potsdam, Germany

T. Mayer-Gürr · A. Kvas
Institute of Geodesy, Technical University of Graz, Graz, Austria

J.-M. Lemoine
Department of Terrestrial and Planetary Geodesy, Centre National d'Etudes Spatiales, Toulouse, France

S. Bourgogne
Stellar Space Studies, Esplanade Compans Caffarelli, Toulouse, France

M. Weigelt
Institut für Erdmessung, University of Hannover, Hannover, Germany

A. Groh
Institut für Planetare Geodäsie, Technische Universität Dresden, Dresden, Germany

1 Introduction

Ultra-precise inter-satellite ranging as performed for more than 15 years by the GRACE mission has been established as the state-of-the-art technique to globally observe mass variations in the system Earth from space (Tapley et al. 2019). Continued meanwhile by its Follow-On mission (GRACE-FO, Flechtner et al. 2013), a growing number of institutions is processing the GRACE/GRACE-FO Level-1B instrument data to derive mass variations on a monthly basis (Level-2

© The Author(s) 2020
J. T. Freymueller, L. Sanchez (eds.), *Beyond 100: The Next Century in Geodesy*,
International Association of Geodesy Symposia 152, https://doi.org/10.1007/1345_2020_109

products). Although each new release of monthly gravity fields represents a significant improvement with respect to earlier releases, the solutions of different institutions usually differ considerably in terms of noise (Jean et al. 2018) and sometimes also in terms of signal (Meyer et al. 2015). In the frame of the European Gravity Service for Improved Emergency Management (EGSIEM) initiative, a prototype of a scientific combination service has been set up to demonstrate that improved solutions may be derived by combining individual solutions which are based on different approaches but agreed-upon processing standards (Jäggi et al. 2019). The Combination Service of Time-variable Gravity Fields (COST-G) continues the activities of the scientific combination prototype service of the EGSIEM initiative to realize a long-awaited standardization of gravity-derived mass transport products under the umbrella of the International Association of Geodesy (IAG). Established at the 2019 General Assembly of the International Union of Geodesy and Geophysics (IUGG) as a new Product Center of IAG's International Gravity Field Service (IGFS) for time-variable gravity fields, COST-G will operationally provide consolidated monthly global gravity models with improved quality, robustness, and reliability both in terms of spherical harmonic (SH) coefficients and thereof derived grids of surface mass changes by combining solutions or normal equations (NEQs) from COST-G analysis centers (ACs). The COST-G ACs are adopting different analysis methods but apply agreed-upon consistent processing standards[1] to deliver time-variable gravity field models, e.g. from GRACE or GRACE-FO low-low satellite-to-satellite tracking (ll-SST) or from non-dedicated data such as GPS high-low satellite-to-satellite tracking (hl-SST) or Satellite Laser Ranging (SLR). In addition COST-G makes use of existing and publicly available solutions or NEQs of Partner Analysis Centers (PCs), who are directly linked with the GRACE and GRACE-Follow On project. PCs are producing quality controlled products following their own project requirements which may not necessarily be in compliance with the COST-G standards.

The article is structured as follows: Sect. 2 provides an overview of the COST-G organization, Sect. 3 describes the COST-G workflow by discussing the first release of COST-G GRACE Level-2 products and Sect. 4 concludes the article with a summary and future perspectives.

2 COST-G Organization

COST-G is organized in close analogy to other IAG services (Drewes et al. 2016). It is steered by a Directing Board, which sets the objectives and the scientific and operational

[1] https://cost-g.org/download/COST_G_STANDARDS.pdf.

goals. The COST-G objectives to derive time-variable gravity fields with improved quality, robustness, and reliability are accomplished by the following components.

2.1 Central Bureau (CB)

The Central Bureau is responsible for all operational activities of the Service. The Central Bureau coordinates COST-G activities, facilitates communications and maintains documentations. The CB is currently located at the Astronomical Institute of the University of Bern (AIUB).

2.2 Analysis Centers (ACs)

The COST-G Analysis Centers produce time-variable gravity field solutions according to the specifications defined by the COST-G Processing Standards defined by the COST-G Directing Board. For the analysis of GRACE/GRACE-FO data the current ACs are, in alphabetical order:

– Astronomical Institute, University of Bern (AIUB)
– Centre National d'Etudes Spatiales (CNES/GRGS)
– German Research Centre for Geosciences (GFZ)
– Institute of Geodesy, Graz University of Technology (IfG/ITSG)

The list of ACs may differ for the processing of non-dedicated satellite data to derive alternative monthly solutions of the Earth's time-variable gravity field, e.g. from Swarm hl-SST data in the frame of an ESA initiative (Teixeira da Encarnação et al. 2019).

The ACs send their solutions to the Analysis Center Coordinator for combination together with a summary describing their processing strategy. Depending on the availability of new or improved AC contributions, new combined solutions are released. GRACE contributions need to cover at least the time period 2003 to mid 2016 to be included in the combination. Shorter periods may be considered for testing. Corresponding rules for GRACE-FO, hl-SST, SLR contributions will be defined when a significantly large number of ACs is available and has consolidated their processing strategies.

2.3 Partner Analysis Centers (PCs)

COST-G will in addition make use of existing and publicly available solutions or NEQs of other processing centers, denoted as Partner Analysis Centers (PC). Currently, these are primarily centers that are part of the GRACE and GRACE-FO project, e.g. the Center for Space Research (CSR) of the University of Texas at Austin or NASA's Jet Propulsion Laboratory (JPL), who are producing quality controlled products following their own project requirements. COST-G retains the right to exclude solutions if they either deviate from the COST-G Processing Standards or do not comply with decisions of the COST-G Directing Board.

2.4 Analysis Center Coordinator (ACC)

The Analysis Center Coordinator, currently located at AIUB, first compares the individual gravity fields for quality control. From the comparison of the solutions passing the quality control the ACC defines empirical weights for the individual contributions. Eventually, he combines the accepted solutions to generate a combined gravity field using the underlying normal equations of the individual ACs (Meyer et al. 2019). If normal equations are not available, combinations may also be performed on the solution level (Jean et al. 2018). The resulting COST-G solutions (Level-2 products) are published at the International Center for Global Earth Models (ICGEM, Ince et al. 2019).

2.5 Validation Center (VC)

Validation of the COST-G products happens at the Validation Center, which is currently operated by the COST-G ACs at CNES and GFZ. This involves the evaluation of the noise of the solutions over dedicated areas of low variability, as well as the evaluation of the solution quality through comparison with external data from altimetry. Eventually the COST-G products are also validated through fits of tracking data of Low Earth Orbiters.

2.6 Product Evaluation Group (PEG)

External expert users not associated with COST-G ACs are forming the COST-G Product Evaluation Group. They assess the COST-G products for studying mass variations related to terrestrial water storage over non-glaciated regions, bottom pressure variations in the oceans and ice mass changes in Antarctica and in Greenland.

2.7 Level-3 Product Center (L3C)

Various corrections and reductions have to be applied to the Level-2 products, resulting in post-processed SH coefficients (Level-2B products), before user-friendly Level-3 products are generated by the Level-3 Product Center, currently located at GFZ. The COST-G Level-3 products are visualized and described at GFZ's Gravity

Information Service (GravIS, http://gravis.gfz-potsdam. de) and will be made available by GFZ's Information System and Data Center (ISDC, https://isdc.gfz-potsdam. de). Additionally, the products can also be visualized at the COST-G Plotter (http://cost-g.org/).

3 COST-G Workflow

The workflow of COST-G is illustrated by Fig. 1. It consists of harmonization and quality control of the individual input solution of the different ACs and PCs, and the combination and validation of the resulting COST-G solution. All aspects of this process are explained and illustrated in more detail in the subsequent subsections by discussing the results of the first release of combined GRACE Level-2 products (GRAC_COSTG_BF01_0100, subsequently addressed as COST-G RL01).

3.1 Harmonization of Input Solutions

The individual input solutions of the different ACs may differ in the underlying constants (Earth's gravity constant, Earth's equatorial radius), tide system, and mean pole convention (see Eq. 21 in Wahr et al. (2015)). Individual solutions may thus need to be re-scaled and individual coefficients (C_{20}, C_{21}, S_{21}) may need to be further corrected to be compliant with the tide system and mean pole convention used by COST-G. Additional background models, such as the atmosphere and ocean dealiasing product (AOD), may also differ between groups and are consolidated before the combination to ensure a consistent signal definition.

3.2 Quality Control of Input Solutions

For quality control of the COST-G products, the signal and noise content of the harmonized AC contributions are first compared. The signal content is analyzed by computing the amplitude of seasonal variations of equivalent water height (EWH) for a large number of river basins and by computing ice mass trends in Greenland and Antarctica. An assessment of the noise is performed by analyzing anomalies, which represent the monthly variability after subtraction of a deterministic model of secular and seasonal variations. The assessment of the seasonal variations of the input solutions for COST-G RL01 is discussed below, whereas the assessment of ice mass trends will be separately discussed in Sect. 3.4.1. The noise assessment is not reproduced here as it largely corresponds to earlier results presented in Jean et al. (2018) and Meyer et al. (2019).

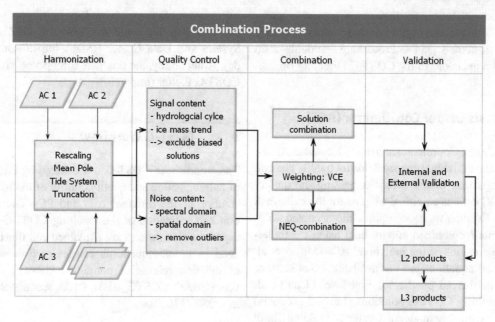

Fig. 1 COST-G workflow

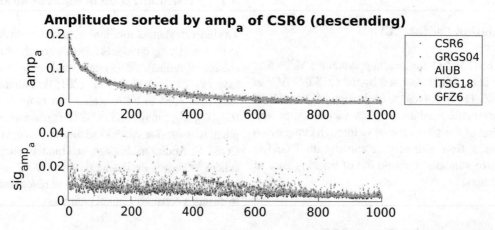

Fig. 2 Amplitude of seasonal variations (top) and formal errors of amplitudes (bottom) in MEWH (m) for major river basins using a 400 km Gauss filter

Figure 2 (top) shows the amplitudes, expressed in mean equivalent water height (MEWH), of the estimated seasonal variations for the 500 largest river basins as derived from all AC and PC solutions used for the COST-G RL01. The underlying river basin masks were taken from the Total Runoff Integrating Pathways (TRIP4[2]) model. The corresponding formal errors of the estimation are shown in Fig. 2 (bottom). The analysis shows that no systematic signal attenuation may be observed for seasonal signals in any of the contributing gravity field time-series. The cluster of outlying larger formal errors visible in Fig. 2 (bottom) is related to regions with small seasonal signals and large non-linear trends as occurring, e.g. for regions with accelerated ice mass loss.

---

[2]http://hydro.iis.u-tokyo.ac.jp/~taikan/TRIPDATA/TRIPDATA.html.

3.3 Combination of Input Solutions

Generally, the planned strategy for future COST-G releases is to provide a combination based on the underlying NEQs of the individual ACs according to the methodology presented in Meyer et al. (2019). As NEQs were not available from all centers, the combination was performed on the solution level for the COST-G RL01 as a weighted combination of the SH coefficients using variance component estimation (VCE) (Jean et al. 2018). The underlying AC und PC solutions are, AIUB RL02 (Meyer et al. 2016), GRGS RL04, GFZ RL06 (Dahle et al. 2019), ITSG-Grace2018 (Kvas et al. 2019), and CSR RL06 (Save et al. 2018).

Apart from AIUB RL02, which is still based on the RL02 of the GRACE Level-1B data, all input solutions are based on

Fig. 3 Weights assigned to input solutions

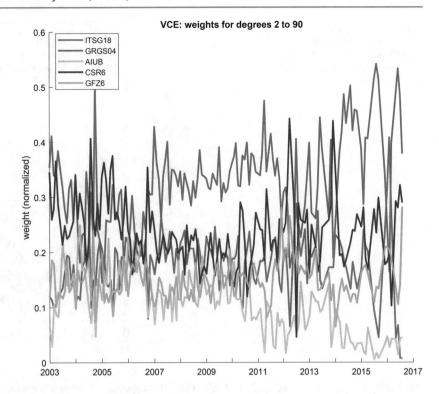

the most recent RL03 of the GRACE Level-1B data (GRACE Level 1B JPL Release 3.0 2018). For the GRGS solution the underlying NEQs were inverted by the ACC to obtain a solution without regularization.

Figure 3 shows the relative weights assigned to the individual input solutions as determined by VCE. The weights can be interpreted as quality indicators of the solutions as they are inversely proportional to the noise levels of the individual contributions. The highest weights are usually assigned to the ITSG-Grace2018 solutions due to their very low noise level. This is largely related to the sophisticated empirical noise modeling of the ITSG-Grace2018 solution (Kvas et al. 2019) and in accordance with analyses based on earlier ITSG releases (Jean et al. 2018). Lowest weights are usually assigned to the AIUB RL02 solutions due to the use of (meanwhile) outdated Level-1B data and, in particular, since active thermal control of the GRACE satellites was switched off in April 2011 (Tapley et al. 2015), which would have required adaptions in the accelerometer parametrization as shown in Klinger and Mayer-Gürr (2016).

Figure 4 shows median degree amplitudes of anomalies for the combined solution as well as for the input solutions that contributed to the combination for the entire GRACE time period with respect to a linear and seasonal model without applying any filtering. The analysis reveals that in the spectral domain the main gain of the combination is in the range of degrees 15–45. When truncating all gravity fields at order 29 to exclude the effect of the noisy higher-order SH coefficients, which are usually attenuated in applications by post-processing filters, e.g. Kusche (2007), the gain of

the combination may even be seen up to about degree 65. The lower noise of the combined solution may also be confirmed in the spatial domain by analyzing the RMS of EWH anomalies over the oceans (not shown).

3.4   Validation of Combined Solution

Internal and external validation is performed to ensure the quality of the COST-G solutions by identifying outlying or systematically deviating input solutions.

3.4.1   Internal Validation

Ice mass loss in the polar regions is of enormous societal relevance (Shepherd et al. 2012). Evaluating GRACE mass change time series for the ice sheets in Antarctica and Greenland, as derived from the individual input solutions and the combined COST-G solution, is thus an essential task of the COST-G product evaluation group to detect potential inconsistencies between the input solutions.

Greenland Ice Sheet

GRACE-derived mass change time series are derived from all input solutions that contributed to the combination for the entire GRACE time period and from the combined COST-G solution for all drainage basins of the Greenland Ice Sheet (GIS[3]) and for the entire Greenland Ice Sheet by adopting the

---

[3]http://icesat4.gsfc.nasa.gov/cryo_data/ant_grn_drainage_systems.php.

Fig. 4 Median degree
amplitudes of residual SH
coefficients. Solid: full spectrum,
dashed: limited to order
m = 0 . . . 29

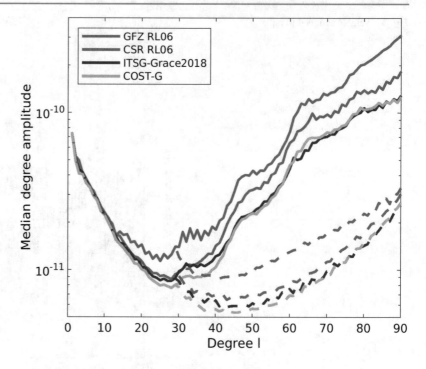

sensitivity kernel approach (Groh et al. 2019 and references in there).

The estimated linear trends of ice mass change for all drainage basins and for the entire Greenland ice sheet agree very well for the different solutions (not shown). Most notably the COST-G time series is characterized by a very favorable noise behavior. Figure 5 shows the noise levels of the mass change time series available over the entire time span in terms of the scaled standard deviation of the derived noise time series for all Greenland Ice Sheet drainage basins and the entire ice sheet following the method of Groh et al. (2019). For the majority of the basins the COST-G time series shows indeed the lowest noise of all solutions used for the combination.

Antarctic Ice Sheet

A similar analysis is performed for the drainage basins of the Antarctic Ice Sheet (AIS, see footnote 3), selected aggregations, and for the entire Antarctic ice sheet. Whereas most of the results of this analysis confirm the level of agreement found for the Greenland Ice Sheet, the linear trends resulting for different gravity field solutions for some drainage basins deserve special attention. Figure 6 compiles for the solutions available over the entire time span the estimated linear trends and error bars in ice mass change for selected drainage basins, the Antarctic Peninsula, East Antarctica, West Antarctica, and for the entire ice sheet. The displayed error bars account for the formal errors of the estimated trends, as well as for leakage errors and errors in the applied model reductions.

Figure 6 shows that the trends resulting from all solutions agree fairly well for West Antarctica. It also reveals, however, that discrepancies are occurring for East Antarctica. Whereas the trends resulting from the CSR RL06 and the ITSG-Grace2018 solutions are in almost perfect agreement, the GFZ RL06 solution suggests a different trend for this entire region. However, all trend estimates agree within their error estimates, since these are clearly dominated by the error of the reduced glacial isostatic adjustment model. As a consequence of the weighting scheme used for the combined solution, the trend of the COST-G solution is in-between the two concurring trends. Due to the larger weights assigned to the CSR / ITSG solution (cf. Fig. 3), it is closer to these solutions. Figure 6 also shows that the different trends over East Antarctica influence the trends resulting for the entire ice sheet. Trend analyses based on shorter time periods common to all solutions show again different trends for the AIUB RL02 and GRGS RL04 solutions (not shown).

3.4.2 External Validation

Currently two methods are realized within COST-G to assess the quality of the solutions by independent data sets. More external validations may follow in the future.

Comparison to Altimetry

Currently two test areas (Caspian Sea, Black Sea) are selected within COST-G for an independent signal assessment. The time series of the time-variable gravity field solutions are compared with the time series of altimetric heights, derived from Hydroweb for the Caspian Sea and

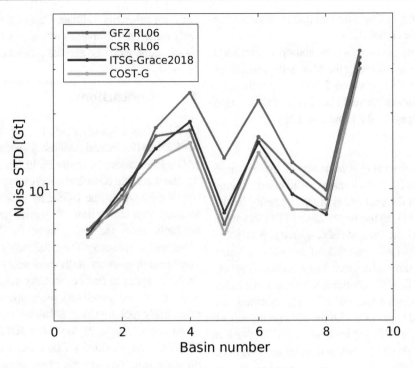

Fig. 5 Noise level of the mass change time series for all GIS basins and the entire GIS (basin no. 9)

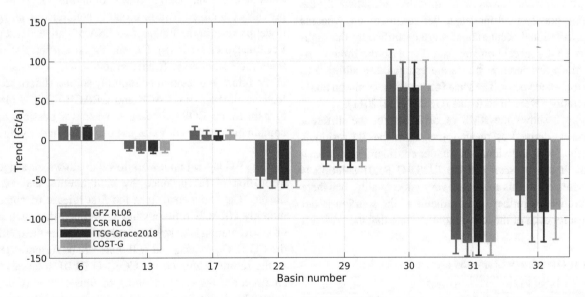

Fig. 6 Linear trend and error bars in ice mass for drainage basins of the AIS (6, 13, 17, 22), selected aggregations (29: Antarctic Peninsula, 30: East Antarctica, 31: West Antarctica) and the entire ice sheet (32)

AVISO+ for the Black Sea. One bias and one scale factor are adjusted to perform the comparison.

Table 1 exemplarily shows the correlation coefficients over the Caspian Sea when filtering the gravity field solutions with different DDK filters (Kusche 2007). It can be seen that the COST-G solution presents in this metric a slight improvement with respect to the input solutions.

Orbital Fits

The long to medium wavelength accuracy of gravity field models can be evaluated through dynamic orbit computations as commonly done for the evaluation of static gravity fields, e.g. Gruber et al. (2011). In the frame of COST-G the same concept is adopted to the time-variable gravity field solutions from the individual ACs and thereof derived combined solution. In order to not suffer from large omission errors of the time-variable GRACE solutions, which are generally only available up to degree and order 90, all solutions are filled up to degree and order 240 with the SH coefficients of the GOCE static model DIR-6 (Förste et al. 2019). For the COST-G RL01 dynamic GOCE orbits with an arc length of 1.25 days were fitted to the GOCE kinematic precise science orbits (Bock et al. 2014) serving as pseudo-observations. The gravitational forces were modeled according to the different gravity field models under consideration, whereas non-gravitational accelerations were described by the high-quality GOCE accelerometer data. For each arc three common mode acceleration biases are estimated in addition to the initial state vector. The scale factors of the common mode accelerations were fixed to one (Gruber et al. 2011).

Table 2 shows the RMS of orbit fits for the different test cases, derived as mean values from the 3D residuals from the 30/31 individual arcs under consideration for each month. It can be seen that the COST-G RL01 solution is also assessed in this metric of very good quality, but there is potential for further improvements as the solution is not yet best for each of the tested months. Note that the orbit test

does not primarily validate C20 of the individual solutions. Only marginal differences would result if C20 is replaced in all solutions with one and the same value (not shown).

4 Conclusions

COST-G has officially started its operational service at the IUGG's 2019 General Assembly as a new Product Center of IAG's International Gravity Field Service (IGFS). It continues the scientific combination prototype service that has been established within the EGSIEM initiative to realize a long-awaited standardization of gravity-derived mass transport products under the umbrella of the IAG. COST-G recognizes and emphasizes the existence and acknowledges the contribution of every individual analysis center and partner analysis center to this community effort. Their participation is a crucial and mandatory prerequisite to the consolidation of monthly global gravity fields within COST-G.

At the occasion of the 2019 IUGG General Assembly COST-G has provided a first release of combined GRACE monthly solutions covering the entire GRACE time period between April 2002 and June 2017 by combining the solutions of five contributing centers. In addition COST-G also provides combined Swarm monthly solutions as an operational product in the frame of an ESA initiative (Teixeira da Encarnação et al. 2019). Depending on the interest of the scientific community, further products may be established in the future, e.g. combined monthly solutions derived from SLR satellites. For GRACE and GRACE-FO it is planned to base future COST-G releases whenever possible on the combination of the normal equations of the underlying input solutions.

COST-G has set up a workflow that allows for a rigorous evaluation of the products by both internal and external means. The experience from the first release of combined monthly GRACE solutions has underlined that such a rigorous evaluation is of key importance to ensure the quality of the COST-G products and will therefore be further extended in the future. Although the COST-G RL01 solutions show an excellent behavior in terms of noise, as it is demonstrated by the various internal and external quality metrics, a discrepancy in the signal has been revealed by the COST-G product evaluation group for the different input solutions when regarding ice mass trends over East Antarctica. As this also affects the trends of the COST-G RL01 for that region, the root cause is currently being further investigated. The COST-G workflow also includes the generation of Level-2B products to derive thereof user-friendly Level-3 products that will enable hydrologists, glaciologists, oceanographers, geodesists and geophysicists to fully profit from one combined, and consolidated monthly GRACE gravity product. Updates will be announced at http://cost-g.org.

Table 1 Correlation with altimetry over the Caspian Sea

Filter	CSR RL06	ITSG-Grace2018	COST-G RL01
DDK5	96.6%	97.0%	97.2%
DDK6	96.3%	96.5%	96.6%

Table 2 RMS of dynamic GOCE orbit fits (cm) for the different gravity fields

Gravity model	2009/11	2009/12	2010/10	2010/11
GFZ RL06	7.4	6.8	6.2	6.2
AIUB RL02	8.7	8.6	7.4	7.2
CSR RL06	6.9	9.0	6.6	6.2
GRGS RL04	5.9	7.3	5.5	5.8
ITSG-Grace2018	5.5	5.1	4.2	4.5
COST-G RL01	5.0	5.5	4.5	4.7

Acknowledgements This article is dedicated to Richard Biancale, who unexpectedly passed away on February 4, 2019. His support and promotion of the activities described in this article are gratefully acknowledged. The study was performed in the framework of an international team, led by Richard Biancale and the first author of this article, that is receiving support from the International Space Science Institute (ISSI) in Bern, Switzerland.

References

Bock H, Jäggi A, Beutler G et al (2014) GOCE: precise orbit determination for the entire mission. J Geod 88(11):1047–1060. https://doi.org/10.1007/s00190-014-0742-8

Dahle C, Murböck M, Flechtner F et al (2019) The GFZ GRACE RL06 monthly gravity field time series: processing details and quality assessment. Remote Sens 11:2116. https://doi.org/10.3390/rs11182116

Drewes H, Kuglitsch F, Ádám J (2016) The geodesist's handbook 2016. J Geod 90(10):907–1205

Flechtner F, Morton P, Watkins M et al (2013) Status of the GRACE follow-on mission. In: Marti U (ed) Gravity, geoid and height systems. IAG symposia, pp 117–121. https://doi.org/10.1007/978-3-319-10837-7_15

Förste C, Abrykosov O, Bruinsma S et al (2019) ESA's Release 6 GOCE gravity field model by means of the direct approach based on improved filtering of the reprocessed gradients of the entire mission. GFZ Data Services. https://doi.org/10.5880/ICGEM.2019.004

GRACE Level 1B JPL Release 3.0 (2018) Ver. 3. PO.DAAC, CA, USA. https://doi.org/10.5067/GRJPL-L1B03

Groh A, Horwath M, Horvath A et al (2019) Evaluating GRACE mass change time series for the Antarctic and Greenland Ice Sheet – methods and results. Geosciences 9:415. https://doi.org/10.3390/geosciences9100415

Gruber T, Visser PNAM, Ackermann C et al (2011) Validation of GOCE gravity field models by means of orbit residuals and geoid comparisons. J Geod 85(11):845–860. https://doi.org/10.1007/s00190-011-0486-7

Ince ES, Barthelmes F, Reissland S et al (2019) ICGEM - 15 years of successful collection and distribution of global gravitational models, associated services, and future plans. Earth Syst Sci Data 11:647–674

Jäggi A, Weigelt M, Flechtner F et al (2019) European Gravity Service for Improved Emergency Management (EGSIEM) – from concept to implementation. Geophys J Int 218(3):1572–1590. https://doi.org/10.1093/gji/ggz238

Jean Y, Meyer U, Jäggi A (2018) Combination of GRACE monthly gravity field solutions from different processing strategies. J Geod 92(11):1313–1328. https://doi.org/10.1007/s00190-018-1123-5

Klinger B, Mayer-Gürr T (2016) The role of accelerometer data calibration within GRACE gravity field recovery: results from ITSG-Grace2016. Adv Space Res 58(9):1597–1609. https://doi.org/10.1016/j.asr.2016.08.007

Kusche J (2007) Approximate decorrelation and non-isotropic smoothing of time-variable GRACE-type gravity field models. J Geod 81(11):733–749. https://doi.org/10.1007/s00190-007-0143-3

Kvas A, Behzadpour S, Ellmer M et al (2019) ITSG-Grace2018: overview and evaluation of a new GRACE-only gravity field time series. J Geophys Res Solid Earth 124:9332. https://doi.org/10.1029/2019JB017415

Meyer U, Jäggi A, Beutler G et al (2015) The impact of common versus separate estimation of orbit parameters on GRACE gravity field solutions. J Geod 89(7):685–696. https://doi.org/10.1007/s00190-015-0807-3

Meyer U, Jäggi A, Jean Y et al (2016) AIUB-RL02: an improved time-series of monthly gravity fields from GRACE data. Geophys J Int 205(2):1196–1207. https://doi.org/10.1093/gji/ggw081

Meyer U, Jean Y, Kvas A et al (2019) Combination of GRACE monthly gravity fields on the normal equation level. J Geod 93:1645–1658. https://doi.org/10.1007/s00190-019-01274-6

Save H, Tapley B, Bettadpur S et al (2018) GRACE RL06 reprocessing and results from CSR, presented at EGU 2018, abstract EGU2018-10697

Shepherd A, Ivins ER, Geruo A et al (2012) A reconciled estimate of ice-sheet mass balance. Science 338:1183–1189. https://doi.org/10.1126/science.1228102

Tapley B, Flechtner F, Watkins M et al (2015) GRACE mission: status and prospects. Presented at: Grace science team meeting, Austin, Texas. https://doi.org/10.1038/s41558-019-0456-2

Tapley BD, Watkins MM, Flechtner F et al (2019) Contributions of GRACE to understanding climate change. Nat Clim Change 9:358–369

Teixeira da Encarnação J, Visser P, Arnold D et al (2019) Multi-approach gravity field models from Swarm GPS data. Earth Syst Sci Data. In press. https://doi.org/10.5194/essd-2019-158

Wahr J, Nerem RS, Bettadpur SV (2015) The pole tide and its effect on GRACE time-variable gravity measurements: implications for estimates of surface mass variations. J Geophys Res Solid Earth 120:4597–4615. https://doi.org/10.1002/2015JB011986

LUH-GRACE2018: A New Time Series of Monthly Gravity Field Solutions from GRACE

Igor Koch, Jakob Flury, Majid Naeimi, and Akbar Shabanloui

Abstract

In this contribution, we present the LUH-GRACE2018 time series of monthly gravity field solutions covering the period January 2003–March 2016. The solutions are obtained from GRACE K-Band Range Rate (KBRR) measurements as main observations. The monthly solutions are computed using the in-house developed GRACE-SIGMA software. The processing is based on dynamic orbit and gravity field determination using variational equations and consists of two main steps. In the first step, 3-hourly orbital arcs of the two satellites and the state transition and sensitivity matrices are dynamically integrated using a modified Gauss-Jackson integrator. In this step, initial state vectors and 3D accelerometer bias parameters are adjusted using GRACE Level-1B reduced-dynamic positions as observations. In the second step, normal equations are accumulated and the normalized spherical harmonic coefficients up to degree and order 80 are estimated along with arc-wise initial states, accelerometer biases and empirical KBRR parameters. Here KBRR measurements are used as main observations and reduced-dynamic positions are introduced to solve for the low frequency coefficients. In terms of error degree standard deviations as well as Equivalent Water Heights (EWH), our gravity field solutions agree well with RL05 solutions of CSR, GFZ and JPL.

Keywords

GRACE · Gravity field recovery · Time-variable gravity field · GRACE Follow-On

1 Introduction

In this paper we report on the status of the determination of monthly Earth gravity field models from orbit positions and K-band satellite-to-satellite tracking measurements of the GRACE and GRACE Follow-On (GRACE-FO) low Earth orbiters at Leibniz Universität's Institut für Erdmessung. We give an overview on the processing approach of the LUH-GRACE2018 time series of monthly solutions from GRACE data. We briefly report on the first monthly solutions

from GRACE-FO data that will be described in detail in an upcoming article.

The first batch of the LUH-GRACE2018 time series, covering 7 years (2003–2009), was presented and published in 2018 (Naeimi et al. 2018). In September 2019, a second batch containing monthly solutions from the years 2010 to 2016 was published. The processing approach for the solutions is the method of dynamic orbit and gravity field determination based on the equations of motion, also often referred to as the variational equations approach, e.g. Montenbruck and Gill (2005), Vallado (2013). We implemented it in a compact all-MATLAB program named GRACE-Satellite orbit Integration and Gravity field analysis in MAtlab (GRACE-SIGMA). The code uses strongly vectorized modules for numerical integration of reference orbits and the contributions to the design matrix (state transition matrix,

I. Koch (✉) · J. Flury · M. Naeimi · A. Shabanloui
Institut für Erdmessung, Leibniz Universität Hannover, Hannover, Germany
e-mail: koch@ife.uni-hannover.de

© The Author(s) 2020
J. T. Freymueller, L. Sanchez (eds.), *Beyond 100: The Next Century in Geodesy*,
International Association of Geodesy Symposia 152, https://doi.org/10.1007/1345_2020_92

sensitivity matrix). To allow for the vectorization, a modified Gauss-Jackson numerical integrator was developed. The least-squares parameter estimation is done in a two-step procedure: in the first step, a suitable reference orbit and accelerometer calibration parameters are estimated as a fit to reduced-dynamic orbit positions. In the second step, K-Band Range Rate (KBRR) observations are added to estimate updates to spherical harmonic gravity field coefficients. In order not to constrain the solutions towards the background force models of the pre-adjusted orbits, arc-specific parameters from step 1 are estimated together with the coefficients of the Earth's time-variable potential (Meyer et al. 2015). To allow for an efficient data handling and processing, observations, parameters and force model data are stored and updated in arc-wise data capsules.

The implementation of the force effects was validated in comparison with computation results from other GRACE ACs within the new COmbination Service for Time-variable Gravity solutions (COST-G) (Meyer et al. 2018). It must be noted that we used RL05 Atmosphere and Ocean De-aliasing (AOD) models to correct for rapid mass variations, as our processing started before the more recent and more accurate RL06 AOD models were available. Also, we used RL02 Level-1B data products as the reprocessed RL03 Level-1B data were not yet available for the first batch of our processing. For the GRACE-FO gravity field solutions RL06 AOD models and RL04 data products were used. For a future time series of monthly GRACE solutions, we plan to use updated data and models.

The LUH-GRACE2018 monthly solutions were in principle estimated up to spherical harmonic degree and order 80. The monthly parameter estimation was done by stacking normal equations from orbital arcs of 3 h each; initial satellite state corrections, 3-axes accelerometer instrument biases and empirical range-rate corrections were estimated arc-wise. Our current stochastic model uses uniform weights for the reduced-dynamic orbit position pseudo-observations on the one hand and the K-band range-rate observations on the other.

Obtained results are compared to the solutions of other ACs in terms of error degree standard deviations that show the level of time-variable signal, modeling errors and random measurement errors with respect to a long-term reference model. In addition, obtained results are compared to those of other ACs in terms of mass variation time series for selected geophysical processes. This evaluation shows that for most months the quality of our results is similar to results from other ACs. For some months with poorer and more heterogeneous data quality, we have obtained a similar quality level by estimating spherical harmonic parameters up to degree 60 only. For other months with an even more difficult data situation, the results are not yet satisfactory and therefore not yet published.

One of the future goals of our group is to test how inhomogeneous data quality as seen, e.g., in post-fit range-rate residuals, propagates in the gravity recovery processing. We expect that changes in the parametrization or modifications of the input sensor data can help to identify and disentangle some of the involved effects.

2   Overview of the Gravity Field Recovery Method

Here, we mention the main elements of dynamic orbit and gravity field determination from GRACE and GRACE-FO sensor data. A paper with more details on our implementation is in preparation. We refer to Fig. 1 for a generalized overview. As the problem is highly nonlinear, a reference orbit that closely matches the true orbit (represented by the GRACE Level-1B data products) is determined using dynamic orbit integration from a priori force models. Then reduced observations are computed. After the linearization, the reduced observations are used to estimate parameter updates.

Orbit integration is based on the equations of motion

$$\ddot{\mathbf{r}} = -\frac{GM_\oplus}{r^3}\mathbf{r} + \ddot{\mathbf{r}}_p \tag{1}$$

with the central (Keplerian) term and a sum of accelerations due to perturbing forces $\ddot{\mathbf{r}}_p$. The perturbing forces include the higher harmonics of the time-variable gravity field, direct and indirect tidal forces, forces from non-tidal mass redistribution, and non-gravitational forces acting on the satellites.

The reference orbit is obtained by numerically integrating the first order ordinary differential equations

$$\left.\begin{aligned} \dot{\mathbf{r}} &= \mathbf{v} \\ \dot{\mathbf{v}} &= -\frac{GM_\oplus}{r^3}\mathbf{r} + \ddot{\mathbf{r}}_p \end{aligned}\right\} . \tag{2}$$

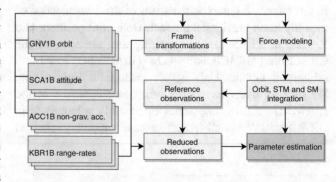

Fig. 1 Simplified procedure of gravity field recovery from satellite-to-satellite tracking data with the variational equations approach. STM: state transition matrix, SM: sensitivity matrix, non-grav. acc: non-gravitational acceleration

For the non-gravitational forces, the accelerometer observations of each satellite are used, considering bias and scale calibration parameters. For the perturbing forces, frame transformations between geocentric and inertial frames or between satellite body frames and inertial frame need to be applied; for the latter, star camera based attitude data are used.

In the first estimation step, the orbit is iteratively fitted to orbit positions from a reduced-dynamic orbit as pseudo-observations, in order to ensure a sufficiently good linearization for the second estimation step. The partial derivatives in the design matrix of the estimation require the integration of the state transition and sensitivity matrices along with the orbit integration. In the second step, K-band range-rate observations are added, and updates to spherical harmonic coefficients are introduced as main unknown parameters.

3   Data

For the GRACE processing, we used the Level-1B data products of the release 2 (RL02) (Case et al. 2010). These data products include K-band range-rate observations (KBR1B), reduced-dynamic positions and velocities (GNV1B), accelerometer measurements representing the non-gravitational accelerations (ACC1B) and satellite attitude (SCA1B). All quantities of the Level-1B data products were used with 5 s sampling during numerical integration. For K-band range-rates, light-time and antenna offset corrections from the KBR1B data products were included. In the processing the standard GNV1B reduced-dynamic orbit positions were used as pseudo-observations. We assume that they allow for a sufficiently good linearization in the first estimation step. Nevertheless, the influence of the orbit type and their characteristics, e.g. dynamic information in reduced-dynamic orbits or high frequency noise in kinematic orbits, on the recovered gravity field solutions may deserve further systematic investigations in the future.

4   Force Modeling

Force models needed for the numerical integration of the satellite trajectory and the state transition and sensitivity matrices are summarized in Table 1. The force models consider gravitational effects of tidal and non-tidal nature as well as non-gravitational effects. The non-gravitational effects are measured by the onboard accelerometers and have to be corrected for scale factors and biases during processing. Except for the acceleration due to the Earth's gravity potential and the non-gravitational acceleration, all effects were pre-computed using the GNV1B reduced-dynamic orbits and were not altered during numerical orbit integration. We

Table 1 Force models applied for the LUH-GRACE2018 time series of monthly gravity field solutions. d/o: indicates the utilized maximum degree/order of the spherical harmonic coefficients

Force	Models and parameters
Gravity field	GIF48 (d/o: 300) [1]
Direct tides	Moon, Sun (DE405) [2]
Solid Earth tides	IERS Conventions 2010 [3]
Ocean tides	EOT11a (d/o: 80) [4], [5]
Relativistic effects	IERS Conventions 2010 [3]
Solid earth pole tides	IERS Conventions 2010 [3]
Ocean pole tides	IERS Conventions 2010 [3], [6]
De-aliasing	AOD1B RL05 (d/o: 100) [7]
Non-gravitational	ACC1B [8]

References: [1]: Ries et al. (2011), [2]: Standish (1998), [3]: Petit and Luzum (2010), [4]: Savchenko and Bosch (2012), [5]: Rieser et al. (2012), [6]: Desai (2002), [7]: Dobslaw et al. (2013), [8]: Case et al. (2010)

assume that future releases of the LUH-GRACE monthly gravity field solutions will benefit from updated ocean tide models such as FES2014 (Carrere et al. 2015) and from the AOD1B RL06 de-aliasing products (Dobslaw et al. 2017).

5   Numerical Integration

An accurate numerical integration of the satellite orbits and the state transition and sensitivity matrices can be regarded as the most time-consuming processing part of gravity field recovery. In order to save computational time while ensuring an integration accuracy suitable for gravity field recovery, we developed a modified version of the widely used predictor-corrector Gauss-Jackson integrator, e.g. Berry and Healy (2004), Montenbruck and Gill (2005), Vallado (2013). The impact of the corrector step on a GRACE-like orbit (near circular with an eccentricity of 0.001 and an altitude of about 500 km) was validated for different integration orders and step sizes. For the typical GRACE integration step size of 5 s and the integration order of 8, the impact of the corrector step is in the order of 10^{-13} m for the position and 10^{-13} m/s for the velocity and thus can be neglected (Naeimi 2018). However, the corrector formulas can be used to simplify the formulation of the predictor step considerably, allowing a straightforward, vectorized and thus efficient implementation. While more details of this integration technique will be covered in an upcoming publication, the modified Gauss-Jackson equations and ancillary sets of coefficients needed for an implementation can be found below. The modified Gauss-Jackson equations for state prediction have the following form:

$$\left.\begin{aligned} \mathbf{r}_{i+1} &= \mathbf{r}_i + h\dot{\mathbf{r}}_i + h^2 \mathbf{pBA} \\ \dot{\mathbf{r}}_{i+1} &= \dot{\mathbf{r}}_i + h\mathbf{qBA} \end{aligned}\right\} \tag{3}$$

Table 2 Parameters $p_j = \delta_{j+1} - \gamma_j^* - \delta_{j+1}^*$ and $q_j = \gamma_j - \gamma_j^*$ for the numerical integration with the modified Gauss-Jackson integrator

j	0	1	2	3	4	5
p_j	0	$\frac{1}{2}$	$\frac{1}{6}$	$\frac{1}{8}$	$\frac{19}{180}$	$\frac{3}{32}$
q_j	0	1	$\frac{1}{2}$	$\frac{5}{12}$	$\frac{3}{8}$	$\frac{251}{720}$

j	6	7	8	9	10	11
p_j	$\frac{33953}{453600}$	$\frac{8183}{115200}$	$\frac{3250433}{47900160}$	$\frac{4671}{71680}$	$\frac{275}{3456}$	$\frac{863}{10080}$
q_j	$\frac{110397}{362880}$	$\frac{8183}{115200}$	$\frac{3250433}{47900160}$	$\frac{4671}{71680}$	$\frac{19087}{60480}$	$\frac{665}{2016}$

These coefficients can be obtained from the Adams-Bashforth coefficients γ_j, the Adams-Moulton coefficients γ_j^* and Stoermer and Cowell coefficients δ_j, δ_j^*

where $\mathbf{r}_{i+1}$, $\dot{\mathbf{r}}_{i+1}$ (row vectors) are the predicted position and velocity vectors; $\mathbf{r}_i$ and $\dot{\mathbf{r}}_i$ (row vectors) are the position and velocity vectors at a current time and h is the integration step size. **pBA** and **qBA** represent the summation parts of the integrator, where the specific integration order is considered. For the computation of the results presented in this study an integration order of 12 was used. In the modified version of the Gauss-Jackson integrator this integration order corresponds to a summation over indices $j = [1, m]$ where m is the integration order decreased by one. The row vectors **p** and **q** contain the coefficients p_j and q_j that are defined in Table 2. Matrix **A** contains row-wise the acceleration vector at the current time stamp i as well as the accelerations of $m - 1$ back points. The result of multiplying matrix **B** with matrix **A** are the acceleration backward differences up to order $m - 1$. The entries of matrix **B** are summarized in Table 3. For initialization a Runge-Kutta integrator of order 4 is used.

6   Parametrization

The parametrization chosen for the two-step approach is summarized in Table 4. The main idea behind the two-

Table 4 Specifics of the two-step approach applied for the LUH-GRACE2018 time series of monthly gravity field solutions

Step 1: Orbit pre-adjustment	
Arc length	3 h
Numerical integrator	Modified Gauss-Jackson
Integration step size	5 s
Observations	GNV1B positions (5 s)
Weighting	Identity matrix
Local parameters	Initial state, acc. biases
Global parameters	No
Constraints	Not applied
Regularization	Not applied
Step 2: Orbit adjustment and gravity field recovery	
Arc length	3 h
Numerical integrator	Modified Gauss-Jackson
Integration step size	5 s
Observations	GNV1B positions (30 s)
	K-band range-rates (5 s)
Weighting	GNV1B positions $\sigma_0 = 0.02$ m
	KBRR $\sigma_0 = 2 \times 10^{-7}$ m/s
Local parameters	Initial state, acc. biases, empirical KBRR parameters
Global parameters	Spherical harmonic coefficients up to degree and order 80 (60)
Constraints	Not applied
Regularization	Not applied

step approach is to reduce the number of iterations during gravity field recovery and therefore to save computational time. This is achieved by estimating appropriate a priori values for the initial satellite states as well as accelerometer calibration parameters using only reduced-dynamic orbit positions as observations in the orbit pre-adjustment (Wang et al. 2015). For this release of monthly gravity field solutions the accelerometer biases are estimated arc-wise while the scale factors are held fixed to a priori values reported in Bettadpur (2009). The satellite orbits are parametrized by an arc length of 3 h. The aim of this arc length is to allow a more precise orbit fit to observations, as inaccuracies, e.g. in force modeling, can be balanced by the very frequent estimation of local arc parameters. Compared to the usual

Table 3 Entries of the matrix **B** for the vectorized computation of backward differences

1	0	0	0	0	0	0	0	0	0	0
1	−1	0	0	0	0	0	0	0	0	0
1	−2	1	0	0	0	0	0	0	0	0
1	−3	3	−1	0	0	0	0	0	0	0
1	−4	6	−4	1	0	0	0	0	0	0
1	−5	10	−10	5	−1	0	0	0	0	0
1	−6	15	−20	15	−6	1	0	0	0	0
1	−7	21	−35	35	−21	7	−1	0	0	0
1	−8	28	−56	70	−56	28	−8	1	0	0
1	−9	36	−84	126	−126	84	−36	9	−1	0
1	−10	45	−120	210	−252	210	−120	45	−10	1

arc length of one day, no constrained dynamic empirical parameters, such as cycle per revolution accelerations or more frequent stochastic parameters, have to be co-estimated in order to achieve a good orbit fit. In the second step the K-band range-rates are used as main observations, along with reduced-dynamic positions. As additional unknowns, kinematic empirical KBRR parameters are introduced. In total 8 empirical parameters consisting of bias and bias-rates as well as periodic bias and bias-rates are estimated per arc (Kim 2000). We solve for the bias and bias-rates every 90 min and for periodic bias and bias-rates every 180 min. Range-rates and reduced-dynamic orbit positions are combined on normal equation level using a technique-specific weighting. The spherical harmonic coefficients of the Earth's gravity potential are obtained together with the arc-specific parameters without any constraints, regularizations or stochastic accelerations. The number of parameters to be estimated in the second step of the procedure equals 6557 global parameters representing the normalized spherical harmonic coefficients of the Earth's potential (for maximum degree 80) as well as 6448 arc-specific parameters if 31 days of observations are considered.

7   GRACE-FO Solutions

In Sect. 8 we will jointly evaluate the LUH-GRACE2018 gravity field solutions (2003–2016) as well as first GRACE-FO results (June 2018–August 2019). For the GRACE-FO results a slightly modified processing strategy was applied. The most recent de-aliasing product AOD1B RL06 (Dobslaw et al. 2017) up to degree and order 180 was used. In addition, atmospheric tides were considered according to Biancale and Bode (2006). Accelerometer scale factors were estimated arc-wise along with the biases. The Earth's gravity potential was estimated up to degree and order 96.

8   Results and Evaluation

The convergence of exemplary gravity field solutions in terms of the error degree standard deviation for March 2006 (good ground track coverage) and September 2004 (sparse ground track coverage) are shown in Fig. 2. Arcs with a length of 3 h are stacked successively, overall leading to a decrease of the error degree standard deviations of the corresponding solutions. Figure 2a, b shows the convergence from the beginning, highlighting the rapid convergence during the first seven days. For the remainder of a month, the convergence rate is much slower as can be seen in Fig. 2c, d. The results demonstrate that satisfactory sub-monthly

gravity field solutions of the time-variable Earth's gravity potential can often be obtained without any constraints, regularizations or a priori information such as solutions from previous months or weeks. Note that the convergence rate can vary significantly depending on sensor data quality and distribution of ground tracks.

The computed monthly gravity field solutions are compared to the solutions of the three official ACs: CSR (Center for Space Research, The University of Texas at Austin), GFZ (German Research Centre for Geosciences) and JPL (Jet Propulsion Laboratory, California Institute of Technology). Since we used AOD1B RL05 products for the correction of rapid mass variations, our solutions are compared to the RL05 previous generation solutions, in order to allow for a fair comparison. The reference solutions CSR RL05 (Bettadpur 2012), GFZ RL05a (Dahle et al. 2012) and JPL RL05 (Watkins and Yuan 2014) are obtained from the International Centre for Global Earth Models (ICGEM) (Ince et al. 2019).

Error degree standard deviations of the monthly solutions are illustrated in Fig. 4. The mean error degree standard deviations are computed for the two periods 2003–2009 and 2010–2016 separately. As a reference field the recent static gravity field model GOCO06s (Kvas et al. 2019) was subtracted from the solutions of all 4 ACs. It can be seen that in general the noise characteristics of the solutions are similar, not considering degree 2. In general the amplitudes of the LUH error degree standard deviations are slightly larger when compared to the other ACs. Degree 4 for both periods as well as degree 3 for the second period show larger error degree standard deviations indicating a larger noise for these degrees. A comparison on the coefficients level showed that the degree 4 deviation is caused by the $\overline{C}_{44}$ and $\overline{S}_{44}$ coefficients. We presume that this deviation is caused by the applied parametrization, since the mean error degree standard deviations of about one year of GRACE-FO processing with a slightly changed parametrization does not show any significant deviations in the low degrees. Nevertheless, further investigations are needed. Error degree standard deviations show effects of orbital resonances (Cheng and Ries 2017) near degrees 31 and 46. The orbital resonance near degree 62 is missing in the second period leading to a smaller noise for the higher degrees 62–80 compared to period 1.

A comparison of exemplary global maps of mass variations in terms of Equivalent Water Height (EWH) can be seen in Fig. 3. Global maps based on CSR RL05 and the LUH-GRACE2018 solutions for every second month of the year 2008 are shown with respect to the static model GOCO06s. The $\overline{C}_{20}$ coefficients of all solutions were replaced by values obtained from Satellite Laser Ranging (SLR) (Cheng and

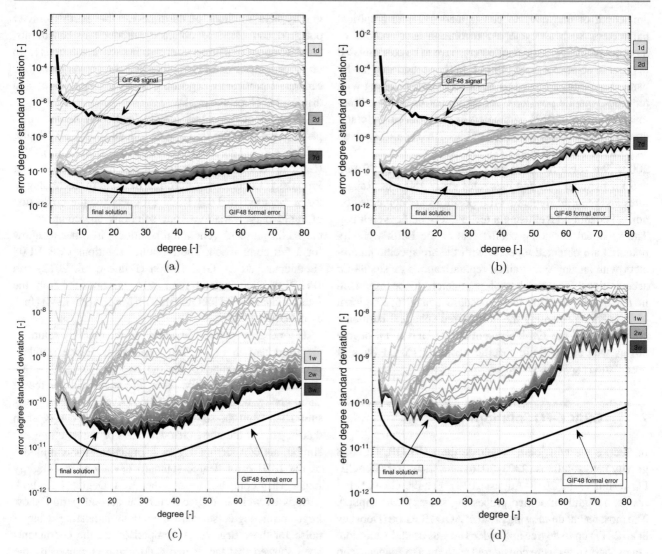

Fig. 2 Exemplary successive stacking of 3 hourly orbital arcs and the error degree standard deviations of the corresponding gravity field solution for March 2006 (good ground track coverage) and September 2004 (sparse ground track coverage) with respect to the gravity field model GIF48. The error degree standard deviations for 1 day, 2 days, 7 days, 2 weeks and 3 weeks of observation data are highlighted. (**a**) March 2006. (**b**) September 2004. (**c**) March 2006 zoom. (**d**) September 2004 zoom

Ries 2017). In order to mitigate the meridional North-South stripes, the spherical harmonic coefficients differences were smoothed using the Gaussian filter (Wahr et al. 1998) with a half width of 400 km before computing the EWH. It can be seen that larger mass variations such as in the Amazon region, central Africa, India, Southeast Asia, Greenland and Northeast Canada can be localized equally well in both CSR and LUH solutions. The meridional stripes in the LUH maps

are slightly stronger, which is consistent with the slightly higher error degree standard deviations in Fig. 4, and their characteristics suggest differences in the analysis noise that should be further studied.

Exemplary Equivalent Water Height time series for Greenland and the Amazon and Ganges basins based on the same processing as applied for the global EWH maps are shown in Fig. 5. The time series cover the GRACE period and

Fig. 3 Global Equivalent Water Height (EWH) for every second month of the year 2008. Left side: CSR RL05 solutions; right side: LUH solutions. The $\overline{C}_{20}$ coefficients were replaced by SLR values in all solutions. As a reference model the static gravity field model GOCO06s (reference epoch: 2010-01-01) was subtracted. The spherical harmonic coefficients differences were smoothed using the Gaussian filter with a half width of 400 km

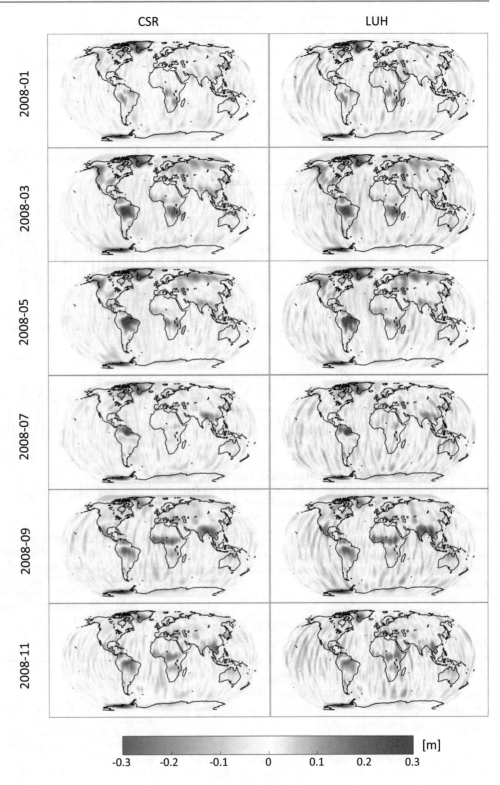

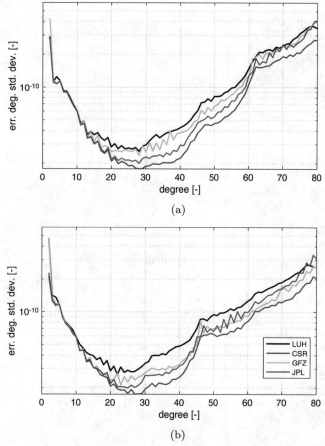

Fig. 4 Mean of error degree standard deviations of the four analysis centers for degrees 2–80 with respect to the static model GOCO06s. The solutions were divided into two periods: 2003–2009 (**a**) and 2010–2016 (**b**). The $\overline{C}_{20}$ coefficients were not replaced. All solutions are zerotide. A specific month is considered when a solution from all 4 centers is available. We define a monthly solution as a solution that is based on satellite data from one calendar month, i.e. solutions combining sensor data of neighboring months are not considered. Periods when regualrizations were applied are excluded

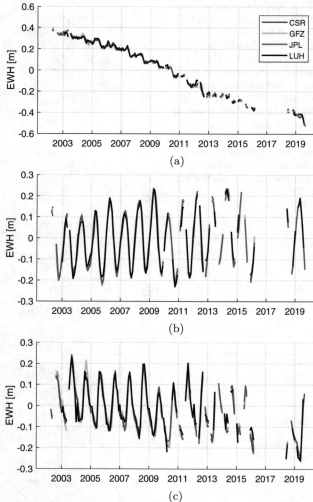

Fig. 5 Equivalent Water Heights (EWH) in three regions: (**a**) Greenland, (**b**) Amazon basin, (**c**) Ganges basin. EWH values are region mean values. The $\overline{C}_{20}$ coefficients in all solutions were replaced by SLR values. As a reference model the static gravity field model GOCO06s (reference epoch: 2010-01-01) was subtracted. The spherical harmonic coefficients differences were smoothed using the Gaussian filter with a half width of 400 km

additionally include the first available GRACE-FO solutions. The EWH signals of all four ACs covering a time span of more than 16 years show a high degree of consistency for all three regions. This indicates that the signal content of the four ACs' solutions does not exhibit large differences. The degradation of the GRACE sensor data manifests as gaps in the second half of the time series.

Acknowledgements We are thankful for the valuable comments of the two anonymous reviewers who helped to improve the manuscript considerably. Part of this work was funded by Deutsche Forschungsgemeinschaft (DFG) (SFB1128 geo-Q: Collaborative Research Centre 1128 "Relativistic Geodesy and Gravimetry with Quantum Sensors").

References

Berry MM, Healy LM (2004) Implementation of Gauss-Jackson integration for orbit propagation. J Astronaut Sci 52(3):331–357

Bettadpur S (2009) Recommendation for a-priori bias & scale parameters for level-1B acc data (version 2), GRACE TN-02. Technical note. Center for Space Research, The University of Texas at Austin

Bettadpur S (2012) UTCSR level-2 processing standards document for level-2 product release 0005. Technical report GRACE, 327–742. Center for Space Research, The University of Texas at Austin

Biancale R, Bode A (2006) Mean annual and seasonal atmospheric tide models based on 3-hourly and 6-hourly ECMWF surface pressure data. Scientific technical report STR06/01. GeoForschungsZentrum Potsdam, Germany

Carrere L, Lyard F, Cancet M, Guillot A (2015) FES2014, a new tidal model on the global ocean with enhanced accuracy in shallow seas and in the Arctic region. Geophys Res Abstr, EGU2015-5481. EGU General Assembly 2015, Vienna, Austria

Case K, Kruizinga G, Wu S-C (2010) GRACE Level 1B data product user handbook (JPL D-22027), Technical report. Jet Propulsion Laboratory, California Institute of Technology, Pasadena, CA, USA

Cheng MK, Ries JC (2017) The unexpected signal in GRACE estimates of C20. J Geod 91(8):897–914. https://doi.org/10.1007/s00190-016-0995-5

Dahle C, Flechtner F, Gruber C, König D, König R, Michalak G, Neumayer, K-H (2012) GFZ GRACE level-2 processing standards document for level-2 product release 0005: revised edition, January 2013, (Scientific Technical Report STR - Data ; 12/02 rev. ed.), Deutsches GeoForschungsZentrum GFZ, 21 p. http://doi.org/10.2312/GFZ.b103-1202-25

Desai SD (2002) Observing the pole tide with satellite altimetry. J Geophys Res 107(C11). https://doi.org/10.1029/2001JC001224

Dobslaw H, Bergmann-Wolf I, Dill R, Poropat L, Thomas M, Dahle C, Esselborn S, König R, Flechtner F (2017) A new high-resolution model of non-tidal atmosphere and ocean mass variability for de-aliasing of satellite gravity observations: AOD1B RL06. Geophys J Int 211(1):263–269. https://doi.org/10.1093/gji/ggx302

Dobslaw H, Flechtner F, Bergmann-Wolf I, Dahle C, Dill R, Esselborn S, Sasgen I, Thomas M (2013) Simulating high-frequency atmosphere-ocean mass variability for dealiasing of satellite gravity observations: AOD1B RL05. J Geophys Res Oceans 118:3704–3711. https://doi.org/10.1002/jgrc.20271

Ince ES, Barthelmes F, Reißland S, Elger K, Förste C, Flechtner F, Schuh H (2019) ICGEM – 15 years of successful collection and distribution of global gravitational models, associated services and future plans. Earth Syst Sci Data 11:647–674. http://doi.org/10.5194/essd-11-647-2019

Kim J (2000) Simulation study of a low-low satellite-to-satellite tracking mission, PhD thesis, The University of Texas at Austin

Kvas A, Mayer-Gürr T, Krauß S, Brockmann JM, Schubert T, Schuh W-D, Pail R, Gruber T, Meyer U, Jäggi A (2019) The satellite-only gravity field model GOCO06s. EGU General Assembly 2019, 7.-12. April 2019, Vienna, Austria. https://doi.org/10.13140/RG.2.2.14101.99047

Meyer U, Jäggi A, Beutler G, Bock H (2015) The impact of common versus separate estimation of orbit parameters on GRACE gravity field solutions. J Geod 89(7):685–696. https://doi.org/10.1007/s00190-015-0807-3

Meyer U, Jenny B, Dahle C, Flechtner F, Save H, Bettadpur S, Landerer F, Boening C, Kvas A, Mayer-Gürr T, Lemoine JM, Bruinsma S, Jäggi A (2018) COST-G: The new international combination service for time-variable gravity field solutions of the IAG/IGFS. GRACE/GRACE-FO Science Team Meeting 2018, 9.-11. October 2018, Potsdam, Germany

Montenbruck O, Gill E (2005) Satellite orbits – Models, Methods and Applications, 3rd edn., Springer, Berlin, Germany, ISBN: 978-3-642-58352-0. http://doi.org/10.1007/978-3-642-58351-3

Naeimi M (2018) A modified Gauss-Jackson method for the numerical integration of the variational equations. EGU General Assembly 2018, 8.-13. April 2018, Vienna, Austria

Naeimi M, Koch I, Khami A, Flury J (2018) IfE monthly gravity field solutions using the variational equations. EGU General Assembly 2018, 8.-13. April 2018, Vienna, Austria. https://doi.org/10.15488/4452

Petit G, Luzum B. (2010) IERS Conventions (2010), IERS Technical Note No. 36. Verlag des Bundesamts für Kartographie, Frankfurt am Main, Germany

Ries JC, Bettadpur S, Poole S, Richter T (2011) Mean Background Gravity Fields for GRACE Processing. GRACE Science Team Meeting 2011, 8.-10. August 2011, Austin, TX, USA

Rieser D, Mayer-Gürr T, Savchenko R, Bosch W, Wünsch J, Dahle C, Flechtner F (2012) The ocean tide model EOT11a in spherical harmonics representation, Technical note

Savchenko R, Bosch W (2012) EOT11a - empirical ocean tide model from multi-mission satellite altimetry. DGFI Report No. 89, Deutsches Geodätisches Forschungsinstitut (DGFI), München, Germany

Standish EM (1998) JPL planetary and lunar ephemerides, DE405/LE405 (JPL Iteroffice Memorandum IOM 312.F-98-048)

Vallado DA (2013) Fundamentals of astrodynamics and applications, 4th edn. Microcosm Press, Hawthorne, CA, USA, ISBN: 978-188188318-0

Wahr J, Molenaar M, Bryan F (1998) Time variability of the Earth's gravity field: Hydrological and oceanic effects and their possible detection using GRACE. J Geophys Res 103(B12):30205–30229. https://doi.org/10.1029/98JB02844

Wang C, Xu H, Zhong M, Feng W (2015) Monthly gravity field recovery from GRACE orbits and K-band measurements using variational equations approach. Geodesy Geodynamics 6(4):253–260. https://doi.org/10.1016/j.geog.2015.05.010

Watkins MM, Yuan D (2014) JPL level-2 processing standards document for level-2 product release 05.1. Technical report GRACE, 327–744, Jet Propulsion Laboratory, California Institute of Technology, Pasadena, CA, USA

A Precise Geoid Model for Africa: AFRgeo2019

Hussein A. Abd-Elmotaal, Norbert Kühtreiber, Kurt Seitz, and Bernhard Heck

Abstract

In the framework of the IAG African Geoid Project, an attempt towards a precise geoid model for Africa is presented in this investigation. The available gravity data set suffers from significantly large data gaps. These data gaps are filled using the EIGEN-6C4 model on a $15' \times 15'$ grid prior to the gravity reduction scheme. The window remove-restore technique (Abd-Elmotaal and Kühtreiber, Phys Chem Earth Pt A 24(1):53–59, 1999; J Geod 77(1–2):77–85, 2003) has been used to generate reduced anomalies having a minimum variance to minimize the interpolation errors, especially at the large data gaps. The EIGEN-6C4 global model, complete to degree and order 2190, has served as the reference model. The reduced anomalies are gridded on a $5' \times 5'$ grid employing an un-equal weight least-squares prediction technique. The reduced gravity anomalies are then used to compute their contribution to the geoid undulation employing Stokes' integral with Meissl (Preparation for the numerical evaluation of second order Molodensky-type formulas. Ohio State University, Department of Geodetic Science and Surveying, Rep 163, 1971) modified kernel for better combination of the different wavelengths of the earth's gravity field. Finally the restore step within the window remove-restore technique took place generating the full gravimetric geoid. In the last step, the computed geoid is fitted to the DIR_R5 GOCE satellite-only model by applying an offset and two tilt parameters. The DIR_R5 model is used because it turned out that it represents the best available global geopotential model approximating the African gravity field. A comparison between the geoid computed within the current investigation and the existing former geoid model AGP2003 (Merry et al., A window on the future of geodesy. International Association of Geodesy Symposia, vol 128, pp 374–379, 2005) for Africa has been carried out.

Keywords

Africa · Geoid determination · Gravity field · Gravity interpolation · Window technique

H. A. Abd-Elmotaal (✉)
Civil Engineering Department, Faculty of Engineering, Minia University, Minia, Egypt
e-mail: hussein.abdelmotaal@gmail.com

N. Kühtreiber
Institute of Geodesy, Graz University of Technology, Graz, Austria
e-mail: norbert.kuehtreiber@tugraz.at

K. Seitz · B. Heck
Geodetic Institute, Karlsruhe Institute of Technology, Karlsruhe, Germany
e-mail: kurt.seitz@kit.edu; bernhard.heck@kit.edu

1 Introduction

The geoid, being the natural mathematical figure of the earth, serves as height reference surface for geodetic, geophysical and many engineering applications. It is directly connected with the theory of equipotential surfaces (Heiskanen and Moritz 1967; Hofmann-Wellenhof and Moritz 2006), and its determination needs sufficient coverage of observation data related to the earth's gravity field, such as gravity

© The Author(s) 2020
J. T. Freymueller, L. Sanchez (eds.), *Beyond 100: The Next Century in Geodesy*,
International Association of Geodesy Symposia 152, https://doi.org/10.1007/1345_2020_122

anomalies. In this investigation, a geoid model for Africa will be determined. The challenge we face here consists in the available data set, which suffers from significantly large gaps, especially on land.

The available data for this investigation is a set of gravity anomalies, both on land and sea. The geoid is computed using the Stokes' integral, which requires interpolating the available data into a regular grid. In order to reduce the interpolation errors, especially in areas of large data gaps, the window remove-restore technique (Abd-Elmotaal and Kühtreiber 1999, 2003) is used. The window technique doesn't suffer from the double consideration of the topographic-isostatic masses in the neighbourhood of the computational point, and accordingly produces un-biased reduced gravity anomalies with minimum variance.

In order to control the gravity interpolation in the large data gaps, these gaps are filled-in, prior to the interpolation process, with an underlying grid employing the EIGEN-6C4 geopotential model (Förste et al. 2014a,b). Hence the interpolation process took place using the unequal weight least-squares prediction technique (Moritz 1980).

Finally, the computed geoid within the current investigation is fitted to the DIR_R5 GOCE satellite-only model by applying an offset and two tilt parameters. This adjustment reduces remaining tilts and a vertical offset in the model. Previous studies (Abd-Elmotaal 2015) have shown that the DIR_R5 GOCE model is best suited for this purpose on the African continent.

The first attempt to compute a geoid model for Africa has been made by Merry (2003) and Merry et al. (2005). A $5' \times 5'$ mean gravity anomaly grid developed at Leeds University was used to compute that geoid model. We regret that this data set has never become available since then again. For the geoid computed by Merry et al. (2005), the remove-restore method, based on the EGM96 geopotential model (Lemoine et al. 1998), was employed. Another geoid model for Africa has been computed by Abd-Elmotaal et al. (2019). This geoid model employed the window remove-restore technique with the EGM2008 geopotential model (Pavlis et al. 2012), up to degree and order 2160, and a tailored reference model (computed through an iterative process), up to degree and order 2160, to fill in the data gaps.

Due to problems with a data set in Morocco, used in the former solution AFRgeo_v1.0 (Abd-Elmotaal et al. 2019), the computed geoid has been compared only to the AGP2003 model (Merry et al. 2005) in the present paper.

2 The Data

2.1 Gravity Data

The available gravity data set for the current investigation comprises data on land and sea. The sea data consists of shipborne point data and altimetry-derived gravity anomalies along tracks. The latter data set was derived from the average of 44 repeated cycles of the satellite altimetry mission GEOSAT by the National Geophysical Data Center NGDC (www.ngdc.noaa.gov) (Abd-Elmotaal and Makhloof 2013, 2014). The goal of the African Geoid Project is the calculation of the geoid on the African continent. Data within the data window which are located on the oceans (shipborne and altimetry data) are used to stabilize the solution at the continental margins to avoid the Gibbs phenomenon.

The land point gravity data, being the most important data set for the geoid at the continent, have passed a laborious gross-error detection process developed by Abd-Elmotaal and Kühtreiber (2014) using the least-squares prediction technique (Moritz 1980). This gross-error detection process estimates the gravity anomaly at the computational point using the neighbour points and defines a possible gross-error by comparing it to the data value. The gross-error detection process deletes the point from the data set if it proves to be a real gross-error after examining its effect to the neighbourhood points. Furthermore, a grid-filtering scheme (Abd-Elmotaal and Kühtreiber 2014) on a grid of $1' \times 1'$ is applied to the land data to improve the behaviour of the empirical covariance function especially near the origin (Kraiger 1988). The statistics of the land free-air gravity anomalies, after the gross-error detection and the grid-filtering, are illustrated in Table 2. Figure 1a shows the distribution of the land gravity data set.

The shipborne and altimetry-derived free-air anomalies have passed a gross-error detection scheme developed by Abd-Elmotaal and Makhloof (2013), also based on the least-squares prediction technique. It estimates the gravity anomaly at the computational point utilizing the neighbourhood points, and defines a possible blunder by comparing it to the data value. The gross-error technique works in an iterative scheme till it reaches 1.5 mgal or better for the discrepancy between the estimated and data values. A combination between the shipborne and altimetry data took place (Abd-Elmotaal and Makhloof 2014). Then a grid-filtering process on a grid of $3' \times 3'$ has been applied to the shipborne and altimetry-derived gravity anomalies to decrease their dominating effect on the gravity data set. The statistics of the shipborne and altimetry-derived free-air anomalies, after the gross-error detection and grid-filtering, are listed in Table 2. The distribution of the shipborne and altimetry data is given in Fig. 1b and c, respectively. More details about the used data sets can be found in Abd-Elmotaal et al. (2018).

2.2 Digital Height Models

If the computation of the topographic reduction is carried out with a software such as TC-program, a fine DTM for the near-zone and a coarse one for the far-zone are required. The TC-software originates from Forsberg (1984). In this investigation a program version was used which was modified by Abd-Elmotaal and Kühtreiber (2003). A set of DTMs for Africa covering the window ($-42° \leq \phi \leq 44°; -22° \leq \lambda \leq 62°$) are available for the current investigation. The AFH16S30 $30'' \times 30''$ and the AFH16M03 $3' \times 3'$ models (Abd-Elmotaal et al. 2017) have been chosen to represent the fine and coarse DTMs, respectively. Figure 2 illustrates the AFH16S30 $30'' \times 30''$ fine DTM for Africa. The heights range between -8291 and 5777 m with an average of -1623 m.

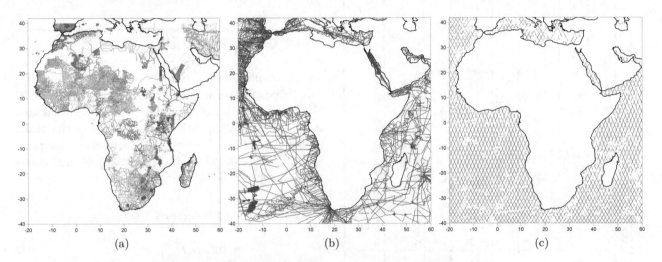

Fig. 1 Distribution of the (**a**) land, (**b**) shipborne and (**c**) altimetry free-air gravity anomaly points for Africa

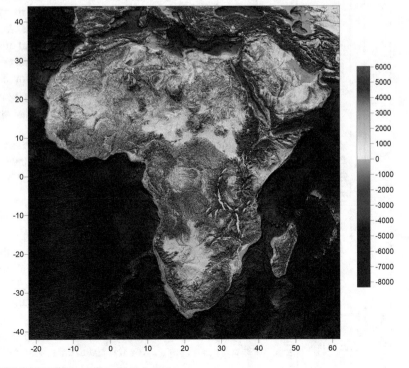

Fig. 2 The $30'' \times 30''$ AFH16S30 DTM for Africa. Units in [m]

Table 1 Used data and techniques in the African Geoid Project

Geoid version	AGP2003	AFRgeo_v1.0	AFRgeo2019
Data base	Leeds University	AFRGDB_v1.0 (Abd-Elmotaal et al. 2015)	AFRGDB_v2.0 and AFRGDB_v2.2 (Abd-Elmotaal et al. 2018, 2020)
Land	5′ × 5′ grided	96,472 (point values)	126,202 (grid filtering)
Shipborne	–	971,945 (point values)	148,674 (grid filtering)
Altimetry	–	119,249 (point values)	70,589 (grid filtering)
Underlying grid	N/A	Tailored model for Africa $N_{max} = 360$ on a 15′ × 15′ unregistered grid	EIGEN-6C4 $N_{max} = 2,190$ on a 15′ × 15′ unregistered grid
Reference model	EGM96 $N_{max} = 360$	Tailored model for Africa through an iterative process $N_{max} = 2,160$	EIGEN-6C4 $N_{max} = 2,190$
De-trended by	–	GOCE DIR_R5 $N_{max} = 280$	GOCE DIR_R5 $N_{max} = 280$
Fine DTM	GLOBE 1′ × 1′ (Hastings and Dunbar 1998)	SRTM30+ 30″ × 30″ (Farr et al. 2007)	AFH16S30 30″ × 30″ (Abd-Elmotaal et al. 2017)
Coarse DTM	GLOBE 5′ × 5′ (Hastings and Dunbar 1998)	SRTM30+ 3′ × 3′ (Farr et al. 2007)	AFH16M03 3′ × 3′ (Abd-Elmotaal et al. 2017)
Reduction technique	Residual Terrain Modelling (RTM) technique	Window remove-restore technique using a tailored reference model for Africa	Window remove-restore technique using EIGEN-6C4 till ultra high degree (2190)
Geoid determination technique	Molodensky integrals with original Stokes kernel, quasigeoid to geoid separation using Rapp (1997)	Stokes integral with original Stokes kernel, 1D-FFT	Stokes integral with Meissl (1971) modified kernel, solution in the space domain

2.3   A Short History of Used Data

The data used to calculate the current geoid solution for Africa have been described in Sects. 2.1 and 2.2. Data acquisition is a continuous tedious task, especially for point gravity values on land. As can be seen in Fig. 1, significant data gaps still need to be closed despite great efforts. In fact, since the first basic calculation of an African geoid by Merry (2003) and Merry et al. (2005), the point gravity data situation is continuously improving, although the original data of Merry et al. (2005) are no longer available. This can be concluded from Table 1. It should be mentioned that no ocean data had been used in the former AGP2003 solution.

3   Gravity Reduction

As stated earlier, in order to get un-biased reduced anomalies with minimum variance, the window remove-restore technique is used. The remove step of the window remove-restore technique when using the EIGEN-6C4 geopotential model (Förste et al. 2014a,b), complete to degree and order 2190, as the reference model can be expressed by (Abd-Elmotaal and Kühtreiber 1999, 2003) (cf. Fig. 3)

$$\Delta g_{win-red} = \Delta g_F - \Delta g_{TI\,win} - \Delta g_{EIGEN\text{-}6C4}\Big|_{n=2}^{n_{max}} +$$
$$+ \Delta g_{wincof}\Big|_{n=2}^{n_{max}} , \tag{1}$$

where $\Delta g_{win-red}$ refers to the window-reduced gravity anomalies, Δg_F refers to the measured free-air gravity anomalies, $\Delta g_{EIGEN\text{-}6C4}$ stands for the contribution of the global reference geopotential model, $\Delta g_{TI\,win}$ is the contribution of the topographic-isostatic masses for the fixed data window, Δg_{wincof} stands for the contribution of the harmonic coefficients of the topographic-isostatic masses of the same data window and n_{max} is the maximum degree ($n_{max} = 2,190$ is used).

For the underlying grid, which is intended to support the boundary values, particularly in areas of data gaps, the free-air gravity anomalies are computed by

$$\Delta g_F = \Delta g_{EIGEN\text{-}6C4}\Big|_{n=2}^{n_{max}} \tag{2}$$

on a 15′ × 15′ grid. This is three times the resolution of the output grid. To avoid identical grid points between the underlying grid and the output grid, the underlying grid is shifted by 2.5′ relative to the output grid. Therefore both grids are called unregistered.

The contribution of the topographic-isostatic masses $\Delta g_{TI\,win}$ for the fixed data window ($-42° \leq \phi \leq 44°$; $-22° \leq \lambda \leq 62°$) is computed using TC-program (Forsberg

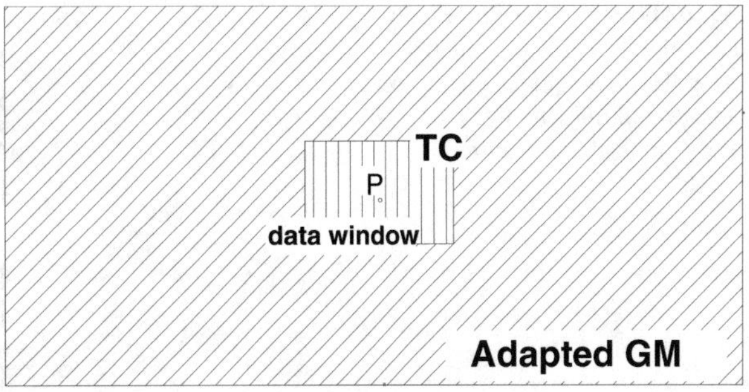

Fig. 3 The window
remove-restore technique

Table 2 Statistics of the free-air
and reduced gravity anomalies

Anomaly type	Category	No. of points	Statistical parameters			
			min	max	Mean	Std
Free-air	Land	126, 202	−163.20	465.50	9.84	40.93
	Shipborne	148, 674	−238.30	354.40	−6.21	34.90
	Altimetry	70, 589	−172.23	156.60	4.09	18.23
	Total	345, 465	−238.30	465.50	1.76	35.44
	Underlying	48, 497	−201.09	500.30	3.45	32.81
Window-reduced	Land	126, 202	−125.26	110.44	−0.19	6.63
	Shipborne	148, 674	−60.24	58.96	−0.88	9.90
	Altimetry	70, 589	−75.26	98.09	6.67	10.14
	Total	345, 465	−125.26	110.44	0.92	9.37
	Underlying	48, 497	−79.94	149.08	0.42	5.85

Units in [mgal]

1984; Abd-Elmotaal and Kühtreiber 2003). The following commonly used parameter set (cf. Kaban et al. 2016; Braitenberg and Ebbing 2009; Heiskanen and Moritz 1967, p. 327) is implemented

$$T_\circ = 30 \, \text{km},$$
$$\rho_\circ = 2.67 \, \text{g/cm}^3, \tag{3}$$
$$\Delta\rho = 0.40 \, \text{g/cm}^3,$$

where $T_\circ$ is the normal crustal thickness, $\rho_\circ$ is the density of the topography and $\Delta\rho$ is the density contrast between the crust and the mantle.

The contribution of the involved harmonic models is computed by the technique developed by Abd-Elmotaal (1998). Alternative techniques can be found, for example, in Rapp (1982) or Tscherning et al. (1994). The potential harmonic coefficients of the topographic-isostatic masses for the data window are computed using the rigorous expressions developed by Abd-Elmotaal and Kühtreiber (2015).

Table 2 illustrates the statistics of the free-air and reduced anomalies for each data category. The great reduction effect using the window remove-restore technique in terms of both the mean and the standard deviation for all data categories is obvious. What is very remarkable is the dramatic drop of the standard deviation of the most important data source, the land gravity data, by about 84%. This indicates that the used reduction technique works quite well. Table 2 also

shows that the underlying grid has a compatible statistical behaviour with the other data categories, which is needed for the interpolation process.

4 Interpolation Technique

An unequal weight least-squares interpolation technique (Moritz 1980) on a $5' \times 5'$ grid covering the African window ($40°S \le \phi \le 42°N$, $20°W \le \lambda \le 60°E$) took place to generate the gridded window-reduced gravity anomalies $\Delta g^G_{win-red}$ from the pointwise window-reduced gravity anomalies $\Delta g_{win-red}$. The following standard deviations have been fixed after some preparatory investigations:

$$\sigma_{land} = 1 \, \text{mgal},$$
$$\sigma_{shipborne} = 3 \, \text{mgal},$$
$$\sigma_{altimetry} = 5 \, \text{mgal}, \tag{4}$$
$$\sigma_{underlying \, grid} = 20 \, \text{mgal}.$$

The generalized covariance model of Hirvonen has been used for which the estimation of the parameter p (related to the curvature of the covariance function near the origin) has been made through the fitting of the empirically determined covariance function by employing a least-squares regression algorithm developed by Abd-Elmotaal and Kühtreiber (2016). A value of $p = 0.364$ has been estimated. The

Fig. 4 Fitting of the empirically determined covariance function using the least-squares regression algorithm developed by Abd-Elmotaal and Kühtreiber (2016)

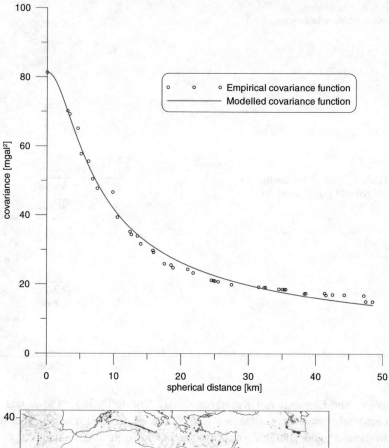

Fig. 5 Interpolated $5' \times 5'$ window-reduced anomalies $\Delta g^G_{win-red}$ for Africa. Units in [mgal]

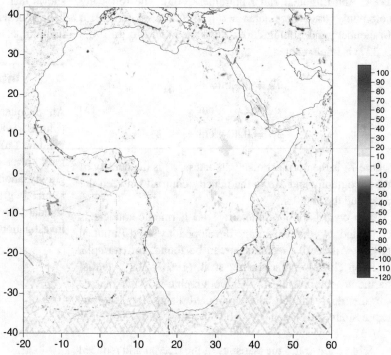

values of the empirically determined variance $C_\circ$ and correlation length ξ for the empirical covariance function are as follows:

$$C_\circ = 81.30 \text{ mgal}^2,$$
$$\xi = 10.38 \text{ km}. \tag{5}$$

Figure 4 shows the excellent fitting of the empirically determined covariance function performed by the above described process.

Figure 5 illustrates the $5' \times 5'$ interpolated window-reduced anomalies $\Delta g^G_{win-red}$ generated using the unequal weight least-squares interpolation technique employing the relative standard deviations described above and specified in Eq. (4). In most areas, the anomalies are less than 10 mgal, indicating that the modelling is appropriate for the remove step. This is particularly evident in the regions on the African mainland and there especially in the areas with large data gaps. Thus it can be concluded that the reduction and interpolation methods, especially developed for this data situation, have not led to any irregularities in the reduced anomalies. The efficiency of the used reduction and interpolation method has been validated by Abd-Elmotaal and Kühtreiber (2019), employing independent point gravity data not used in the interpolation process; this validation proved an external precision of about 7 mgal over various test areas on the African continent indicating the good feasibility of the applied approach.

5   Geoid Determination

For a better combination of the different wavelengths of the earth's gravity field (e.g., Featherstone et al. 1998; Abd-Elmotaal and Kühtreiber 2008), the contribution of the reduced gridded gravity anomalies $\Delta g^G_{win\text{-}red}$ to the geoid $N_{\Delta g_{win\text{-}red}}$ is determined on a $5' \times 5'$ grid covering the African window using Stokes' integral employing Meissl (1971) modified kernel, i.e.,

$$N_{\Delta g_{win\text{-}red}} = \frac{R}{4\pi\gamma} \iint_\sigma \Delta g^G_{win\text{-}red} \, K^M(\psi) \, d\sigma \,, \quad (6)$$

where $K^M(\psi)$ is the Meissl modified kernel, given by

$$K^M(\psi) = \begin{cases} S(\psi) - S(\psi_\circ) & \text{for } 0 < \psi \leq \psi_\circ \\ 0 & \text{for } \psi > \psi_\circ \end{cases} \,. \quad (7)$$

A value of the cap size $\psi_\circ = 3°$ has been used. $S(\cdot)$ is the original Stokes function. The choice of the Meissl modified kernel has been made because it proved to give good results (cf. Featherstone et al. 1998; Abd-Elmotaal and Kühtreiber 2008).

The full geoid restore expression for the window technique reads (Abd-Elmotaal and Kühtreiber 1999, 2003)

$$N = N_{\Delta g_{win\text{-}red}} + N_{TI\,win} + \zeta_{EIGEN\text{-}6C4} \Big|_{n=2}^{n_{max}} -$$
$$- \zeta_{wincof} \Big|_{n=2}^{n_{max}} + (N - \zeta)_{win} \,, \quad (8)$$

where $N_{TI\,win}$ gives the contribution of the topographic-isostatic masses (the indirect effect) for the same fixed data window as used for the remove step, $\zeta_{EIGEN\text{-}6C4}$ gives the contribution of the EIGEN-6C4 geopotential model, ζ_{wincof} stands for the contribution of the dimensionless harmonic coefficients of the topographic-isostatic masses of the data

window, and $(N - \zeta)_{win}$ is the conversion from quasi-geoid to geoid for the terms related to the quasi-geoid, i.e., $\zeta_{EIGEN\text{-}6C4}$ and ζ_{wincof}. The term $(N - \zeta)_{win}$ can be determined by applying the quasi-geoid to geoid conversion given by Heiskanen and Moritz (1967, p. 327) (see also Eq. (11)). This gives immediately

$$(N - \zeta)_{win} = \frac{H}{\bar{\gamma}} \left(\Delta g_{EIGEN\text{-}6C4} - \Delta g_{wincof} \right) \,, \quad (9)$$

where $\Delta g_{EIGEN\text{-}6C4}$ and Δg_{wincof} are the free-air gravity anomaly contributions of the EIGEN-6C4 geopotential model and the harmonic coefficients of the topographic-isostatic masses of the data window, respectively, and $\bar{\gamma}$ is a mean value of the normal gravity.

In order to fit the gravimetric geoid model for Africa to the individual height systems of the African countries, one needs some GNSS stations with known orthometric height covering the continental area. Unfortunately, despite our hard efforts, this data is still not available to the authors. As an alternative, the computed geoid is embedded using the GOCE DIR_R5 satellite-only model (Bruinsma et al. 2014), which is complete to degree and order 300. It represents the best available global geopotential model approximating the gravity field in Africa; this has been investigated by Abd-Elmotaal (2015). In the present application, the DIR_R5 model was evaluated up to d/o 280, since the signal-to-noise ratio for higher degrees is greater than one, and thus the coefficients of higher degrees are not considered. The general discrepancies between the GOCE DIR_R5 geoid and our calculated geoid solution have been represented by a trend model consisting of a vertical offset and two tilt parameters. These parameters have been estimated through a least-squares regression technique from the residuals between the two geoid solutions. This parametric model has been used to remove the trend which may be present in the computed geoid within the current investigation. This trend may be caused by errors in the long-wavelength components of the used reference model EIGEN-6C4 or the point gravity data. The Dir_R5 geoid undulations N_{Dir_R5} can be computed by

$$N_{Dir\_R5} = \zeta_{Dir\_R5} + (N - \zeta) \,, \quad (10)$$

where ζ_{Dir_R5} refers to the contribution of the Dir_R5 geopotential model, and the term $(N - \zeta)$ is computed by (Heiskanen and Moritz 1967, p. 327)

$$(N - \zeta) = \frac{\Delta g_{Dir\_R5} - 2\pi\rho_\circ GH}{\bar{\gamma}} H \,, \quad (11)$$

where Δg_{Dir_R5} refers to the free-air gravity anomalies computed by using the Dir_R5 geopotential model, G is Newton's gravitational constant, and $\rho_\circ$ is the density of the topography, given by Eq. (3).

Fig. 6 The AFRgeo2019
African de-trended geoid model.
Contour interval: 2 m

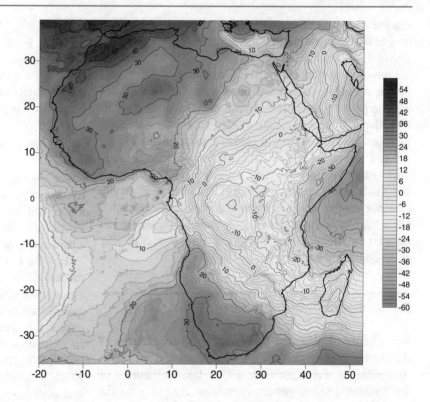

Figure 6 shows the AFRgeo2019 African de-trended geoid as stated above. The values of the AFRgeo2019 African geoid range between −55.34 and 57.34 m with an average of 11.73 m.

6 Geoid Comparison

As stated earlier, the first attempt to determine a geoid model for Africa "AGP2003" has been carried out by Merry (2003) and Merry et al. (2005). Since then, the data base has been further enhanced. In particular, the calculation method, statistical combination of the various types of gravity anomalies, has been revised and further developed. This has led to a significant improvement of the African geoid model. Figure 7 shows the difference between the de-trended AFRgeo2019 and the AGP2003 geoid models. The light yellow pattern in Fig. 7 indicates differences below 1 m in magnitude. Figure 7 shows that the differences between the two geoids amount to several meters in the continental area, especially in East Africa. The large differences over the Atlantic Ocean arise from the fact that the AGP2003 didn't include ocean data in the solution. Figure 7 shows some edge effects, which are again a direct consequence of using no data outside the African continent in the AGP2003 solution.

As the AFRGDB_v1.0, which has been the basis for computing the AFRgeo_v1.0, has been greatly influenced by a wrong data set in Morocco (cf. Abd-Elmotaal et al. 2015,

2019), it has been decided to skip the comparison between AFRgeo_v1.0 and the current geoid model.

7 Conclusions

In this paper, we successfully computed an updated version of the African geoid model. The computed geoid model is based on the window remove-restore technique (Abd-Elmotaal and Kühtreiber 2003), which gives very small and smooth reduced gravity anomalies. This helped to minimize the interpolation errors, especially in the areas of large data gaps. Filling these data gaps with synthesized gravity anomalies using the EIGEN-6C4 geopotential model, complete to degree and order 2190, has stabilized the interpolation process at the data gaps.

The reduced gravity anomalies employed for the AFRgeo2019 geoid model show a very good statistical behaviour (especially on land) because they are centered, smooth and have relatively small range (cf. Fig. 5 and Table 2). The smoothness of the residuals indicates that the interpolation technique proposed by Abd-Elmotaal and Kühtreiber (2019) did not induce aliasing effects, especially in the areas with point data gaps. Hence, they give less interpolation errors, especially in the large gravity data gaps. The reduced gravity data were interpolated using an unequal least-squares interpolation technique, giving the land data the highest precision, the sea data a moderate precision and the underlying grid the lowest precision.

Fig. 7 Difference between the de-trended AFRgeo2019 and the AGP2003 geoid models. Contour interval: 1 m

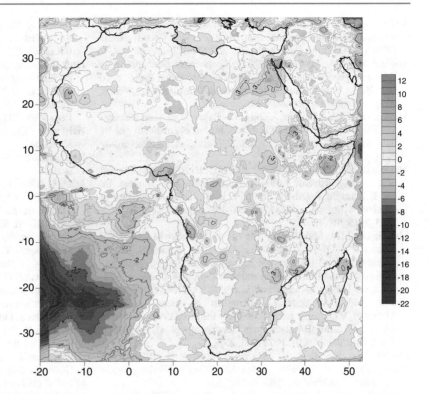

In order to optimally combine the spectral components in the remove-compute-restore technique, the Stokes function in the Stokes integral is replaced by a modified kernel function. In the geoid solution presented, the modification according to Meissl (1971) was used. Alternative modifications have been discussed by Wong and Gore (1969), Jekeli (1980), Wenzel (1982), Heck and Grüninger (1982), Featherstone et al. (1998) or Sjöberg (2003).

Finally, the computed geoid model for Africa has been de-trended by the use of the DIR_R5 GOCE model. In comparison with the previous model AGP2003, the progress made in determining the African height reference surface becomes visible.

Unfortunately, despite of strong efforts, extended precise GNSS positioning data over the African continent have not been made available to the authors. Thus, a rigorous comparison of the presented geoid model with an independent data set can only be made with further international efforts.

Acknowledgements The support by the International Association of Geodesy (IAG) and the International Union of Geodesy and Geophysics (IUGG) is kindly acknowledged. The authors would like to thank Dr. Sylvain Bonvalot, Director of the Bureau Gravimétrique International (BGI), for providing part of the used data set for Africa. The authors would like to thank the editor of this paper, Professor Roland Pail, and two anonymous reviewers for their useful suggestions and critical comments.

References

Abd-Elmotaal HA (1998) An alternative capable technique for the evaluation of geopotential from spherical harmonic expansions. Boll Geodesia Sci Affin 57(1):25–38

Abd-Elmotaal HA (2015) Validation of GOCE models in Africa. Newton's Bull 5:149–162. http://www.isgeoid.polimi.it/Newton/Newton_5/11_Hussein_149_162.pdf

Abd-Elmotaal HA, Kühtreiber N (1999) Improving the geoid accuracy by adapting the reference field. Phys Chem Earth Pt A 24(1):53–59. https://doi.org/10.1016/S1464-1895(98)00010-6

Abd-Elmotaal HA, Kühtreiber N (2003) Geoid determination using adapted reference field, seismic Moho depths and variable density contrast. J Geod 77(1–2):77–85. https://doi.org/10.1007/s00190-002-0300-7

Abd-Elmotaal HA, Kühtreiber N (2008) An attempt towards an optimum combination of gravity field wavelengths in geoid computation. In: Sideris MG (ed) Observing our changing earth. International Association of Geodesy Symposia, vol 133, pp 203–209. https://doi.org/10.1007/978-3-540-85426-5_24

Abd-Elmotaal HA, Kühtreiber N (2014) Automated gross-error detection technique applied to the gravity database of Africa. Geophysical Research Abstracts, vol 16, EGU General Assembly 2014:92. http://meetingorganizer.copernicus.org/EGU2014/EGU2014-92.pdf

Abd-Elmotaal HA, Kühtreiber N (2015) On the computation of the ultra-high harmonic coefficients of the topographic-isostatic masses within the data window. Geophysical Research Abstracts, vol 17, EGU General Assembly 2015:355. http://meetingorganizer.copernicus.org/EGU2015/EGU2015-355.pdf

Abd-Elmotaal HA, Kühtreiber N (2016) Effect of the curvature parameter on least-squares prediction within poor data coverage: case study for Africa. Geophysical Research Abstracts, vol 18, EGU General Assembly 2016:271. http://meetingorganizer.copernicus.org/EGU2016/EGU2016-271.pdf

Abd-Elmotaal HA, Kühtreiber N (2019) Suitable gravity interpolation technique for large data gaps in Africa. Stud Geophys Geod 63(3):418–435. https://doi.org/10.1007/s11200-017-0545-5

Abd-Elmotaal HA, Makhloof A (2013) Gross-errors detection in the shipborne gravity data set for Africa. Geodetic Week, Essen, 8–10 Oct 2013. www.uni-stuttgart.de/gi/research/Geodaetische_Woche/2013/session02/Abd-Elmotaal-Makhloof.pdf

Abd-Elmotaal HA, Makhloof A (2014) Combination between altimetry and shipborne gravity data for Africa. In: 3rd international gravity field service (IGFS) General Assembly, Shanghai, 30 June–6 July 2014

Abd-Elmotaal HA, Seitz K, Kühtreiber N, Heck B (2015) Establishment of the gravity database AFRGDB_V1.0 for the African geoid. In: Jin S, Barzaghi R (eds) IGFS 2014. International Association of Geodesy Symposia, vol 144, pp 131–138. https://doi.org/10.1007/1345_2015_51

Abd-Elmotaal HA, Makhloof A, Abd-Elbaky M, Ashry M (2017) The African 3″ × 3″ DTM and its validation. In: Vergos GS, Pail R, Barzaghi R (eds) International symposium on gravity, geoid and height systems 2016. International Association of Geodesy Symposia, vol 148, pp 79–85. https://doi.org/10.1007/1345_2017_19

Abd-Elmotaal HA, Seitz K, Kühtreiber N, Heck B (2018) AFRGDB_V2.0: the gravity database for the geoid determination in Africa. International Association of Geodesy Symposia, vol 149, pp 61–70. https://doi.org/10.1007/1345_2018_29

Abd-Elmotaal HA, Seitz K, Kühtreiber N, Heck B (2019) AFRgeo_v1.0: a geoid model for Africa. KIT Scientific Working Papers 125. https://doi.org/10.5445/IR/1000097013

Abd-Elmotaal HA, Kühtreiber N, Seitz K, Heck B (2020) The new AFRGDB_v2.2 gravity database for Africa. Pure Appl Geophys 177. https://doi.org/10.1007/s00024-020-02481-5

Braitenberg C, Ebbing J (2009) New insights into the basement structure of the West Siberian Basin from forward and inverse modeling of GRACE satellite gravity data. J Geophys Res 114(B06402):1–15. https://doi.org/10.1029/2008JB005799

Bruinsma S, Förste C, Abrikosov O, Lemoine JM, Marty JC, Mulet S, Rio MH, Bonvalot S (2014) ESA's satellite-only gravity field model via the direct approach based on all GOCE data. Geophys Res Lett 41(21):7508–7514. https://doi.org/10.1002/2014GL062045

Farr TG, Rosen PA, Caro E, Crippen R, Duren R, Hensley S, Kobrick M, Paller M, Rodriguez E, Roth L, Seal D, Shaffer S, Shimada J, Umland J, Werner M, Oskin M, Burbank D, Alsdorf D (2007) The shuttle radar topography mission. Rev Geophys 45(RG2004):1–33. https://doi.org/10.1029/2005RG000183

Featherstone WE, Evans JD, Olliver JG (1998) A Meissl-modified Vaníček and Kleusberg kernel to reduce the truncation error in geoid computations. J Geod 72:154–160. https://doi.org/10.1007/s001900050157

Forsberg R (1984) A study of terrain reductions, density anomalies and geophysical inversion methods in gravity field modelling. Ohio State University, Department of Geodetic Science and Surveying, Rep 355

Förste C, Bruinsma S, Abrikosov O, Lemoine JM, Schaller T, Götze HJ, Ebbing J, Marty JC, Flechtner F, Balmino G, Biancale R (2014a) EIGEN-6C4 the latest combined global gravity field model including GOCE data up to degree and order 2190 of GFZ Potsdam and GRGS Toulouse. 5th GOCE User Workshop, Paris, 25–28 Nov 2014

Förste C, Bruinsma S, Sean L, Abrikosov O, Lemoine JM, Marty JC, Flechtner F, Balmino G, Barthelmes F, Biancale R (2014b) EIGEN-6C4 the latest combined global gravity field model including GOCE data up to degree and order 2190 of GFZ Potsdam and GRGS Toulouse. GFZ Data Services. http://doi.org/10.5880/icgem.2015.1

Hastings D, Dunbar P (1998) Development and assessment of the global land one-km base elevation digital elevation model (GLOBE). ISPRS Arch 32(4):218–221

Heck B, Grüninger W (1982) Modification of Stokes's integral formula by combining two classical approaches. In: Proceedings of the IAG symposia, XIXth general assembly of the international union of geodesy and geophysics, Vancouver/Canada, 10–22 August 1987, Paris 1988, Tome II pp 319–337

Heiskanen WA, Moritz H (1967) Physical geodesy. Freeman, San Francisco

Hofmann-Wellenhof B, Moritz H (2006) Physical geodesy. Springer, Berlin

Jekeli C (1980) Reducing the error of geoid undulation computations by modifying Stokes' function. Report 301, Department of Geodetic Science and Surveying, The Ohio State University, Columbus

Kaban MK, El Khrepy S, Al-Arifi N (2016) Isostatic model and isostatic gravity anomalies of the Arabian plate and surroundings. Pure Appl Geophys 173:1211–1221. https://doi.org/10.1007/s00024-015-1164-0

Kraiger G (1988) Influence of the curvature parameter on least-squares prediction. Manuscr Geod 13(3):164–171

Lemoine F, Kenyon S, Factor J, Trimmer R, Pavlis N, Chinn D, Cox C, Klosko S, Luthcke S, Torrence M, Wang Y, Williamson R, Pavlis E, Rapp R, Olson T (1998) The development of the joint NASA GSFC and the National Imagery and Mapping Agency (NIMA) geopotential model EGM96. NASA/TP-1998-206861, NASA Goddard Space Flight Center, Greenbelt, Maryland

Meissl P (1971) Preparation for the numerical evaluation of second order Molodensky-type formulas. Ohio State University, Department of Geodetic Science and Surveying, Rep 163

Merry C (2003) The African geoid project and its relevance to the unification of African vertical reference frames. 2nd FIG Regional Conference Marrakech, Morocco, 2–5 Dec 2003

Merry CL, Blitzkow D, Abd-Elmotaal HA, Fashir H, John S, Podmore F, Fairhead J (2005) A preliminary geoid model for Africa. In: Sansò F (ed) A window on the future of geodesy. International Association of Geodesy Symposia, vol 128, pp 374–379. https://doi.org/10.1007/3-540-27432-4_64

Moritz H (1980) Advanced physical geodesy. Wichmann, Karlsruhe

Pavlis N, Holmes S, Kenyon S, Factor J (2012) The development and evaluation of the earth gravitational model 2008 (EGM2008). J Geophys Res 117(B04406). https://doi.org/10.1029/2011JB008916

Rapp RH (1982) A Fortran program for the computation of gravimetric quantities from high degree spherical harmonic expansions. Ohio State University, Department of Geodetic Science and Surveying, Rep 334

Rapp RH (1997) Use of potential coefficient models for geoid undulation determinations using a spherical harmonic representation of the height anomaly/geoid undulation difference. J Geod 71(5):282–289. https://doi.org/10.1007/s001900050096

Sjöberg LE (2003) A general model for modifying Stokes' formula and its least-squares solution. J Geod 77:459–464. https://doi.org/10.1007/s00190-003-0346-1

Tscherning CC, Knudsen P, Forsberg R (1994) Description of the GRAVSOFT package. Geophysical Institute, University of Copenhagen Technical Report

Wenzel HG (1982) Geoid computation by least-squares spectral combination using integration kernels. In: Proceedings of IAG general meeting, Tokyo, The Geodetic Society of Japan, pp 438–453

Wong L, Gore R (1969) Accuracy of geoid heights from modified Stokes kernels. Geophys J Int 18(1):81–91. https://doi.org/10.1111/j.1365-246X.1969.tb00264.x

A First Assessment of the Corrections for the Consistency of the IAU2000 and IAU2006 Precession-Nutation Models

José M. Ferrándiz, Dhygham Al Koudsi, Alberto Escapa, Santiago Belda, Sadegh Modiri, Robert Heinkelmann, and Harald Schuh

Abstract

The Earth precession-nutation model endorsed by resolutions of each the International Astronomical Union and the International Union of Geodesy and Geophysics is composed of two theories developed independently, namely IAU2006 precession and IAU2000A nutation. The IAU2006 precession was adopted to supersede the precession part of the IAU 2000A precession-nutation model and tried to get the new precession theory dynamically consistent with the IAU2000A nutation.

However, full consistency was not reached, and slight adjustments of the IAU2000A nutation amplitudes at the micro arcsecond level were required to ensure consistency. The first set of formulae for these corrections derived by Capitaine et al. (Astrophys 432(1):355–367, 2005), which was not included in IAU2006 but provided in some standards and software for computing nutations. Later, Escapa et al. showed that a few additional terms of the same order of magnitude have to be added to the 2005 expressions to get complete dynamical consistency between the official precession and nutation models. In 2018 Escapa and Capitaine made a joint review of the problem and proposed three alternative ways of nutation model and its parameters to achieve consistency to certain different extents, although no estimation of their respective effects could be worked out to illustrate the proposals. Here we present some preliminary results on the assessment of the effects of each of the three sets of corrections suggested by Escapa and Capitaine (Proceedings of the Journées, des Systémes de Référence et de la Rotation Terrestre: Furthering our Knowledge of Earth Rotation, Alicante, 2018) by testing them in conjunction with the conventional celestial pole offsets given in the IERS EOP14C04 time series.

Keywords

Earth orientation parameters · Earth rotation models · Earth rotation theory · Precession-nutation

Electronic supplementary material The online version of this chapter (https://doi.org/10.1007/1345_2020_90) contains supplementary material, which is available to authorized users.

J. M. Ferrándiz · D. Al Koudsi
UAVAC, University of Alicante, Alicante, Spain
e-mail: jm.ferrandiz@ua.es

A. Escapa
UAVAC, University of Alicante, Alicante, Spain

Department of Aerospace Engineering, University of León, León, Spain

S. Belda
UAVAC, University of Alicante, Alicante, Spain

Image Processing Laboratory (IPL) - Laboratory of Earth Observation (LEO), University of Valencia, Valencia, Spain

S. Modiri (✉) · H. Schuh
Technische Universität Berlin, Institute for Geodesy and Geoinformation Science, Berlin, Germany

GFZ German Research Centre for Geosciences, Potsdam, Germany
e-mail: sadegh@gfz-potsdam.de

R. Heinkelmann
GFZ German Research Centre for Geosciences, Potsdam, Germany

J. T. Freymueller, L. Sanchez (eds.), *Beyond 100: The Next Century in Geodesy*,
International Association of Geodesy Symposia 152, https://doi.org/10.1007/1345_2020_90

1   Introduction

In 2000, the Resolution B1.6 of the XXIV General Assembly (GA) of the International Astronomical Union (IAU) endorsed the IAU2000A nutation theory, which entered in force on January 1, 2003. Resolution B1 of the XXVI IAU GA held in 2006 adopted the IAU2006 precession model based on the P03 theory by Capitaine et al. (2003, 2005), following the recommendations made by the IAU Division I Working Group (WG) "On Precession and the Ecliptic" (Hilton et al. 2006). Both models were then adopted by resolutions of the International Union of Geodesy and Geophysics (IUGG) taken in 2003 and 2007. As pointed up by Escapa and Capitaine (2018a), strictly speaking Resolution B1.6 approved the "IAU 2000A precession-nutation model". However, its precession component was just a set of empirical, small corrections to the offsets and rates of its and obliquity precession. Therefore, this model was not intended to supersede the former IAU1976 precession theory (Lieske et al. 1977), and Resolution B1.6 itself encouraged the development of new expressions for precession consistent with the IAU2000A model. Both models IAU2000 and IAU2006 were required to be dynamically consistent, but before the approval of the second it was already known that they were not (Capitaine et al. 2005), but some small corrections with amplitudes of few microarcseconds (μas) had to be added to the nutation model. However, the WG in charge considered that nutations were out of the scope of its task, and the fact was not mentioned in the IAU resolution. Later, Escapa et al. (2014, 2016, 2017) showed that a few additional terms of the same order of magnitude have to be added to the 2005 expressions to get complete dynamical consistency between the official precession and nutation models. The issue was discussed in several occasions, particularly inside the IAU/ International Association of Geodesy (IAG) Joint Working Group on Theory of Earth rotation and validation (JWG TERV), and the main authors were invited to propose actions. Escapa and Capitaine (2018b) made a joint review of the problem and proposed three alternative ways of correcting nutations to achieve consistency to certain different extents, although no estimation of their respective effects could be worked out to illustrate the proposals. The document was subject to a wide consultation extended to all the members of the Sub-WG 1, precession and nutation, of the IAU/IAG JWG TERV, as well as to many other experts, including current and past officers of IAU and IAG. Given the short time available to take a solid decision based on the actual impact of each option, it was agreed not to propose any resolution on that direction to IAU before getting more

insight into the interrelationship between precession and nutation theories and analyzing the practical implications of the different possibilities. The origin of those corrections is twofold:

1. IAU2006 included a mean constant rate for J_2, proportional to the dynamical ellipticity H_d, which is a factor of all the nutation amplitudes and the rate of the precession in the longitude of the equator, at the first order of approximation;
2. IAU2006 adopted different values than IAU2000A for other important parameters, namely the constant term ϵ_0 of the obliquity and the longitude rate, at the reference epoch J2000.0.

To give more insight into the implications of both facts, let us recall that the IAU2000 nutation amplitude for each frequency was derived by applying the MHB2000 transfer function (Mathews et al. 2002) to multiply the corresponding rigid-Earth amplitude of REN2000 (Souchay et al. 1999). The latter amplitudes are implicitly factorized by J_2 through H_d or the $K_{S,M}$ Kinoshita's constants (Kinoshita and Souchay 1990), and besides they depend on several circular functions of the ϵ_0 obliquity. Therefore, the total induced variations of the non-rigid Earth amplitudes cannot be got by simply making a rescaling associated only to J_2 (Escapa et al. 2014, 2016, 2017; Escapa and Capitaine 2018a,b).

2   Fundamentals and Methodology

Escapa and Capitaine (2018b) cast the components of those corrections in three groups according to their origin:

(a) A geometrical effect due to the impact of the IAU2000-to-IAU2006 change in the obliquity value on the projection of the CIP motion in space onto the ecliptic (i.e., nutation in longitude); it keeps unchanged the amplitudes of the IAU2000A nutation referred to the IAU 2000 ecliptic.
(b) The J_2 rate effect (a dynamical effect) due to the introduction of that rate into the IAU2000 expressions for nutation.
(c) The so-called ΔPP effect (a dynamical effect) due to the IAU2000-to-IAU2006 changes of the formerly said Precession Parameters (PP).

A detailed explanation appears in that reference.

2.1 Equations of Models

The three models proposed for their consideration were labeled as (a), (b) and (c), and their expressions in terms of celestial pole offsets (CPO) dX, dY, are:

(a)

$$-dXa = +18.8t \sin \Omega + 1.4t \sin(2F - 2D + 2\Omega) - 0.8t^2 \cos \Omega,$$

$$-dYa = -24.6t \cos \Omega - 1.6t \cos(2F - 2D + 2\Omega) - 0.6t^2 \sin \Omega,$$

(b)

$$-dXb = +15.4t \sin \Omega + 1.4t \sin(2F - 2D + 2\Omega) - 0.6t^2 \cos \Omega,$$

$$-dYb = -25.4t \cos \Omega - 1.8t \cos(2F - 2D + 2\Omega) - 0.3t^2 \sin \Omega,$$

(c)

$$-dXc = (-6.2 + 15.4t) \sin \Omega + 1.4t \sin(2F - 2D + 2\Omega)$$
$$+ (-0.8 - 0.3t^2) \sin \Omega,$$

$$-dYc = (0.8 - 25.4t) \cos \Omega + (0.3 - 1.8t) \cos(2F - 2D + 2\Omega)$$
$$+ (-0.8 - 0.3t^2) \sin \Omega,$$

gathering terms with amplitudes above the μas level.

Subindexes identify the relevant model, coefficients units are μas, time t is measured in Julian centuries since J2000.0 and the arguments are certain linear combinations of the Delaunay ones. In all models the dominant term is that of period 18.6 years and the other arguments is semiannual. Notice that the corrections are presented with reversed sign like in Escapa et al. (2017), so that the right hand sides should be added to the CPO instead of being subtracted.

3 Methodology

Looking at the small magnitude of the terms, the application of any of those corrections would not likely produce a significant reduction of the WRMS (weighted root mean square) of the observed CPO series, particularly if the time t is not far from the origin J2000.0. Therefore, we decided to perform the tests with a twofold purpose:

1. Test the hypothesis of potential intercourses between the nutation corrections and the coefficients of the precession polynomials at short time intervals.
2. Checking the accuracy of the precession formulae after more than a decade, not only the effect of the corrections on the "residuals" (or unexplained by theory) CPO.

Concerning the first objective, the idea behind is that a polynomial of low degree is able of providing good approximations of long period oscillations when the time interval is short enough, but not when it exceeds certain length. Because of that, for each of the proposed correction models we computed time series of daily CPO generated from their respective formulae, and fitted to them polynomials of degrees 1 to 5, the highest degree present in the current precession model (Belda et al. 2017a,b). We used a least squares method, either un-weighted or with weights derived from the IERS EOP14C04 series (Bizouard et al. 2019) in the usual way of most EOP data analyses (AlKoudsi 2019). Different time spans were tested, paying special attention to the period with VLBI observations available when IAU2006 was derived, presumably extended not beyond 2003.

4 Results

We can only present some results addressing the first of the former two purposes, due to the length constraints. The full set of results will be presented in a forthcoming paper. First we consider the time interval 1984–2003. The first numeric row of Table 1 displays the WRMS (weight root mean square) of the time series for dX containing only the correction (b), which was one the preferred because it contains the full set of secular-mixed (or Poisson) terms needed to rend IAU2000 consistent with IAU2006. Next rows display the WRMS after fitting polynomials of degrees 1–5, and the coefficients of them. It can be seen that the polynomial of degree 5 provides a very accurate approximation of the corrections values. That fact can be easily visualized in Figs. 1 and 2. The upper graphics in those figures show the correction (b) values in green together with the respective fit polynomials of degrees 1 and 5. The last polynomial almost reproduces only its low frequency variability. The respective residuals are shown in the lower graphics. Lower plot of Fig. 2 suggests that the main secular-mixed term of pseudo-period 18.6 years has been almost perfectly reproduced by the polynomial, and the remaining mostly semiannual oscillation is visible in the residuals.

Table 1 Polynomial approximation of daily values of correction (b) for dX in the interval 1984–2003

Degree	WRMS	Offset	Trend	t^2	t^3	t^4	t^5
No fit	0.47						
1	0.35	−0.18	0.05				
2	0.34	−0.17	0.03	−0.00			
3	0.12	0.08	−0.05	−0.06	−0.004		
4	0.09	0.02	−0.07	−0.04	−0.000	0.0002	
5	0.04	0.01	−0.14	−0.04	0.008	0.0015	0.0000

No. points = 6,941. Units: μas and years

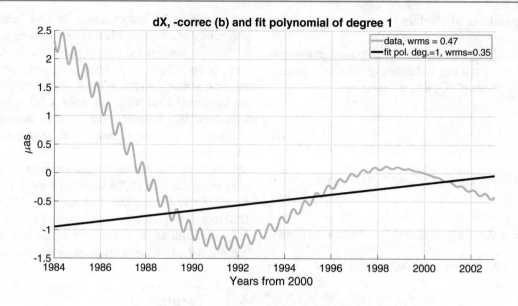

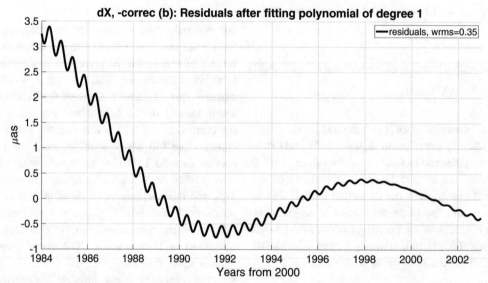

Fig. 1 Upper plot: Daily values of dX correction (b) in the interval 1984–2003 and fit polynomial of degree 1. Lower plot: Residuals. Units μas and years

Next, we present the results for dX correction (b) for a much longer time interval, the two centuries 1900–2100. Table 2 shows that the signal can hardly be reproduced by any polynomial only to a minimum extent. A quick look at Fig. 3, similar to Fig. 2, allows to visualize the reason: Any low degree polynomial, even the fifth that was excellent in 1984–2003, can not reproduce the input pattern, made of many quasi-periodic cycles with amplitude increasing far from the time origin, set at year 2000.

Finally, we present a case corresponding to a time interval ending in September 2018. Table 3 is similar to the former ones. This time the WRMS decreases, but not so much as on Table 1. Figure 4 helps to intuit why: The data curve bends too many times to be reproduced with accuracy below 1 μas by a polynomial up to degree 5, and a long period oscillation is still visible in the residuals plot shown in the lower part of the figure.

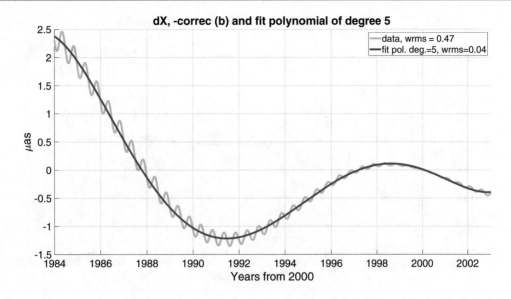

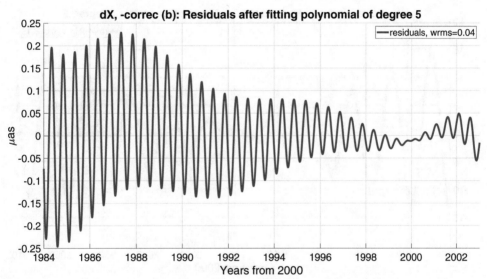

Fig. 2 Upper plot: Daily values of dX correction (b) in the interval 1984–2003 and fit polynomial of degree 5. Lower plot: Residuals. Units µas and years

Table 2 Polynomial approximation of daily values of correction (b) for dX in the interval 1900–2100

Degree	WRMS	Offset	Trend	t^2	t^3	t^4	t^5
No fit	6.27						
1	6.25	−0.20	−0.01				
2	6.23	0.33	−0.01	−0.00			
3	6.21	0.33	0.01	−0.00	−0.000		
4	6.17	−0.47	0.01	0.00	−0.000	−0.0000	
5	6.16	−0.47	−0.01	0.00	0.0008	−0.0000	−0.0000

No. points = 73,050. Units: µas and years

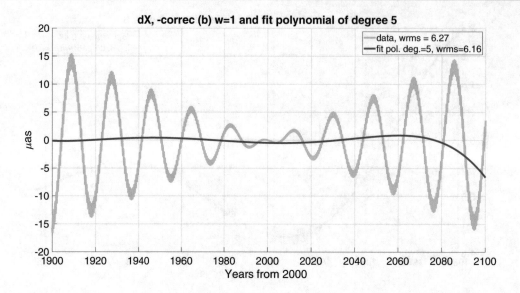

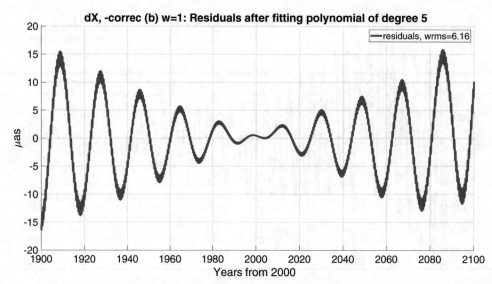

Fig. 3 Upper plot: Daily values of dX correction (b) in the interval 1900–2100 and fit polynomial of degree 5. Lower plot: Residuals. Units μas and years

Table 3 Polynomial approximation of daily values of correction (b) for dX in the interval 1984–2018

Degree	WRMS	Offset	Trend	t^2	t^3	t^4	t^5
No fit	0.79						
1	0.76	−0.13	0.03				
2	0.71	−0.00	0.08	−0.00			
3	0.56	−0.32	0.15	0.01	−0.001		
4	0.45	−0.42	0.06	0.02	0.000	−0.0001	
5	0.31	−0.18	0.01	−0.00	0.002	0.0000	−0.0000

No. points = 12,680. Units: μas and years

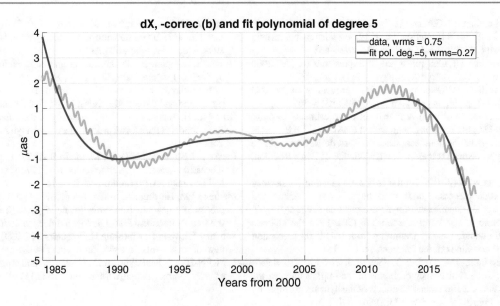

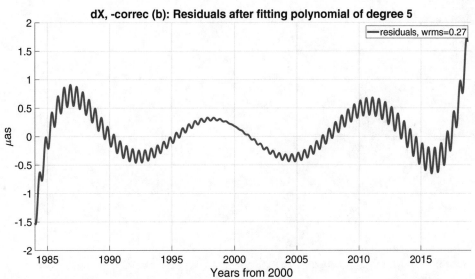

Fig. 4 Upper plot: Daily values of dX correction (b) in the interval 1984–2018 and fit polynomial of degree 5. Lower plot: Residuals. Units μas and years

5   Conclusions

The results show that the lack of application of the correction making IAU2000 and IAU2006 consistent with each other can be masked in the period 1984–2003 by a fifth degree polynomial capable of absorbing more than 90% of the variance due to the additional terms that contain the nutation corrections. That fact implies that the coefficients of the IAU2006 reference polynomials include a small spurious contribution that has no physical origin but replace the effect of the absent nutation corrections.

Acknowledgement The four first authors were partially supported by Spanish Project AYA2016-79775-P (AEI/FEDER, UE).

Conflict of Interest The authors declare that they have no conflict of interest.

References

AlKoudsi D (2019) Análisis de los efectos de la insuficiente consistencia dinámica de las teorias de precesión y nutación IAU2006 e IAU2000 (Spanish language). PhD thesis, University of Alicante

Belda S, Heinkelmann R, Ferrándiz JM, Karbon M, Nilsson T, Schuh H (2017a) An improved empirical harmonic model of the celestial intermediate pole offsets from a global VLBI solution. Astron J 154(4):166

Belda S, Heinkelmann R, Ferrándiz JM, Nilsson T, Schuh H (2017b) On the consistency of the current conventional EOP series and the celestial and terrestrial reference frames. J Geodesy 91(2):135–149

Bizouard C, Lambert S, Gattano C, Becker O, Richard JY (2019) The IERS EOP 14C04 solution for Earth orientation parameters consistent with ITRF 2014. J Geodesy 93(5):621–633

Capitaine N, Wallace PT, Chapront J (2003) Expressions for IAU 2000 precession quantities. Astron Astrophys 412(2):567–586

Capitaine N, Wallace P, Chapront J (2005) Improvement of the IAU 2000 precession model. Astron Astrophys 432(1):355–367

Escapa A, Capitaine N (2018a) A global set of adjustments to make the IAU 2000a nutation consistent with the IAU 2006 precession. In: Proceedings of the Journées, des Systémes de Référence et de la Rotation Terrestre: Furthering our Knowledge of Earth Rotation, Alicante

Escapa A, Capitaine N (2018b) On the IAU 2000 nutation consistency with IAU 2006 precession,draft note. https://web.ua.es/es/wgterv/documentos/other-documents/draft-note-escapa-capitaine-2018.pdf

Escapa A, Getino J, Ferrándiz J, Baenas T (2014) On the changes of IAU 2000 nutation theory stemming from IAU 2006 precession theory. In: Proceedings of the Journées, pp 148–151

Escapa A, Ferrándiz JM, Baenas T, Getino J, Navarro JF, Belda-Palazón S (2016) Consistency problems in the improvement of the IAU precession–nutation theories: effects of the dynamical ellipticity differences. Pure Appl Geophys 173(3):861–870

Escapa A, Getino J, Ferrándiz J, Baenas T (2017) Dynamical adjustments in IAU 2000a nutation series arising from IAU 2006 precession. Astron Astrophys 604:A92

Hilton JL, Capitaine N, Chapront J, Ferrandiz JM, Fienga A, Fukushima T, Getino J, Mathews P, Simon JL, Soffel M et al (2006) Report of the international astronomical union division i working group on precession and the ecliptic. Celest Mech Dyn Astron 94(3):351–367

Kinoshita H, Souchay J (1990) The theory of the nutation for the rigid Earth model at the second order. Celest Mech Dyn Astron 48(3):187–265

Lieske J, Lederle T, Fricke W, Morando B (1977) Expressions for the precession quantities based upon the IAU/1976/system of astronomical constants. Astron Astrophys 58:1–16

Mathews PM, Herring TA, Buffett BA (2002) Modeling of nutation and precession: new nutation series for nonrigid earth and insights into the Earth's interior. J Geophys Res Solid Earth 107(B4):ETG–3

Souchay J, Loysel B, Kinoshita H, Folgueira M (1999) Corrections and new developments in rigid Earth nutation theory-III. Final tables "REN-2000" including crossed-nutation and spin-orbit coupling effects. Astron Astrophys Suppl Ser 135(1):111–131

Report of the IAU/IAG Joint Working Group on Theory of Earth Rotation and Validation

José M. Ferrándiz, Richard S. Gross, Alberto Escapa, Juan Getino, Aleksander Brzeziński, and Robert Heinkelmann

Abstract

This report focuses on some selected scientific outcomes of the activities developed by the IAU/IAG Joint Working Group on Theory of Earth rotation and validation along the term 2015–2019. It is based on its end-of-term report to the IAG Commission 3 published in the Travaux de l'IAG 2015–2019, which in its turn updates previous reports to the IAG and IAU, particularly the triennial report 2015–2018 to the IAU Commission A2, and the medium term report to the IAG Commission 3 (2015–2017). The content of the report has served as a basis for the IAG General Assembly to adopt Resolution 5 on Improvement of Earth rotation theories and models.

Keywords

Earth rotation theory · Earth rotation models · Earth orientation parameters · Precession–nutation · Polar motion · UT1

1   Introduction

In 2015 the IAG, jointly with the International Astronomical Union (IAU), established the IAU/IAG Joint Working Group on Theory of Earth rotation and validation (IAU/IAG JWG

J. M. Ferrándiz (✉)
University of Alicante VLBI Analysis Centre, Alicante, Spain
e-mail: jm.ferrandiz@ua.es

R. S. Gross
Jet Propulsion Laboratory, California Institute of Technology, Pasadena, CA, USA

A. Escapa
University of Alicante VLBI Analysis Centre, Alicante, Spain

Department of Aerospace Engineering, University of León, León, Spain

J. Getino
Department of Applied Mathematics, University of Valladolid, Valladolid, Spain

A. Brzeziński
Warsaw University of Technology, Warsaw, Poland

R. Heinkelmann
GFZ German Research Centre for Geosciences, Potsdam, Germany

TERV, or only JWG for short) that continued the former IAU/IAG JWG on Theory of Earth rotation (ThER), which operated in 2013–2015. This JWG had the purpose of promoting the development of theories of Earth rotation fully consistent and in agreement with observations, useful for providing predictions of the Earth orientation parameters (EOP) with the accuracy required to meet the needs of the near future as recommended by GGOS, the IAG Global Geodetic Observing System. The accuracy and stability goals are very stringent, since the benchmarks set by the JWG are 30 μas and 3 μa/y in terms of geocentric angles; those figures arise from the requirements to the Terrestrial Reference Frames (TRF) accuracy and stability that are necessary for monitoring the sea level rise properly and adopting the policies suitable to act against global change and minimize its prejudicial effects.

The JWG addressed the whole set of five Earth orientation parameters (EOP), since there are interrelations among them and consistency was a main goal besides of accuracy. Because of that the JWG had a complex structure, with a Vice-chair (Richard Gross) and three partially overlapped sub-working groups (SWG) that operated independently but in coordination:

© The Author(s) 2020
J. T. Freymueller, L. Sanchez (eds.), *Beyond 100: The Next Century in Geodesy*,
International Association of Geodesy Symposia 152, https://doi.org/10.1007/1345_2020_103

(1) Precession/Nutation, chaired by Juan Getino and Alberto Escapa,
(2) Polar Motion and UT1, chaired by Aleksander Brzeziński, and
(3) Numerical Solutions and Validation, chaired by Robert Heinkelmann.

The complete terms of reference appear in The Geodesist's Handbook 2016 (Drewes et al. 2016), and a website with further information is hosted by the University of Alicante (UA) (https://web.ua.es/en/wgterv).

Coordination among the SWGs and with other IAG components (in particular GGOS and IERS) was facilitated by the existence of common members (including correspondents) affiliated to the JWG and e.g. to the IERS Earth Orientation Centre, Rapid Service/Prediction Centre, Conventions Centre, and Central Bureau, the IERS Analysis Coordinator and the GGOS Scientific Panel, Bureau of Products and Standards, and Committee on Essential Geodetic Variables. Coordination with the IAU was also guaranteed by a majority of JWG members in the Organizing Committee of IAU Commission A2, Rotation of the Earth, the IAU body to which the JWG reported. More details on that can be found on the Travaux 2015–2019 Commission 3 report (Drewes and Kuglitsch 2019), which also details the organization of splinter meetings or sessions at large conferences. Additional information appears in the precedent JWG reports, like those to the IAU in 2018 and to the IAG in 2017 (Drewes and Kuglitsch 2017)

The scientific outcomes and findings can be cast according to their level of maturity in

– Advances or findings on topics that can be considered scientifically solved, and
– Advances showing remarkable improvement of knowledge but still on progress

Next section emphasizes on outcomes of the first kind. Many of them are related to precession and nutation, since those parameters are a main object of theoretical developments due to its origin, astronomical forcing in the main, and subsequent better predictability.

2 Selected Outcomes

Several papers published in recent years by JWG members in the main unveil that a noticeable part of the unexplained variance of the determined EOP series can be attributed mainly to:

– Systematic errors, e.g. in conventional or background models,
– Inconsistencies, either internal to theories or among components of them, and

– Need of updating some specific components after 20 years of their derivation

That happens particularly for the Celestial pole offsets (CPO) that provide the deviations of the precession-nutation parameters with respect to the conventional models adopted by the IAU and IAG/IUGG. Some of the main outcomes related to the CPO are:

1. The amplitudes of the main nutation terms have to be updated after almost 20 years of use. This is particularly important for the 21 frequencies used to fit the nutation theory IAU2000, at a time in which the amplitude formal errors were not better than $5 \mu as$ (Herring et al. 2002) and may exceed some tens. Currently the number of separable frequencies has increased drastically up to several tens, and the uncertainties of the fitted amplitudes are reduced to about 2–$3 \mu as$. The existence of amplitude inaccuracies that can reach several tens μas has been confirmed by many different independent results, e.g. Malkin (2014), Gattano et al. (2016), Belda et al. (2017a), Schuh et al. (2017), Zhu et al. (2017). Methodologies are varied; for instance, Malkin (2014) fitted a reduced set of amplitudes to various single and combined CPO time series, in different time intervals. Gattano et al. (2016) also used different single and combined series, whereas Belda et al. (2017a) used a global VLBI solution derived from 2990 sessions ranging from 1990 to 2010 (the last year used for the ICRF2 realization) to fit the widest set of amplitudes, 179. To give an idea of the magnitude of the potential improvement, the decrease of the WRMS of the CPO residuals is roughly around $15 \mu as$ when the 14 major amplitudes and precession offsets and trends are corrected according to that fit.

2. Also for nutation theory, it has been found that the two independent blocks that compose the IAU2000 series, namely lunisolar and planetary, are inconsistent with each other (Ferrándiz et al. 2018). In fact, the MHB2000 transfer function was not applied to the amplitudes of the 687 planetary terms, either direct or indirect; instead, those terms were taken without change from an early version of the rigid-Earth theory REN2000. Besides, the planetary terms are nutations of the angular momentum vector, whereas the 678 lunisolar terms are nutations of the figure axis of a non-rigid three-layer Earth. That surprising fact was clearly reported by Herring et al. (2002), and a likely cause might be that the effect of the transfer function application on individual amplitudes was assumed to be negligible and less than $5 \mu as$, the threshold for truncation that IAU recommended at that time for the renewal of the nutation theory. It is not really the case, since the magnitude of this effect has been proved to reach near $20 \mu as$ in single amplitudes, a value much larger than the joint contribution

Table 1 Largest Oppolzer terms of planetary origin for the Earth's figure axis

L_{Ve}	L_E	L_{Ma}	L_J	p_A	Period (days)	dX_{in} (sin)	dX_{out} (cos)	dY_{in} (cos)	dY_{out} (sin)	Origin code
0	1	0	−1	0	398.884	0.2	0	0.2	0	0
2	−4	0	0	2	−487.638	−1.8	−0.1	1.7	−0.1	1
0	1	0	−2	0	439.332	1.1	−14	1.3	14.3	1
0	3	−4	0	0	418.266	−3.2	0.5	−3.1	−0.4	1
3	−4	0	0	0	416.688	−2.9	8.2	−2.9	−7.9	1
0	1	0	−1	0	**398.884**	**−18.8**	−3	**−17.2**	3	1
0	2	−2	0	0	389.968	−4.3	−0.4	−3.8	0.4	1
2	−4	0	0	2	−487.638	−5.6	−0.5	5.6	−0.5	2
0	1	0	−1	0	398.884	1.6	0.2	1.5	−0.2	5

Units: amplitudes in μas, periods in mean solar days
Effect origin code: 0 indirect Moon; 1 indirect Sun; 2 direct Venus; 5 direct Jupiter

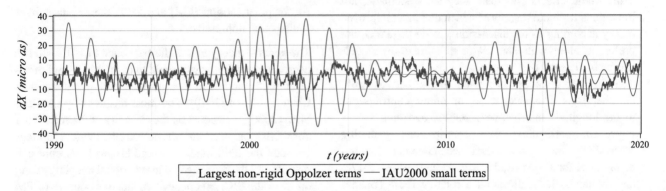

Fig. 1 dX: Comparison of the accumulated effects of the six leading non-rigid planetary Oppolzer terms (in red) and the 533 planetary terms of IAU2000 with amplitude <1 μas (in blue). Period 1990–2020

of several hundreds of small planetary terms included in the IAU2000 model. The joint effect of the neglected terms can be above the GGOS threshold. Those facts are illustrated in Table 1 (an abridged version of Table 1 in Ferrándiz et al. 2018) and Fig. 1. The largest amplitudes and its associated period are marked in bold in Table 1.

3. The background geophysical models of IAU2000, particularly those corresponding to the ocean mass and currents effects, were among the best ones available before 2000, but since then those models have become obsolete. For instance, the computation of oceanic effects is reported as based on the GOT94 model (Chao et al. 1996). Outdating of background models poses a new source of inconsistency, since they are different from the corresponding models currently used to process the observation data for determining the EOP, either separately or jointly with a terrestrial reference frame (TRF). The impact of this fact on the accuracy of those IAU2000 components has not been assessed in most cases. Besides, the update needed to improve consistency and assessing accuracy is not straightforward, since the final MHB2000 nutation series were computed numerically from the dynamical equations and not from the simpler resonance

formulae, as described in 6.1 of Mathews et al. (2002); besides, the full set of either oceanic or anelastic contributions was never published separately and only a few sample terms were displayed on the cited paper.

4. Regarding the mutual consistency of the conventional precession and nutation models, it has been proved that the precession theory IAU2006 is not fully dynamically consistent with the nutation theory IAU2000 (MHB2000 by Mathews et al. 2002), though dynamical consistency was required by the 2006 IAU Resolution B1 endorsing P03 (Capitaine et al. 2003). Inconsistencies arise from the fact that IAU2006 considers J_2 as a linear function instead of a constant like IAU2000, and besides uses different values, at J2000.0, for the obliquity and the rate of longitude (*precession constant*) than those of IAU2000. Making the two theories consistent requires applying certain corrections to the nutation part, as already noticed by Capitaine et al. (2005) although no correction was recommended by the IAU WG in charge (Hilton et al. 2006) nor included in the text of the Resolution. The set of corrections already recommended in the IERS Conventions (2010) (Petit and Luzum 2010) has been found to be incomplete, but full consistency can be achieved by applying to the IAU2000 series a recently determined set of small corrections that

include a few so-called Poisson or secular-mixed terms, whose amplitudes are factorized by the time and thus increase as it departs from J2000.0 in either sense (Escapa et al. 2017a,b; Escapa and Capitaine 2018). Using μas and Julian centuries as units, the complete set is given by

$$
\begin{aligned}
(-d\,\Delta\psi) = {} & -15.6\sin\Omega - 1.4\cos\Omega - 0.5\cos l_s \\
& +39.8t\sin\Omega - 0.6t\sin 2\Omega, \\
(-d\,\Delta\epsilon) = {} & +0.8\cos\Omega - 0.8\sin\Omega - 25.1t\cos\Omega \\
& -1.7t\cos(2F - 2D + 2\Omega).
\end{aligned}
\tag{1}
$$

While these effects are small, they are systematic, not random, and should therefore be included in an improved theory according to the discussions inside the JWG, but preferably along with other major updating of the models for the final users' convenience.

5. The precession model has been re-assessed as well. On the theoretical side, a set of minor contributions to the longitude rate has been revised and their values improved, particularly two contributions gathering respectively the mathematical second order solution component for a non-rigid Earth (Baenas et al. 2017a) and the anelasticity effects on a rotating Earth (Baenas et al. 2017b), the latter effect named as *non-linear* in the Mathews et al. (2002) terminology. Besides, those findings imply that the value of the Earth's dynamical ellipticity, H_d, must be adjusted since the observed precession rate is of course unchanged. The H_d variation is of some ppm and the resulting corrections to nutations, or *indirect* effects, are non-negligible since they approach near $100\,\mu$as for certain terms (Baenas et al. 2019).

6. From a more practical perspective, the accuracy of the precession polynomial has been checked by several authors, e.g. Malkin (2014), Liu and Capitaine (2017), Gattano et al. (2016), Belda et al. (2017a). The most recent results show that the offsets and trends of dX and dY deviate from 0 slightly, but significantly, and reach the μas and μas/y levels—i.e. the current precession model may be not 100% accurate although not at a worrying extent. In general, the offsets of dX and dY are $>30\,\mu$as, the target accuracy recommended by the IAG Global Geodetic Observing System (GGOS). In contrast, the uncertainties of rates are rather compliant with the GGOS goals.

7. The free core nutation (FCN) is a major source of unexplained variance of the CPO. FCN models have never been included in the IAU and IAG/IUGG theories of Earth rotation, since their excitation mechanism is closer to that of polar motion than to the astronomically forced nutations. Its modeling has been addressed by different approaches; some of them are new, like convolution (Chao

and Hsieh 2015) and numerical integration with excitation functions that may include geomagnetic jerks (GMJ) (Vondrák and Ron 2015, 2019). Besides, new accurate empirical models have been derived by Belda et al. (2016) using a sliding window approach with high temporal resolution, therefore closer to Malkin's methods (2013, 2014, 2016) than to Lambert's (2007). Furthermore, FCN models are dependent on the EOP solutions used in their derivation at the current accuracy level (Malkin 2017). Summarizing, the WRMS of the CPO residuals can be reduced near the vicinity of $100\,\mu$as by using suitable correction models.

As for polar motion (PM) and UT1, the advances have been also quite impressive.

1. The general theory that is the backbone of IAU2000 has been extended from symmetric to triaxial two- and three-layer Earth models in a series of papers that starts with Chen and Shen (2010). The most recent is by Guo and Shen (2020), who consider the elasticity of the solid inner core (SIC) as well as viscoelectromagnetic couplings between the fluid outer core and elastic inner core and mantle, and the pressure and gravitational coupling acting on inner the SIC. Frequency-dependent responses of PM to excitations have been further investigated by Chen et al. (2013a,b).

2. The S1 signal, considered anomalous for long, has been further explained. Schindelegger et al. (2016, 2017) showed that the S1 LOD estimate ($6\,\mu$s) determined from VLBI is in agreement with atmosphere-ocean excitation estimates.

3. The analyses of PM and UT1 benefit to a remarkable extent from the improvement of their excitation function models. For instance, we can cite the continuous archive of the NCEP-CAR reanalyses for excitations of PM and UT1 at the IERS Special Bureau (SB) for the Atmosphere, applying the methodology introduced by Salstein et al. (1993) and Zhou et al. (2006) to the new input data; the new release of the oceanic ECCO model by Quinn et al. (2019), available from the IERS SB for Oceans; the time series of operational effective angular momentum (AM) functions provided by the ESM (Advanced Earth system modelling capacity) at GFZ in Potsdam (e.g. Dill and Dobslaw 2019; Dill et al. 2019), which include 3h atmospheric (AAM), 3h oceanic (OAM), 24h hydrologic (HAM) and 24h sea-level (SLAM) contributions, etc.

4. The numerical integration of Brzeziński's broad-band Liouville equations (Brzeziński 1994) with geophysical excitations like OAM, AAM from several sources has shown that all the EOP are sensitive to those excitations. The agreement with observations is improved when GMJ are considered (Vondrák and Ron 2016). The method has

been applied to derive new estimates of the periods and Q of the Chandler Wobble (CW) and FCN (Vondrák et al. 2017, 2019).

5. More insight into the hydrological effects on PM, taking into account time-varying gravity, has been provided by the research performed at the Polish Space Research Centre (e.g. Wińska and Śliwińska 2019; Śliwińska and Nastula 2019).

Regarding the EOP determination, the advances have been also illuminating. A few of them are:

1. In VLBI data analysis, the usual session-wise solutions can be complemented with global solutions (Belda et al. 2017a) and with the simultaneous determination of "quasi instantaneous" terrestrial reference frames (TRFs) and EOP by Kalman filter and more sophisticated methods (Abbondanza et al. 2017; Soja et al. 2016a,b, 2018a).

2. In the search for potential sources of discrepancies between theory and observations, several experiments have assessed the impacts of the variations of reference frames or processing strategies. It has been shown that different realizations of TRFs or data processing strategies can give rise to not negligible differences in the EOP determination at the GGOS level of accuracy (see e.g. Wielgosz et al. 2016; Heinkelmann et al. 2017, Belda et al. 2017b; Soja et al. 2018b). This is not irrelevant from the theoretical perspective since theory must explain observations and help predictions, but it does not accommodate to the actual observational environment as tightly as desirable in some aspects. For instance, the reference systems used in the derivation different sets of fundamental equations that the EOP must satisfy, with IAU2000 among them, have been never realized (Chen and Shen 2010; Ferrándiz et al. 2015) although the analysis of observations is carried out for specific realizations of reference frames— e.g. the current conventional EOP 14C04 series (Bizouard et al. 2019) links the ITRF14 (Altamimi et al. 2016) and the ICRF2-ext2 (Fey et al. 2015), to be superseded by the ICRF3 recently approved by IAU and IAG/IUGG.

3. There is also a wide agreement on the need of improving the consistency between the terrestrial and celestial (CRF) reference frames and the rotation relating them given by the EOP. This is a major challenge since TRFs and CRFs and realized independently and using data that cover different time spans. Better consistency can be expected from simultaneous realization of those three elements (Heinkelmann et al. 2017), but that is not easy and a deeper insight into the meaning of the realizations resulting from each procedure is required; it is well-known by theory that the definitions of EOP and TRFs

are intrinsically related, but can be done in infinitely many ways (Munk and McDonald 1960). In the term 2019–2023 a dedicated IAG/IAU/IERS JWG on Consistent realization of TRF, CRF, and EOP will tackle the problem.

Besides those findings, there are many valuable research works still in progress; a non-comprehensive list was included in the said JWG final report. Because of their theoretical interest, let us comment only some work related to the improvement of the Earth's interior modeling. For instance, the evaluation of the ellipticity of the inner layers, and the theoretical estimates of the free periods, particularly Chandler's, have been brought closer to their observed values (Huang et al. 2019; Liu et al. 2019); more insight has been got into effects related to the inner gravitational interactions among the Earth's components (Chao 2017; Rochester et al. 2018).

3   Conclusion and Outlook

From all those findings and research in progress, it is possible to conclude that at least a partial update of the Earth rotation theory is needed and feasible within a reasonable time span. Not only accuracy but also consistency among EOP, ICRF, and ITRF has to be improved. The extent of the renewal is to be determined in the forthcoming years, since neither any complete new theory nor any integrated set of corrections aimed at improving the theories in force have been published or proposed so far. Future potential candidates should be thoroughly validated with observations and compared to the current theories regarding accuracy and consistency before taking decisions on the update.

Those conclusions were the basis of the Resolution 5, on *Improvement of the Earth's Rotation Theories and Models*, approved by the 2019 IAG General Assembly. The IAG resolved:

- To encourage a prompt improvement of the Earth rotation theory regarding its accuracy, consistency, and ability to model and predict the essential EOP,
- That the definition of all the EOP, and related theories, equations, and ancillary models governing their time evolution, must be consistent with the reference frames and the resolutions, conventional models, products, and standards adopted by the IAG and its components,
- That the new models should be closer to the dynamically time-varying, actual Earth, and adaptable as much as possible to future updating of the reference frames and standards.

Finally, a new working group was created by the IAG to help in the implementation of these recommendations.

Acknowledgements JMF, AE, and JG were partially supported by Spanish Project AYA2016-79775-P (AEI/FEDER, UE). The work of RSG described in this paper was performed at the Jet Propulsion Laboratory, California Institute of Technology, under contract with the National Aeronautics and Space Administration. Support for that work was provided by the Earth Surface and Interior Focus Area of NASA's Science Mission Directorate.

Conflict of Interest

The authors declare that they have no conflict of interest.

References

Abbondanza C, Chin TM, Gross RS, Heflin MB, Parker JW, Soja BS, van Dam T, Wu X (2017) JTRF2014, the JPL Kalman filter and smoother realization of the international terrestrial reference system. J Geophys Res 122:8474–8510. https://doi.org/10.1002/2017/JB014360

Altamimi Z, Rebischung P, Métivier L, Collilieux X (2016) ITRF2014: a new release of the international terrestrial reference frame modeling non-linear station motions. J Geophys Res Solid Earth 121. https://doi.org/10.1002/2016JB013098

Baenas T, Ferrándiz JM, Escapa A, Getino J, Navarro JF (2017a) Contributions of the elasticity to the precession of a two-layer Earth model. Astron J 153:79. https://doi.org/10.3847/1538-3881/153/2/79

Baenas T, Escapa A, Ferrándiz JM, Getino J (2017b) Application of first-order canonical perturbation method with dissipative Hori-like kernel. Int J Non-Linear Mech 90:11–20. https://doi.org/10.1016/j.ijnonlinmec.2016.12.017

Baenas T, Escapa A, Ferrándiz JM (2019) Precession of the non-rigid Earth: effect of the mass redistribution. Astron Astrophys. https://doi.org/10.1051/0004-6361/201935472

Belda S, Ferrándiz JM, Heinkelmann R, Nilsson T, Schuh H (2016) Testing a new free core nutation empirical model. J Geodyn 94:59–67. https://doi.org/10.1016/j.jog.2016.02.002

Belda S, Heinkelmann R, Ferrándiz JM, Karbon M, Nilsson T, Schuh H (2017a) An improved empirical harmonic model of the Celestial intermediate pole offsets from a global VLBI solution. Astron J 154:166. https://doi.org/10.3847/1538-3881/aa8869

Belda S, Heinkelmann R, Ferrándiz JM, Nilsson T, Schuh H (2017b) On the consistency of the current conventional EOP series and the celestial and terrestrial reference frames. J Geod 91:135–149. https://doi.org/10.1007/s00190-016-0944-3

Bizouard C, Lambert S, Gattano C, Becker O, Richard J (2019) The IERS EOP 14C04 solution for Earth orientation parameters consistent with ITRF 2014. J Geod 93:621–633. https://doi.org/10.1007/s00190-018-1186-3

Brzeziński A (1994) Polar motion excitation by variations of the effective angular momentum function: II. Extended Model. Manuscr Geodaet 19:157–171

Capitaine N, Wallace PT, Chapront J (2003) Expressions for IAU 2000 precession quantities. Astron Astrophys 412:567–586. https://doi.org/10.1051/0004-6361:20031539

Capitaine N, Wallace PT, Chapront J (2005) Improvement of the IAU 2000 precession model. Astron Astrophys 432:355–367. https://doi.org/10.1051/0004--6361:20041908

Chao BF (2017) Dynamics of the inner core wobble under mantle-inner core gravitational interactions. J Geophys Res Solid Earth 122:7437–7448. https://doi.org/10.1002/2017JB014405

Chao BF, Hsieh Y (2015) The Earth's free core nutation: Formulation of dynamics and estimation of eigenperiod from the very-long-baseline interferometry data. Earth Planet Sci Lett 432:483–492. https://doi.org/10.1016/j.epsl.2015.10.010

Chao BF, Ray RD, Gipson JM, Egbert GD, Ma C (1996) Diurnal/semidiurnal polar motion excited by oceanic tidal angular momentum. J Geophys Res 101:20,151–20,163

Chen W, Shen WB (2010) New estimates of the inertia tensor and rotation of the triaxial nonrigid Earth. J Geophys Res 115:B12419. https://doi.org/10.1029/2009JB007094

Chen W, Ray J, Li J, Huang CL, Shen WB (2013a) Polar motion excitations for an Earth model with frequency-dependent responses: 1. A refined theory with insight into the Earth's rheology and core-mantle coupling. J Geophys Res Solid Earth 118:4975–4994. https://doi.org/10.1002/jgrb.50314

Chen W, Ray J, Shen WB, Huang CL (2013b) Polar motion excitations for an Earth model with frequency-dependent responses: 2. Numerical tests of the meteorological excitations. J Geophys Res Solid Earth 118. https://doi.org/10.1002/jgrb.50313

Dill R, Dobslaw H (2019) Seasonal variations in global mean sea-level and consequences on the excitation of length-of-day changes. Geophys J Int 218:801–816. http://doi.org/10.1093/gji/ggz201

Dill R, Dobslaw H, Thomas M (2019) Improved 90-day EOP predictions from angular momentum forecasts of atmosphere, ocean, and terrestrial hydrosphere. J Geodesy 93:287–295. http://doi.org/10.1007/s00190-018-1158-7

Drewes H, Kuglitsch F (eds) (2017) Travaux de l'IAG 2015–2017/ IAG Reports, vol 40. Available at https://iag.dgfi.tum.de/en/iag-publications-position-papers/iag-reports-2017-online/ Accessed 17 Jan 2020

Drewes H, Kuglitsch F (eds) (2019) Travaux de l'IAG 2015–2019/ IAG Reports, vol 41. Available at https://iag.dgfi.tum.de/en/iag-publications-position-papers/iag-reports-2019-online/ Accessed 17 Jan 2020

Drewes H, Kuglitsch F, Adám J3, Szabolcs Rózsa S (eds) (2016) The geodesist handbook 2016. J Geodesy 90:907–1205. https://doi.org/10.1007/s00190-016-0948-z

Escapa A, Capitaine, N (2018) A global set of adjustments to make the IAU 2000A nutation consistent with the IAU 2006 precession. In Proc. Journées 2017, des Systèmes de Référence et de la rotation Terrestre: furthering our knowledge of Earth rotation Alicante, Spain, 2017 (in press)

Escapa A, Ferrándiz JM, Baenas T, Getino J, Navarro JF, Belda-Palazón S (2017a) Consistency problems in the improvement of the IAU precession-nutation theories: effects of the dynamical ellipticity differences. Pure Appl Geophys 173:861–870. https://doi.org/10.1007/s00024-015-1154-2

Escapa A, Getino J, Ferrándiz JM, Baenas T (2017b) Dynamical adjustments in IAU 2000A nutation series arising from IAU 2006 precession. Astron Astrophys. https://doi.org/10.1051/0004-6361/201730490.

Ferrándiz JM, Belda S, Heinkelmann R, Getino J, Schuh H, Escapa A (2015) Reference frames in Earth rotation theories. Geophys Res Abstr 17:EGU2015-11566

Ferrándiz JM, Navarro JF, Martínez-Belda MC, Escapa A, Getino J (2018) Limitations of the IAU2000 nutation model accuracy due to the lack of Oppolzer terms of planetary origin Astron Astrophys 618:A69. https://doi.org/10.1051/0004-6361/201730840

Fey AL, Gordon D, Jacobs CS, Ma C, Gaume RA, Arias EF, Bianco G, Boboltz DA, Boeckmann S, Bolotin S, Charlot P, Collioud A, Engelhardt G, Gipson J, Gontier AM, Heinkelmann R, Ojha R,

Skurikhina E, Sokolova J, Souchay J, Sovers OJ, Tesmer V, Titov O, Wang G, Zharov V (2015) The second realization of the international celestial reference frame by very long baseline interferometry. Astron J 150:58. https://doi.org/10.1088/0004-6256/150/2/58

Gattano C, Lambert S, Bizouard C (2016) Observation of the Earth's nutation by the VLBI: how accurate is the geophysical signal. J Geod. https://doi.org/10.1007/s00190-016-0940-7

Guo Z, Shen WB (2020) Formulation of a triaxial three-layered Earth rotation: theory and rotational normal mode solutions. J Geophys Res: Solid Earth. https://doi.org/10.1029/2019JB018571

Heinkelmann R, Belda, S, Ferrándiz JM, Schuh H (2017) How consistent are the current conventional celestial and terrestrial reference frames and the conventional earth orientation parameters? In: International Association of Geodesy Symposia series. https://doi.org/10.1007/1345_2015_149

Herring TA, Mathews PM, Buffett BA (2002) Modeling of nutation - precession: very long baseline interferometry results. J Geophys Res 107:2069. https://doi.org/10.1029/2001JB000165

Hilton, JL, Capitaine N, Chapront J, Ferrándiz JM, Fienga A, Fukushima T, Getino J, Mathews P, Simon JL, Soffel M, Vondrak J, Wallace P, Williams J (2006) Report of the international astronomical union division I working group on precession and the ecliptic. Celes Mech Dyn Astron 94:351–367. https://doi.org/10.1007/s10569-006-0001-2

Huang C, Liu Y, Liu C, Zhang M (2019) A generalized theory of the figure of the Earth: formulae. J Geodesy 93:297–317. https://doi.org/10.1007/s00190-018-1159-6

Lambert S (2007) Empirical model of the retrograde free core nutation. Technical Note, Available at ftp://hpiers.obspm.fr/eop-pc/models/fcn/notice.pdf. Accessed 17 Jan 2020

Liu FC, Capitaine N (2017) Evaluation of a possible upgrade of the IAU 2006 precession. Astron Astrophys 597:A83, 12 pp. https://doi.org/10.1051/0004-6361/201628717

Liu C, Huang C, Liu Y, Zhang M. (2019) A generalized theory of the figure of the Earth: on the global dynamical flattening. J Geodesy 93:19–331. https://doi.org/10.1007/s00190-018-1163-x

Malkin Z (2013) Free core nutation and geomagnetic jerks. J Geod 72:53–58. https://doi.org/10.1016/j.jog.2013.06.001

Malkin ZM (2014) On the accuracy of the theory of precession and nutation. Astron Rep 58:415–425. https://doi.org/10.1134/S1063772914060043

Malkin Z (2016) Free core nutation: new large disturbance and connection evidence with geomagnetic jerks. arXiv160303176M

Malkin Z (2017) Joint analysis of celestial pole offset and free core nutation series. J Geod 91. https://doi.org/839-84810.1007/s00190-016-0966-x

Mathews PM, Herring TA, Buffett BA (2002) Modeling of nutation and precession: new nutation series for nonrigid Earth and insights into the Earth's interior. J Geophys Res 107:2068. https://doi.org/10.1029/2001JB000390

Munk WH, McDonald GJF (1960) The rotation of the Earth. Cambridge University Press, Cambridge

Petit G, Luzum B (2010) IERS Conventions (2010). IERS Technical Note 36, vol 179. Verlag des Bundesamts für Kartographie und Geodësie, Frankfurt am Main. ISBN:3-89888-989-6 (2010)

Quinn KJ, Ponte RM, Heimbach P, Fukumori I, Campin JM (2019) Ocean angular momentum from a recent global state estimate, with assessment of uncertainties. Geophys J Internat 216:584–597. https://doi.org/10.1093/gji/ggy452

Rochester MG, Crossley D, Chao BF (2018) On the physics of the inner core wobble: corrections to "Dynamics of the inner-core wobble under mantle- inner core gravitational interactions" by B. F. Chao. J Geophys Res. https://doi:10.1029/2018JB016506

Salstein DA, Kann DM, Miller AJ, Rosen RD (1993) The sub-bureau for atmospheric angular momentum of the International Earth Rotation Service: a meteorological data center with geodetic applications. Bull Am Meteorol Soc 74:67–80.

Schindelegger M, Einšpigel D, Salstein D, Böhm J (2016) The global S_1 tide in Earth's nutation. Surv Geophys, 37(3):643–680. https://doi.org/10.1007/s10712-016-9365-3

Schindelegger M, Salstein D, Einšpigel D, Mayerhofer C (2017) Diurnal atmosphere-ocean signals in Earth's rotation rate and a possible modulation through ENSO. Geophys Res Lett 44:2755–2762. https://doi.org/10.1002/2017GL072633

Schuh H, Ferrándiz JM, Belda S. Heinkelmann R, Karbon M, Nilsson T (2017) Empirical corrections to nutation amplitudes and precession computed from a global VLBI solution. American Geophysical Union Fall Meeting 2017, abstract #G11A-0695

Śliwińska J, Nastula J (2019) Determining and evaluating the hydrological signal in polar motion excitation from gravity field models obtained from kinematic orbits of LEO satellites. Remote Sens 11:1784. https://doi.org/10.3390/rs11151784

Soja B, Nilsson T, Balidakis K, Glaser S, Heinkelmann R, Schuh H (2016a) Determination of a terrestrial reference frame via Kalman filtering of very long baseline interferometry data. J Geod 90:1311–1327. https://doi.org/10.1007/s00190-016-0924-7

Soja B, Nilsson T, Balidakis K, Glaser S, Heinkelmann R, Schuh H (2016b) Erratum to: determination of a terrestrial reference frame via Kalman filtering of very long baseline interferometry data. J Geod 90:1329–1329. https://doi.org/1010.1007/s00190-016-0953-2

Soja B, Gross RS, Abbondanza C, Chin TM, Heflin MB, Parker JW, Wu X, Balidakis K, Nilsson T, Glaser S, Karbon M, Heinkelmann R, Schuh H (2018a) Application of time-variable process noise in terrestrial reference frames determined from VLBI data. Adv Space Res 61:2418–2425. https://doi.org/10.1016/j.asr.2018.02.023

Soja B, Gross RS, Abbondanza C, Chin TM, Heflin MB, Parker JW, Wu X, Nilsson T, Glaser S, Balidakis K, Heinkelmann R, Schuh H (2018b) On the long-term stability of terrestrial reference frame solutions based on Kalman filtering. J Geod 92:1063–1077. https://doi.org/10.1007/s00190-018-1160-0

Vondrák J, Ron C (2015) Earth orientation and its excitations by atmosphere, oceans, and geomagnetic jerks. Serb Astron J 191:59–66. https://doi.org/10.2298/SAJ1591059V

Vondrák J, Ron C (2016) Geophysical fluids from different data sources, geomagnetic jerks, and their impact on Earth's orientation. Acta Geodyn Geomater 13:241–247. https://doi.org/10.13168/AGG.2016.0005

Vondrák J, Ron C (2019) New GFZ effective angular momentum excitation functions and their impact on nutation. Acta Geodyn Geomater 16:151–155. https://doi.org/10.13168/AGG.2019.0012

Vondrák J, Ron C, Chapanov Y (2017) New determination of period and quality factor of Chandler wobble, considering geophysical excitations. Adv Space Res 59(5):1395–1407. https://doi.org/10.1016/j.asr.2016.12.001

Wielgosz A, Tercjak M, Brzeziński A (2016) Testing impact of the strategy of VLBI data analysis on the estimation of Earth orientation parameters and station coordinates. Rep Geod Geoinform 101:1–15. https://doi.org/:10.1515/rgg-2016-0017

Wińska M, Śliwińska J (2019) Assessing hydrological signal in polar motion from observations and geophysical models. Stud Geophys Geod 63. https://doi.org/10.1007/s11200-018-1028-z

Zhou YH, Salstein DA, Chen JL (2006) Revised atmospheric excitation function series related to Earth variable rotation under consideration of surface topography. J Geophys Res 111:D12108. https://doi.org/10.1029/2005JD006608

Zhu P, Rivoldini A, Koot L, Dehant V (2017) Basic Earth's Parameters as estimated from VLBI observations. Geod Geodyn 8:427–432. http://dx.doi.org/10.1016/j.geog.2017.04.007

Achievements of the First 4 Years of the International Geodynamics and Earth Tide Service (IGETS) 2015–2019

Jean-Paul Boy, Jean-Pierre Barriot, Christoph Förste, Christian Voigt, and Hartmut Wziontek

Abstract

We present the activities and improvements of the International Geodynamics and Earth Tide Service (IGETS) over the last four years. IGETS collects, archives and distributes long time series from geodynamic sensor, in particular superconducting gravimeter data currently from more than 40 stations and 60 different sensors. In addition to the raw 1-s and 1-min gravity and atmospheric pressure data (Level 1), IGETS produces end-user products on different levels. These include gravity and atmospheric pressure data corrected for major instrumental perturbations and ready for tidal analysis (Level 2). Since 2019, IGETS provides gravity residuals corrected for most geophysical contributions (Level 3) which can be used directly for geophysical applications without any expert knowledge in the processing of gravimetric time series.

Keywords

IAG and IGFS service · Superconducting gravimeters

1 Introduction

The International Geodynamics and Earth Tide Service (IGETS) was established at the 2015 IUGG meeting in Prague as an official service of the IAG within the IGFS. Its main objective is to provide a service to monitor temporal variations of the Earth gravity field through long-term records from ground gravimeters, including superconducting gravimeters (SG) and other geodynamic sensors, such as tiltmeters, extensometers, etc. (Voigt et al. 2016). IGETS continues the activities of the Global Geodynamic Project

J.-P. Boy (✉)
IPGS (UMR 7516), Université de Strasbourg/EOST, CNRS, Strasbourg Cedex, France
e-mail: jeanpaul.boy@unistra.fr

J.-P. Barriot
Observatoire Géodésique de Tahiti, University of French Polynesia (UPF), Tahiti, French Polynesia

C. Förste · C. Voigt
GFZ German Research Centre for Geosciences, Potsdam, Germany

H. Wziontek
Federal Agency for Cartography and Geodesy (BKG), Leipzig, Germany

(Crossley and Hinderer 2009) to provide support to geodetic and geophysical research activities using SG data within the context of an international network. IGETS also continues the activities of the International Center for Earth Tides, in particular, in collecting, archiving and distributing Earth tide records from long series of gravimeters, and other geodynamic sensors.

We present here the status and progress of IGETS and its different products after four years of existence for the support of geophysical research activities. We first document the observation network, with 42 existing stations equipped by 60 different instruments, including seven double-sphere SGs. We then describe the three levels of data products, from the raw 1-s or 1-min gravity and atmospheric pressure records (Level 1), to the pre-processed datasets ready for tidal analysis (Level 2) and the gravity residuals corrected for major geophysical contributions (Level 3).

2 Observation Network

In 2016, the IGETS data base was set up at GFZ with GGP data from 32 different stations and has increased its number of stations by 25–42% according to the sta-

© The Author(s) 2020
J. T. Freymueller, L. Sanchez (eds.), *Beyond 100: The Next Century in Geodesy*,
International Association of Geodesy Symposia 152, https://doi.org/10.1007/1345_2020_94

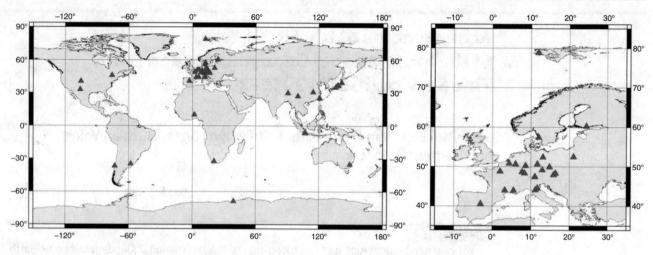

Fig. 1 Map of the 42 stations contributing to the IGETS data base

tus from summer 2019 (Fig. 1). IGETS is interested in collecting more long time series of geodynamic sensors. Benefits for station operators and data providers are the long-term storage of their data sets, the provision of different processing levels and an optional DOI assignment by the research repository of GFZ Data Services. This guarantees best practice with regard to open data policy. The visibility and usage of IGETS has also been increased with a steadily increasing number of almost 500 registered users of the data base.

3 Data Products

A detailed documentation of the IGETS data products and the data base is provided by Voigt et al. (2016). Three different products of the SG time series are stored in monthly files:

- Raw gravity and local atmospheric pressure records sampled at 1 or 2 s, in addition to the same records decimated at 1-min samples (Level 1 products). These are uploaded by each station operator.
- Gravity and atmospheric pressure data at 1-min and 1-h sampling corrected for instrumental perturbations, ready for tidal analysis. This product is derived from the previous datasets and is available in two versions computed by UPF and EOST (Level 2 products).
- Gravity residuals in 1 min sampling after particular geophysical corrections (including solid Earth tides, polar motion, tidal and non-tidal loading effects). This product is derived from the previous dataset of EOST and is also computed by EOST (Level 3 products).

3.1 Level 1

Available raw gravity and atmospheric pressure data from the station operators are shown in Table 1 for the different IGETS stations and sensors. As raw gravity data are not calibrated, each station operator is asked to provide amplitude and phase calibrations and their changes over time if existing in a separate calibration file. Otherwise this information is still provided in the Level 1 header files. In addition, log files are provided by the station operators documenting periods of instrumental disturbances inducing steps, gaps and large spikes with regard to a subsequent processing.

3.2 Level 2

Level 2 data products are currently computed by two different analysis centers: primary at UPF (University of French Polynesia, Tahiti) and secondary also at EOST (Ecole et Observatoire des Sciences de la Terre, Strasbourg, France). This pre-processing aims to remove instrumental perturbations from the Level 1 raw gravity and atmospheric pressure data, producing time series ready for tidal analysis. Level 2 products by UPF are available for all stations and sensors providing Level 1 data and are updated on a regular basis. Level 2 products by EOST were initially computed for 26 stations and 36 sensors with long time series.

While the UPF processing follows the well-known approaches of GGP and ICET, the EOST processing was introduced in 2019 and is described in the following. Raw 1-min gravity and atmospheric pressure (Level 1 data) are first calibrated using the available calibration files. First, interpolated hourly surface pressure from MERRA2 (Gelaro

Table 1 Temporal coverage of Level 1 data provided to IGETS from station operators

Station	Sensor	1989	90	91	92	93	94	95	96	97	98	99	2000	01	02	03	04	05	06	07	08	09	10	11	12	13	14	15	16	17	18	19
Apache Point	ap046																													■	■	■
Bad Homburg	bh030-1													■	■	■	■	■	■													
Bad Homburg	bh030-2													■	■	■	■	■	■	■												
Bad Homburg	bh044																				■	■	■	■	■	■	■	■	■	■	■	
Bandung	ba009									■	■	■	■	■	■	■	■	■	■	■												
Borowa Gora	bg027																													■		
Borowa Gora	bg1036																															■
Boulder	bo024							■	■	■	■	■	■	■	■	■																
Brasimone	br015							■	■																							
Brussels	be003																															
Canberra	cb031																															
Cantley	ca012	■	■	■	■	■	■			■	■	■	■	■	■	■	■	■	■	■												
Cibinong	ci022																				■	■	■	■	■	■	■	■	■	■	■	■
Concepcion	tc038																■	■	■	■	■	■	■	■	■	■						
Conrad	co025																											■	■	■	■	■
Djougou	dj060																						■	■								
Esashi	es007									■	■	■	■	■	■	■	■	■	■	■	■	■	■	■	■	■	■	■	■	■	■	■
Hsinchu	hs048																					■	■	■	■							
Kamioka	ka016																					■	■	■	■	■	■	■	■	■	■	■
Kyoto	ky009									■	■	■	■	■	■	■																
Larzac	la002																							■	■	■	■	■	■	■	■	■
La Plata	lp038																						■	■	■	■	■	■	■	■	■	■
Lhasa	lh057																										■	■	■	■	■	■
Lijiang	li066																					■	■	■	■	■						
Matsushiro	ma011									■	■	■	■	■	■	■	■	■	■	■	■	■	■	■	■	■	■	■	■	■	■	■
Medicina	mc023								■	■	■	■																				
Membach	mb021							■	■	■	■	■	■	■	■	■	■	■	■	■	■	■	■	■	■	■	■	■	■	■	■	■
Metsahovi	me013																															
Metsahovi	me020						■	■	■	■	■	■	■	■	■	■	■	■	■	■	■	■										
Metsahovi	me022																															
Metsahovi	me073-1																											■	■			
Metsahovi	me073-2																													■	■	■
Mizusawa	mi007																															
Moxa	mo034-1												■	■	■	■	■	■	■	■	■	■	■	■								
Moxa	mo034-2																								■	■	■	■	■	■	■	■
Ny-Alesund	ny039												■	■	■	■	■	■	■	■	■	■	■	■	■	■	■	■	■	■	■	■
Onsala	os054																								■	■	■	■	■	■	■	■
Pecny	pe050																			■	■	■	■	■	■	■	■	■	■	■	■	■
Potsdam	po018				■	■	■	■	■	■																						
Rustrel	ru024																											■	■			
Schiltach	bf056-1																				■	■	■	■	■	■	■	■				
Schiltach	bf056-2																											■	■	■	■	■
Strasbourg	st023																						■	■	■	■	■	■				
Strasbourg	st026								■	■	■	■	■	■	■	■	■	■	■	■	■	■	■	■	■	■	■	■	■	■	■	
Sutherland	su037-1									■	■	■	■	■	■	■	■	■	■													
Sutherland	su037-2																				■		■	■								
Sutherland	su052																					■	■	■	■	■	■	■	■	■	■	■
Syowa	sy016									■	■	■	■	■	■	■	■	■	■	■	■	■	■	■	■	■	■	■	■	■	■	■
Trappes	tr005																										■	■	■	■	■	■
Vienna	vi025									■	■	■	■	■	■	■	■	■	■	■	■	■	■	■	■	■	■	■	■	■	■	■
Wettzell	we006									■	■	■	■	■	■	■	■	■	■	■	■	■	■	■	■	■	■	■	■	■	■	■
Wettzell	we029-1																					■	■	■	■	■						
Wettzell	we029-2																									■	■	■	■	■	■	■
Wettzell	we030-1																															
Wettzell	we030-2																															
Wettzell	we103							■	■	■	■																					
Wuhan	wu004									■	■	■	■	■	■	■	■	■	■	■	■	■	■	■	■	■						
Wuhan	wu065																							■	■	■	■	■	■	■		
Yebes	ys064																								■	■	■	■	■	■	■	
Zugspitze	zu052																															

et al. 2017) reanalysis model is removed from the pressure data. Then these residuals are manually corrected for eventual offsets, and gaps are filled by linear interpolation. The de-gapped series is then corrected for the remaining perturbations (spikes) using a threshold on its derivative, following the procedure of Crossley et al. (1993). The full pressure is then restored by adding back the surface pressure from MERRA2.

For gravity data, the methodology is similar: The calibrated gravity values are corrected for a local tidal model as well as polar motion and local air pressure effects. Offsets are then manually corrected, gaps are filled by linear interpolation, and remaining perturbations (spikes, earthquakes) are corrected using a threshold on the derivative of the gravity residuals. The corrected gravity signal is then restored by adding back the modeled tidal signal, polar motion and air pressure effects. A major advance of this processing is that corrected offsets and filled gaps are documented in separate channels.

Differences between the Level 2 data sets have been detected at some stations, which are dominated by discrepancies in the step corrections, reaching up to 100 nm/s^2. Different strategies for spike reduction and gap filling causing differences of a few tens of nm/s^2 (Hinderer et al. 2002). Also differences in air pressure may reach up to several hPa in sections where gaps were filled. Currently, the processing strategies of UPF and EOST are evaluated, therefore no recommendation for a specific Level 2 data set is given. IGETS strives for standardization of the Level 2 products in the near future.

3.3   Level 3

Level 3 data are currently computed from the Level 2 data processed by EOST. Gravity residuals with 1-min resolution are computed after subtracting from the Level 2 data:

- Solid Earth tides and ocean tidal loading,
- Atmospheric loading,
- Polar motion and length-of-day induced gravity changes,
- Instrumental drift.

Tidal gravity variations are computed differently for the long-period tides and for the diurnal and sub-diurnal bands:

- At high frequency, a local tidal model, adjusted by least-squares, is used.
- At low frequency, we model the tidal signal using the DDW99 gravimetric factor (Dehant et al. 1999) and HW95 tidal potential (Hartmann and Wenzel 1995) for the Solid Earth tides, and FES2014b (Carrère et al. 2016) for the ocean tidal loading using its seven different constituents (Sa, Ssa, Mm, Msf, Mf, Mtm and Msqm).

Atmospheric loading is computed according to Boy et al. (2002), using MERRA2 (Gelaro et al. 2017) hourly surface pressure, and assuming an inverted barometer ocean response to pressure. The MERRA2 pressure is replaced by the 1-min local pressure record for angular distance less than 0.10° to the station. The polar motion and length-of-day induced gravity variations are modeled using the IERS EOPC04 daily series (Wahr 1985), and assuming a δ_2 factor of 1.16. We also model ocean pole tide as a self-consistent equilibrium response (Agnew and Farrell 1978). Depending on the sensor, the instrumental drift is generally modeled as a polynomial or an exponential function (Van Camp and Francis 2007). When available, we use time series from absolute gravimeters for the adjustment.

The Level 3 gravity residuals can be used directly by scientists without any expert knowledge in the processing and reduction of gravimetric time series for specific applications e.g. in geodesy, geophysics or hydrology. In addition, the provision of the effect from various reduction models in separate channels easily allows restoring the signal of interest back to the gravity residuals.

4   Scientific Applications of IGETS Data

Since its establishment in 2015, IGETS data has been used as the basis for a large number of scientific studies in various disciplines that were only possible by the global integration of the individual stations. In the following we can only give a brief selection. Mikolaj et al. (2019) provide uncertainty estimates of SG reductions for various IGETS stations especially with regard to subsequent hydrological signal separation. Bogusz et al. (2018) do inter-comparisons of nearby GPS and SG time series at several IGETS stations for an improved understanding of the structure, dynamics and evolution of the system Earth. Ziegler et al. (2016) and Gruszczynska et al. (2017) have estimated the gravimetric pole tide from long SG time series. A recent study by Cui et al. (2018) analyzes the time variability of free core nutation (FCN) period based on several IGETS stations, while Sun et al. (2019) re-analyze the Earth's background free oscillations using various SG data sets. Xu et al. (2019) investigated the Earth toroidal modes after Sumatra-Andaman earthquakes using various SGs. Karkowska and Wilde-Piórko (2019) have used raw gravity data from IGETS stations to study long-period surface waves. Very recently, first attempts have been made to apply the IGETS network for detection of dark matter (Horowitz and Widmer-Schnidrig 2020; McNally and Zelevinsky 2020; Hu et al. 2019).

Full studies of the only equatorial station, Djougou in Niger (Africa), were performed by Hinderer et al. (2019, 2020). Antokoletz et al. (2019) conducted a preliminary study of the South American station of La Plata (Argentina)

5　Conclusions

Since its establishment in 2015, IGETS has collected, archived and distributed superconducting gravimeter data from more than 40 stations and 60 different instruments. In addition to the raw 1-s and 1-min gravity and atmospheric pressure data (Level 1), IGETS is producing end-user products, including gravity and pressure data corrected for major instrumental perturbations (Level 2), ready for tidal analysis, and gravity residuals corrected for geophysical contributions (Level 3). The Level 2 data are produced by the two analysis centers located at the University of Polynesia and EOST, Strasbourg, France; the level 3 data are produced by EOST. The IGETS data base including all data sets is hosted by GFZ. The access for data users is free after a mandatory registration. Interested operators of geodynamic sensors are invited to join IGETS as data producers.

Acknowledgements We thank all station operators for providing their valuable data to IGETS.

References

Agnew DC, Farrell WE (1978) Self-consistent equilibrium ocean tides. Geophys J R Astr Soc 55:171–181

Antokoletz ED, Wziontek H, Tocho C (2019) First six months of superconducting gravimetry in Argentina. In: International symposium on gravity, geoid and height systems 2016, Thessaloniki, 19–23 September 2016, pp 111–118 https://doi.org/10.1007/1345_2017_13

Bogusz J, Rosat S, Klos A, Lenczuk A (2018) On the noise characteristics of time series recorded with nearby located GPS receivers and superconducting gravity meters. Acta Geod Geophys 53:201–220. https://doi.org/10.1007/s40328-018-0212-5

Boy J-P, Gegout P, Hinderer J (2002) Reduction of surface gravity data from global atmospheric pressure loading. Geophys J Int 149:534–545

Carrère L, Lyard F, Cancet M, Guillot A, Picot N (2016) FES 2014, a new tidal model – validation results and perspectives for improvements. In: ESA living planet conference, Prague, 9–13 May 2016

Crossley D, Hinderer J (2009) A review of the GGP network and scientific challenges. J Geodyn 48:299–304

Crossley D, Hinderer J, Jensen O, Xu HH (1993) A slew rate detection criterion applied to SG data processing. Bull d'Inf Marées Terr 117:8675–8704

Cui X, Sun H, Xu J, Zhou J, Chen X (2018) Detection of free core nutation resonance variation in Earth tide from global superconducting gravimeter observations. Earth Planets Space 70:199. https://doi.org/10.1186/s40623-018-0971-9

Dehant V, Defraigne P, Wahr J (1999) Tides for a convective earth. J Geophys Res 104(B1):1035–1058. https://doi.org/10.1029/1998JB900051

Gelaro R, McCarty W, Suarez MJ et al (2017) The modern-era retrospective analysis for research and applications, version-2 (MERRA-2). J Clim 30:5419–5454

Gruszczynska M, Rosat S, Klos A, Janusz Bogusz J (2017) Providing long-term trend and gravimetric factor at Chandler period from superconducting gravimeter records by using singular spectrum analysis along with its multivariate extension. In: AGU Fall Meeting, New Orleans, 11–15 December 2017, G11C-0423

Hartmann T, Wenzel HG (1995) The HW95 tidal potential catalogue. Geophys Res Lett 22(24):3553–3556. https://doi.org/10.1029/95GL03324

Hinderer J, Rosat S, Crossley D, Amalvict M, Boy J-P, Gégout P (2002) Influence of different processing methods on retrieval of gravity signal from GGP data. Bull Infor Marèes Terr 135:10653–10667

Hinderer J, Riccardi U, Rosat S, Boy J-P, Hector B, Calvo M, Littel F, Bernard J-D (2019) A study of the solid Earth tides, ocean and atmospheric loadings using an 8-year record (2010-2018) from superconducting gravimeter OSG-060 at Djougou (Benin, West Africa). J Geodyn 134:101692. https://doi.org/10.1016/j.jog.2019.101692

Hinderer J, Hector B, Riccardi U, Rosat S, Boy J-P, Calvo M, Littel F, Bernard J-D (2020) A study of the monsoonal hydrology contribution using a 8 year record (2010-2018) from superconducting gravimeter OSG-060 at Djougou (Benin, West Africa). Geophys J Int 22(1):431–439. https://doi.org/10.1093/gji/ggaa027

Horowitz CJ, Widmer-Schnidrig R (2020) Gravimeter search for compact dark matter objects moving in the Earth. Phys Rev Lett 124(5), 051102. https://doi.org/10.1103/PhysRevLett.124.051102

Hu W, Lawson M, Budker D, Figueroa NL, Jackson Kimball DF, Mills Jr. AP, Voigt C (2019) A network of superconducting gravimeters as a detector of matter with feeble non-gravitational coupling. arXiv:1912.01900v1 (submitted to EPJD in 2020)

Karkowska K, Wilde-Piórko M (2019) How can gravimeters improve recordings of earthquakes? Geophysical Research Abstracts 21, EGU2019-427

McNally RL, Zelevinsky T (2020) Constraining domain wall dark matter with a network of superconducting gravimeters and LIGO. arXiv:1912.06703v1 (submitted to EJPD in 2020)

Mikolaj M, Reich M, Güntner A (2019) Resolving geophysical signals by terrestrial gravimetry: a time domain assessment of the correction-induced uncertainty. JGR: Solid Earth 124(2):2153–2165. https://doi.org/10.1029/2018JB016682

Sun H, Zhang M, Xu J, Chen X (2019) Reanalysis of background free oscillations using recent SG data. Terr Atmos Ocean Sci 30:757–763. https://doi.org/10.3319/TAO.2019.03.14.03

Van Camp M, Francis O (2007) Is the instrumental drift of superconducting gravimeters a linear or exponential function of time? J Geod 81(5):337–344. https://doi.org/10.1007/s00190-006-0110-4

Voigt C, Förste C, Wziontek H, Crossley D, Meurers B, Pálinkáš V, Hinderer J, Boy J-P, Barriot J-P, Sun HP (2016) Report on the data base of the international geodynamics and earth tide service (IGETS), GFZ German Research Centre for Geosciences, https://doi.org/10.2312/gfz.b103-16087

Wahr JM (1985) Deformation induced by polar motion. J Geophys Res 90(B11):9363–9368. https://doi.org/10.1029/JB090iB11p09363

Xu C, Yu H, Jian G, Deng S, Zhou B, Wu Y (2019) Low-degree toroidal modes from the Sumatra-Andaman event observed by superconducting gravimeters. Geod Geodyn 10(6):477–484. https://doi.org/10.1016/j.geog.2019.07.002

Ziegler Y, Hinderer J, Rogister Y, Rosat S (2016) Estimation of the gravimetric pole tide by stacking long time-series of GGP superconducting gravimeters. Geophys J Int 205(1):77–88. https://doi.org/10.1093/gji/ggw007

Inter-Comparison of Ground Gravity and Vertical Height Measurements at Collocated IGETS Stations

Severine Rosat, Jean-Paul Boy, Janusz Bogusz, and Anna Klos

Abstract

Vertical displacements and time-varying gravity fluctuations are representative of various deformation mechanisms of the Earth occurring at different spatial and temporal scales. The inter-comparison of ground-gravity measurements with vertical surface displacements enables to estimate the transfer function of the Earth at various time-scales related to the rheological properties of the Earth. In this paper, we estimate the gravity-to-height changes ratio at seasonal time-scales due mostly to hydrological mass variabilities. We investigate this ratio at nine sites where Global Navigation Satellite System (GNSS) and Superconducting Gravimeter continuous measurements are collocated. Predicted gravity-to-height change ratios for a hydrological model are around -2 nm/s^2/mm when there is no local mass effect. This is in agreement with theoretical modeling for an elastic Earth's model. Spectral analysis of vertical displacement and surface gravimetric time-series show a coherency larger than 50% at seasonal time-scales at most sites. The obtained gravity-to-height change ratios range between -5 and -2 nm/s^2/mm for stations Lhasa, Metsahovi, Ny-Alesund, Onsala, Wettzell and Yebes. At Canberra and Sutherland, this ratio is close to zero. Finally, at Strasbourg site the coherency is low and the ratio is positive because of local mass effects affecting gravimetric records.

Keywords

Earth's transfer function · GNSS · Hydrological loading · Superconducting gravimeters · Surface gravity variations · Vertical deformation

1 Introduction

Vertical displacements and time-varying gravity are representative of various deformation mechanisms of the Earth occurring at different spatial and temporal scales. We can quote for instance post-glacial rebound, tidal deformation, surficial loading, co- and post- seismic as well as volcanic deformations. The involved temporal scales range from seconds to years and the spatial scales range from millimeters to continental dimension. Daily Global Navigation Satellite System (GNSS) solutions precisely monitor local deformation while sub-daily gravimetric measurements integrate Newtonian mass redistribution and deformation at a site. The inter-comparison of the ground-gravity measurements with vertical surface displacements enables to estimate the transfer function of the Earth at various time-scales related to the elastic and visco-elastic properties of the Earth. We can hence achieve a separation of the contribution of mass redistribution from surface deformation.

In this paper, we estimate the gravity-to-height change ratios, later on denoted dg/du, at seasonal time-scales. We will focus on the ratio obtained with GNSS and Super-conducting Gravimeter (SG) observations. We consider sta-

S. Rosat (✉) · J.-P. Boy
Institut de Physique du Globe de Strasbourg; UMR 7516, Université de Strasbourg/EOST, CNRS, Strasbourg Cedex, France
e-mail: Severine.Rosat@unistra.fr

J. Bogusz · A. Klos
Faculty of Civil Engineering and Geodesy, Military University of Technology, Warsaw, Poland

© The Author(s) 2020
J. T. Freymueller, L. Sanchez (eds.), *Beyond 100: The Next Century in Geodesy*,
International Association of Geodesy Symposia 152, https://doi.org/10.1007/1345_2020_117

tions contributing to the IGETS (International Geodynamics and Earth Tide Service) where both techniques are available with records longer than several years i.e. Canberra (Australia), Lhasa (Tibet, China), Metsahovi (Finland), Ny-Alesund (Norway), Onsala (Sweden), Strasbourg (France, Boy et al. 2017), Sutherland (South Africa), Wettzell (Germany) and Yebes (Spain). In the past, de Linage et al. (2007, 2009) computed this ratio for surface loading models (hydrological, atmospheric and oceanic). Rosat et al. (2009) already performed a comparison of GNSS and SG measurements with hydrological model at the Strasbourg (France) station, but they have computed neither the spectral coherency nor the gravity-to-height changes ratio. In the present paper, using collocated gravity and displacement measurements, we will estimate the gravity (g) to height (u) change ratio dg/du that we compare with theoretical predictions from the MERRA2 (Gelaro et al. 2017) model. In the following, we first explain the methods used to compute the transfer function and the gravity-to-height change ratio. We then remind quickly about the data processing. Finally, we show the results for the hydrological surface loading predictions and for GNSS and SG observations at nine IGETS sites.

2   Methodology and Data Processing

2.1   Spectral Coherency and Gravity-to-Height Ratio

We compute the magnitude squared coherence estimate of the GNSS and SG time-records C_{xy} given by

$$C_{xy} = \left| P_{xy} \right|^2 / \left(P_{xx} P_{yy} \right), \qquad (1)$$

where P_{xx} and P_{yy} are the Power Spectral Density (PSD) estimates of x denoting the height changes (du) and y denoting the gravity variations (dg), respectively, and P_{xy} is the cross-PSD between x and y. The PSD is obtained using the Welch's averaged, modified periodogram method, i.e. the signals x and y are divided into sections of 4 years with 75% overlap and tapered with a Hamming window. For each section, a modified periodogram is computed and the eight periodograms are averaged.

The transfer function between the surface gravity changes and vertical height changes is computed in the same way using

$$T_{xy}(f) = P_{yx}(f) / P_{xx}(f). \qquad (2)$$

The ratio dg/du corresponds to the real part of the transfer function. Please note that in case the coherency is low, computing the transfer function between both datasets would be meaningless.

2.2   GNSS Data Processing

We compute daily displacements from a global set of 117 worldwide stations using the GAMIT/GLOBK (v10.6) software (Herring et al. 2015). We use the latest tropospheric mapping function (VMF1; Böhm et al. 2006) with a priori values of zenith hydrostatic delay derived from the ECMWF meteorological reanalysis fields and residual Zenith Wet Delays (ZTDs) estimated at 2-h intervals with two gradients per day. Standard solid Earth tide, ocean tidal loading (using FES 2014a; Carrère et al. 2016) and pole tide corrections are applied to follow standards recommended by the International Earth Rotation and Reference Systems Service convention (IERS 2010) (Petit and Luzum 2010). We slightly modify the GAMIT software to include in our processing 3-h atmospheric and non-tidal oceanic loading effects at the observation level using ECMWF surface pressure field and TUGO-m (Carrère and Lyard 2003) barotropic ocean model forced by air pressure and winds (see Boy and Lyard 2008; Gegout et al. 2010).

2.3   SG Data Processing

Superconducting Gravimeters (SG) are Level-3 products from IGETS (http://igets.u-strasbg.fr/data_products.php; Voigt et al. 2016) that is to say they are 1-min gravity residuals after cleaning for gaps, large steps and spikes and after particular geophysical corrections (including solid Earth tides, polar motion, tidal and non-tidal loading effects) as described on the IGETS website. Hydrological loading effects are hence remaining. We have applied a low-pass filter and decimated the gravity residuals to 1 h, and then we have performed a simple moving average to have daily solutions. We have selected records to avoid major instrumental trouble or changes of instruments. As for GNSS solutions, we have finally decimated SG time-series to 10 days.

3   Results

3.1   Gravity-to-Height Ratio from Surface Loading Computations

Surface loading computations are available from the EOST loading service at http://loading.u-strasbg.fr/. We consider the hydrological loading computed from MERRA2 model in terms of surface gravity and vertical height changes between January 1980 and April 2018. The spatial and temporal resolutions of MERRA2 are, respectively, 0.625° (~50 km) in latitude and longitude and 1 h. The loading computation was performed up to degree 72. The complete hydrological loading model for surface gravity changes is the sum of

Fig. 1 Non-local surface loading gravity and vertical changes predicted at Strasbourg (France) from MERRA2 hydrological model. (**a**) dg (black line) and du (dashed blue line) surface loading time-series; (**b**) magnitude squared coherence; (**c**) real part of the transfer function between dg and du. Vertical dashed red lines indicate the annual and semi-annual periods

Fig. 2 Local surface loading gravity and vertical changes predicted at Strasbourg (France) from MERRA2 hydrological model. (**a**) dg (black line) and du (dashed blue line) surface loading time-series; (**b**) magnitude squared coherence; (**c**) real part of the transfer function between dg and du. Vertical dashed red lines indicate the annual and semi-annual periods

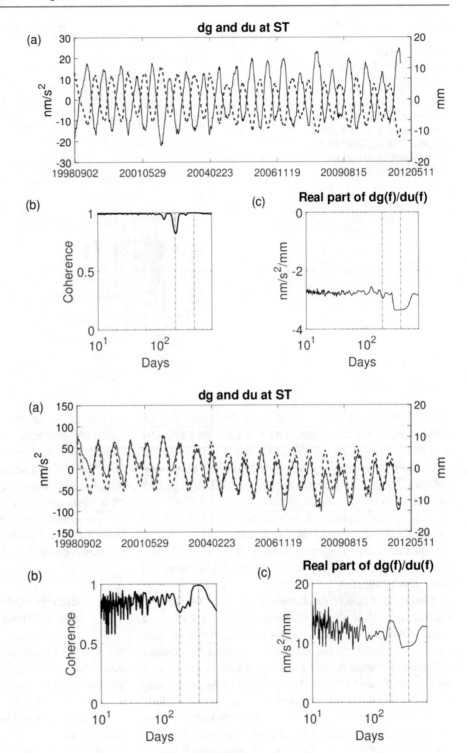

a local contribution (simple Bouguer approximation using an admittance of −4.2677 nm/s²/cm assuming all local mass changes are located above the sensor) and a "non-local" contribution computed using Green's functions (Farrell 1972) as described in Boy et al. (2002) and at the EOST loading website. The vertical displacement due to elastic hydrology loading is also computed using Green's functions.

In Figs. 1, 2 and 3 we have plotted an example at the underground Strasbourg (France) station respectively for the non-local, local and total contributions of the hydrological loading. The time-series of dg and du are in subplots (a), the magnitude squared coherence in (b) and the real part of the transfer function is in (c). de Linage et al. (2009) have shown that theoretically the ratio dg/du should tend

Fig. 3 Total surface loading gravity and vertical changes predicted at Strasbourg (France) from MERRA2 hydrological model. (**a**) dg (black line) and du (dashed blue line) surface loading time-series; (**b**) magnitude squared coherence; (**c**) real part of the transfer function between dg and du. Vertical dashed red lines indicate the annual and semi-annual periods

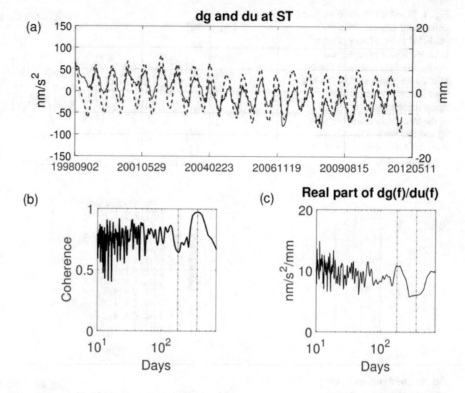

to -2.6 nm/s^2/mm for large spherical harmonic degrees when there is no local load. The non-local part provides a ratio dg/du of -3.3 nm/s^2/mm at annual periods and of -2.7 nm/s^2/mm (Fig. 1c) at shorter periods (below semi-annual). There is an agreement between MERRA2 predictions and theoretical modeling by de Linage et al. (2009). At the semi-annual period, we have a drop of coherence between dg and du. At semi-annual time-scales, hydrological models exhibit rather high-frequency content on most continental area like Europe resulting in almost no displacement but some gravity signal.

Gravimetric effects are known to be more sensitive to local masses while vertical changes, which correspond to a displacement (so a double integration of gravity), are less sensitive to local effects. Figure 2 illustrates the smaller coherence between du and local dg effects. In the case of underground station in Strasbourg the ratio dg/du becomes positive and large as already noted by de Linage et al. (2009). Figure 3 represents the total gravimetric loading effect versus the vertical displacement. The obtained dg/du ratio at annual period is 6 nm/s^2/mm. This ratio is even larger around 10 nm/s^2/mm at periods smaller than 100 days (cf. Fig. 3c).

We provide another example for station Lhasa (Tibet, China) which is above ground. In Figs. 4, 5 and 6 we respectively compare the non-local, local, and total gravimetric contributions to vertical loading displacement. For this station, the annual coherency is close to one. The hydrological signal is below the station and the ratio dg/du

is close to -7 nm/s^2/mm at the annual period and around -10 nm/s^2/mm at the semi-annual period.

Hydrological signal is coherent between vertical displacement and surface gravity loading predictions from MERRA2 model at annual and semi-annual periods for all the stations we have considered. We however do not show the plots since they are similar to the ones already shown for Strasbourg and Lhasa stations. We will now use SG and GNSS observations to retrieve the dg/du ratio.

3.2 Gravity-to-Height Ratio from GNSS and SG Time-Series

We show the transfer functions between vertical surface displacement and surface gravity measurements at Strasbourg (France) and Lhasa (China) in Figs. 7 and 8. We can see that at the Strasbourg station the coherency is close to 50% at the annual period, while at Lhasa, the seasonal coherency is close to 100%.

Table 1 summarizes values of coherence and of dg/du ratios at seasonal time-scales for the nine stations considered here. We have indicated the period at which the coherency is maximum and its corresponding dg/du value. At sites where the local hydrological system is more complicated than a varying underground water table, the coherence is less than 50%. That is the case of Strasbourg for instance. At the other sites, the coherence at annual time-scales is larger than 50%.

Fig. 4 Non-local surface loading gravity and vertical changes predicted at Lhasa (Tibet) from MERRA2 hydrological model. (**a**) dg (black line) and du (dashed blue line) surface loading time-series; (**b**) magnitude squared coherence; (**c**) real part of the transfer function between dg and du. Vertical dashed red lines indicate the annual and semi-annual periods

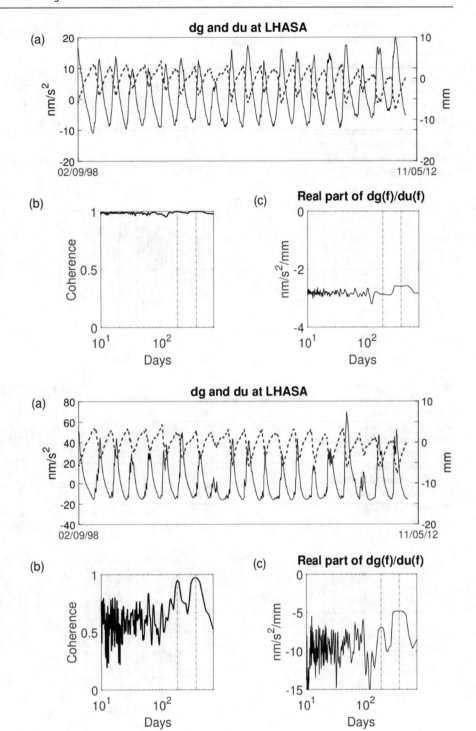

Fig. 5 Local surface loading gravity and vertical changes predicted at Lhasa (Tibet) from MERRA2 hydrological model. (**a**) dg (black line) and du (dashed blue line) surface loading time-series; (**b**) magnitude squared coherence; (**c**) real part of the transfer function between dg and du. Vertical dashed red lines indicate the annual and semi-annual periods

Please note that for Wettzell station, we used the GNSS time-series from GRAZ station, which is located 300 km away from the SG instrument, explaining a coherence as low as 50%. At Lhasa, the coherence is also close to 100% at the semi-annual period.

Table 1 shows that we can distinguish three groups of stations: Lhasa, Metsahovi, Ny-Alesund, Onsala, Wettzell and Yebes for which the ratio dg/du is negative and roughly lies between −2 and −5 nm/s^2/mm. Hydrological models have most of their spectral energy below degree 10 corre-

Fig. 6 Total surface loading gravity and vertical changes predicted at Strasbourg (France) from MERRA2 hydrological model. (**a**) dg (black line) and du (dashed blue line) surface loading time-series; (**b**) magnitude squared coherence; (**c**) real part of the transfer function between dg and du. Vertical dashed red lines indicate the annual and semi-annual periods

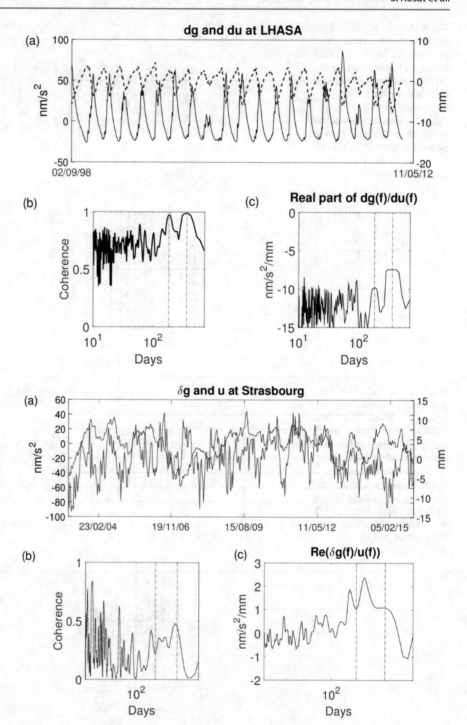

Fig. 7 Surface gravity and vertical changes recorded at Strasbourg (France). (**a**) dg (black line) and du (dashed blue line) time-series; (**b**) magnitude squared coherence; (**c**) real part of the transfer function between dg and du. Vertical dashed red lines indicate the annual and semi-annual periods

sponding to theoretical dg/du values closer to −3 than to −2.6 nm/s^2/mm elastic values (de Linage et al. 2009). So the obtained negative values agree rather well with theoretical predictions. The second group is Strasbourg station for which the ratio is positive illustrating the strong influence of local hydrological masses above the gravimeter. The third group contains Canberra and Sutherland for which the ratio is close to zero. Hydrological loading signals in gravity are indeed smaller at these two sites while vertical displacement is similar.

When comparing the dg/du ratio from MERRA2 hydrological predictions and the dg/du ratio for real observations, we can see that despite the good coherence at the annual period for most sites, the values are quite different. We should further investigate the impact of data pre-processing and the influence of local hydrological masses.

Fig. 8 Surface gravity and vertical changes recorded at Lhasa (Tibet, China). (**a**) dg (black line) and du (dashed blue line) time-series; (**b**) magnitude squared coherence; (**c**) real part of the transfer function between dg and du. Vertical dashed red lines indicate the annual and semi-annual periods

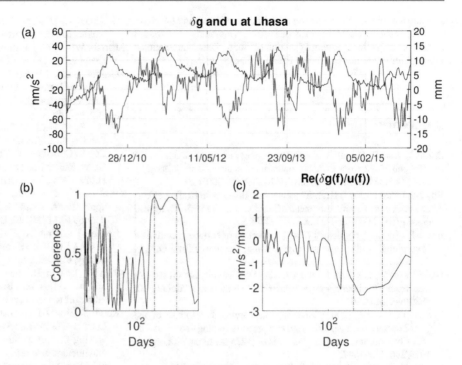

Table 1 Time spans of analyzed GNSS and Superconducting Gravimeter time-series and estimated ratios dg/du at the period of their maximum coherence

| Station name | IGETS instrument code | Time span of series | dg/du at seasonal period from MERRA2 predictions (nm/s²/mm) | | Maximum coherence at seasonal periods | | dg/du at seasonal period (nm/s²/mm) |
			Local hydro below	Local hydro above	%	Period of maximum coherence	
Canberra (Australia)	cb031	2003/01-2015/12	−23	16	60	320 days	−0.5
Lhasa (Tibet, China)	lh057	2009/12-2015/11	−7	2	97	Semi-annual & annual	−2
Metsahovi (Finland)	me020	2003/01-2015/04	−13	7	60	Annual	−1.8
Ny-Alesund (Norway)	ny039	2003/01-2012/06	−39	32	85	Annual	−4.3
Onsala (Sweden)	os054	2009/07-2015/12	−15	9	95	Annual	−3
Strasbourg (France)	st026	2003/01-2015/12	−8	6	48	Annual	1.1
Sutherland (South Africa)	su037-1	2010/05-2015/12	−26	20	59	320 days	−0.4
Wettzell (Germany)	we029-1	2003/01-2015/12	−12	6	58	320 days	−1.7
Yebes (Spain)	ys064	2011/12-2015/12	−16	9	98	Annual	−5.0

Maximum coherence of MERRA2 predictions is at the annual period. For predicted MERRA2 dg/du ratios, we have considered both cases of local hydrological masses below and above the instrument

4   Summary

We have verified the spectral coherency at seasonal time-scales between surface deformation and surface gravity changes recorded at nine collocated sites. The ratio of surface gravity perturbations over vertical height changes strongly depends on the local masses fluctuations. We have shown that the seasonal signal is coherent at eight sites over the nine studied here. The obtained dg/du ratios at the annual period are different between real observations and hydrological loading predictions. Influence of data pre-processing and local hydrological masses should be further investigated. Interpreting the obtained dg/du ratios in terms of rheological properties of the Earth would be the next step.

Acknowledgments We thank two anonymous reviewers and the editor for their suggestions to improve this paper. This study was funded under the International Emerging Action (n°8056, former international program for scientific cooperation PICS) of CNRS.

References

Böhm J, Werl B, Schuh H (2006) Troposphere mapping functions for GPS and VLBI from ECMWF operational analysis data. J Geophys Res 111:B02406. https://doi.org/10.1029/2005JB003629

Boy J-P, Lyard F (2008) High-frequency non-tidal ocean loading effects on surface gravity measurements. Geophys J Int 175:35–45. https://doi.org/10.1111/j.1365-246X.2008.03895.x

Boy J-P, Gegout P, Hinderer J (2002) Reduction of surface gravity data from global atmospheric pressure loading. Geophys J Int 149:534–545

Boy JP, Rosat S, Hinderer J, Littel F (2017) Superconducting gravimeter data from Strasbourg – level 1. GFZ Data Serv. https://doi.org/10.5880/igets.st.ll.001

Carrère L, Lyard F (2003) Modeling the barotropic response of the global ocean to atmospheric wind and pressure forcing - comparisons with observations. Geophys Res Lett 30(1275):4. https://doi.org/10.1029/2002GL01647

Carrère L, Lyard F, Cancet M, Guillot A, Picot N (2016) FES 2014, a new tidal model - Validation results and perspectives for improvements, ESA Living Planet Conference, Prague

de Linage C, Hinderer J, Rogister Y (2007) A search for the ratio between gravity variation and vertical displacement due to a surface load. Geophys J Int 171:986–994

de Linage C, Hinderer J, Boy J-P (2009) Variability of the gravity-to-height ratio due to surface loads. Pure Appl Geophys 166:1217–1245

Farrell WE (1972) Deformation of the earth by surface loads. Rev Geophys Space Phys 10:761–797

Gegout P, Boy J-P, Hinderer J, Ferhat G (2010) Modeling and observation of loading contribution to time-variable GPS sites positions. IAG Symposia 135(8):651–659. https://doi.org/10.1007/978-3-642-10634-7_86

Gelaro R, McCarty W, Suarez MJ, Todling R, Molod AM, Takacs LL, Randles C, Darmenov A, Bosilovich MG, Reichle RH, Wargan K, Coy L, Cullather RI, Akella SR, Bachard V, Conaty AL, da Silva A, Gu W, Koster RD, Lucchesi RA, Merkova D, Partyka GS, Pawson S, Putman WM, Rienecker MM, Schubert SD, Sienkiewicz ME, Zhao B (2017) The modern-era retrospective analysis for research and applications, version-2 (MERRA-2). J Clim 30:5419–5454. https://doi.org/10.1175/JCLI-D-16-0758.1

Herring TA, King RW, Floyd MA, McCluskey SC (2015) Introduction to GAMIT/GLOBK, release 10.6. Massachusetts Institute of Technology, Cambridge

Petit G, Luzum B, IERS Conventions (2010) IERS technical note no. 36. Verlag des Bundesamts für Kartographie und Geodäsie, Frankfurt am Main, p 179

Rosat S, Boy J-P, Ferhat G, Hinderer J, Amalvict M, Gegout P, Luck B (2009) Analysis of a ten-year (1997-2007) record of time-varying gravity in Strasbourg using absolute and superconducting gravimeters: new results on the calibration and comparison with GPS height changes and hydrology. J Geodyn 48:360–365

Voigt C, Förste C, Wziontek H, Crossley D, Meurers B, Palinkas V, Hinderer J, Boy J-P, Barriot J-P, Sun H (2016) Report on the data base of the international geodynamics and earth tide service (IGETS). Scientific technical report STR Potsdam, GFZ German research. Centre for Geosciences, Potsdam, 24 p. https://doi.org/10.2312/gfz.b103-16087

A Benchmarking Measurement Campaign to Support Ubiquitous Localization in GNSS Denied and Indoor Environments

Guenther Retscher, Allison Kealy, Vassilis Gikas, Jelena Gabela, Salil Goel, Yan Li, Andrea Masiero, Charles K. Toth, Harris Perakis, Wioleta Błaszczak-Bąk, Zoltan Koppanyi, and Dorota Grejner-Brzezinska

Abstract

Localization in GNSS-denied/challenged indoor/outdoor and transitional environments represents a challenging research problem. As part of the joint IAG/FIG Working Groups 4.1.1 and 5.5 on Multi-sensor Systems, a benchmarking measurement campaign was conducted at The Ohio State University. Initial experiments have demonstrated that Cooperative Localization (CL) is extremely useful for positioning and navigation of platforms navigating in swarms or networks. In the data acquisition campaign, multiple sensor platforms, including vehicles, bicyclists and pedestrians were equipped with combinations of GNSS, Ultra-wide Band (UWB), Wireless Fidelity (Wi-Fi), Raspberry Pi units, cameras, Light Detection and Ranging (LiDAR) and inertial sensors for CL. Pedestrians wore a specially designed helmet equipped with some of these sensors. An overview of the experimental configurations, test scenarios, characteristics and sensor specifications is given. It has been demonstrated that all involved sensor platforms in the different test scenarios have gained a significant increase in positioning accuracy by using ubiquitous user localization. For example,

G. Retscher (✉)
Department of Geodesy and Geoinformation, TU Wien – Vienna University of Technology, Vienna, Austria
e-mail: guenther.retscher@tuwien.ac.at

A. Kealy
Department of Geospatial Science, RMIT University, Melbourne, VIC, Australia
e-mail: allison.kealy@rmit.edu.au

V. Gikas · H. Perakis
School of Rural and Surveying Engineering, National Technical University of Athens, Athens, Greece
e-mail: vgikas@central.ntua.gr; hperakis@central.ntua.gr

J. Gabela
Department of Electrical and Electronic Engineering, The University of Melbourne, Melbourne, VIC, Australia
e-mail: jgabela@student.unimelb.edu.au

S. Goel
Department of Civil Engineering, Indian Institute of Technology, Kanpur, Uttar Pradesh, India
e-mail: sgoel@iitk.ac.in

Y. Li
SMART Infrastructure Facility, University of Wollongong, Wollongong, NSW, Australia
e-mail: liyan@uow.edu.au

A. Masiero
Interdepartmental Research Center of Geomatics, University of Padova, Padova, Italy
e-mail: masiero@dei.unipd.it

C. K. Toth
Department of Civil, Environmental and Geodetic Engineering, The Ohio State University, Columbus, OH, USA
e-mail: toth.2@osu.edu

W. Błaszczak-Bąk
Institute of Geodesy of the University of Warmia and Mazury, Olsztyn, Poland
e-mail: wioleta.blaszczak@uwm.edu.pl

Z. Koppanyi
Leica Geosystems, Heerbrugg, Switzerland
e-mail: zoltan.koppanyi@gmail.com

D. Grejner-Brzezinska
College of Engineering, The Ohio State University, Columbus, OH, USA
e-mail: grejner-brzezinska.1@osu.edu

© The Author(s) 2020
J. T. Freymueller, L. Sanchez (eds.), *Beyond 100: The Next Century in Geodesy*,
International Association of Geodesy Symposia 152, https://doi.org/10.1007/1345_2020_102

in the indoor environment, success rates of approximately 97% were obtained using Wi-Fi fingerprinting for correctly detecting the room-level location of the user. Using UWB, decimeter-level positioning accuracy is demonstrable achievable under certain conditions. The full sets of data is being made available to the wider research community through the WG on request.

Keywords

Cooperative localization (CL) · GNSS-denied environments · Indoor positioning · Positioning · Navigation and Timing (PNT) · Relative ranging · Ultra-wideband (UWB) · Wireless Fidelity (Wi-Fi)

1   Introduction and Objectives

In GNSS challenged environments, an augmentation with other emerging positioning technologies is an unremitting requirement. This requirement led to the development of multi-sensor systems and their integration using sensor fusion. Thus, for ubiquitous positioning solutions several technologies are researched and further developed. One strategy is to use so-called wireless signals-of-opportunity which were originally not intended for positioning, such as Wireless Fidelity (Wi-Fi). Moreover, designated technologies based on pre-deployed signal transmission infrastructure as well as technologies not based on signals are developed and enhanced in the research conducted by the IAG Sub-Commission 4.1 in the last years. In the first category fall systems using infrared or ultrasonic signals, Ultra-wide Band (UWB), ZigBee, Radio Frequency Identification (RFID), Bluetooth, Light Emitting Diodes (LED), Dedicated Short Range Communication (DSRC) or other radio frequency (RF) based systems. Vision/camera systems as well as inertial sensors, such as accelerometers, gyroscopes, magnetometers, employed for dead reckoning belong to the second category. Also, a typical application field is smartphone positioning which plays an important role in the interdisciplinary research conducted under the umbrella of the SC 4.1. Furthermore, the Sub-Commission lays an emphasis on multi-sensor cooperative systems which employ all aforementioned variety of sensors on different platforms for sharing their absolute and relative locations. Platforms include mobile vehicles, robots as well as pedestrians and most recently UAS (Unmanned Aerial Systems). Their land and airborne navigation applications range from transportation, personal mobility, industrial and indoor positioning applications and to a lesser extent environmental monitoring. Thus, the major key objective of the SC is to examine the potential and capabilities of low-cost sensors including GNSS systems and smartphone navigation sensors. Primary sensors of interest include inertial and wireless technologies as well as vision-based systems and laser scanning for improving the navigation performance. Furthermore, other objectives include to contribute in research that depends on big data handling, sensor synchronization, data fusion, real-time processing as well as to support standardization activities and to study and monitor the progress of new multi-sensor applications, as well as, to support and promote knowledge exchange and reporting on the development trends, possibilities and limitations of emerging positioning technologies. Thereby the development of new measurement integration algorithms based around innovative modeling techniques in other research domains, such as machine learning and genetic algorithms, spatial cognition etc., plays also an important role.

2   PNT Application Requirements

Figure 1 provides an overview about the PNT (Positioning, Navigation and Timing) user requirements listed in a 'fish plot'. These requirements can be categorized in four different classes which are positioning, cost, security and legal as well as interface requirements. Thereby the most relevant positioning requirements in our view are apart from positioning accuracy also integrity, availability and coverage, latency and continuity as well as sampling and update rate. The other three requirements, however, also must not to be ignored. Operational and maintenance costs, for instance, are very important too when designing a low-cost positioning system. The GNSS Market Report of the European Commission in 2017 (GNSS Market Report 2017) identified the key GNSS requirements and performance parameters. Here they are also applied for alternative positioning technologies and techniques, such as UWB and Wi-Fi, as they are valid for any other PNT applications not involving only GNSS but also other sensors and technologies which are additionally and independently used (Retscher et al. 2020b). Regarding availability the number of transmitters (UWB stationary

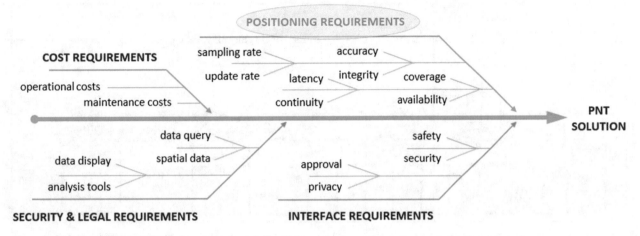

Fig. 1 Overview of PNT user requirements

transmitters or Wi-Fi Access Points) replaces the number of satellites. Especially integrity is often neglected and not paid full attention. It can be seen as a very important key parameter. The way that integrity is ensured and assessed, and the means of delivering integrity related information to the user are highly application dependent. Time-to-first-fix (TTFF) in the case of Wi-Fi positioning is highly correlated to the received signal strength (RSS) scan duration of a certain mobile device. This is especially important in kinematic positioning. As seen by the authors in Retscher and Leb (2019) the appearing scan durations can vary significantly for different smartphones which results in a different level of achievable positioning accuracy in dependence of the walking speed in the case of pedestrian navigation. For different users robustness may have a different meaning, such as the ability of the solution to respond following a server shadowing event. Here, robustness is defined as the ability of the solution to mitigate interference. Other requirements and performance parameters are power consumption, resiliency, connectivity, interoperability and traceability. Especially in the case of mobile devices power consumption is still very critical to provide a long-term solution possibility. Resiliency is the ability to prepare for and adapt to changing conditions, such as it is the case for Wi-Fi RSSI (Received Signal Strength Indicator) signal variations and fluctuations. To encounter for their influence new robust schemes are necessary and need to be developed.

3 Field Campaign at The Ohio State University

A benchmarking measurement campaign dealing with cooperative localization (CL) of different mobile sensor platforms navigating within a neighbourhood was conducted in October 2017 (Retscher et al. 2020a). Pedestrians as well as

vehicle test were carried out. In the case of pedestrian CL, four pedestrians with a specially designed helmet equipped with GNSS, two UWB systems (i.e., from TimeDomain and Pozyx), Raspberry Pi, Wi-Fi and smartphone camera were moving around jointly, with the objective of achieving precise positioning in indoor environments, as well as providing a seamless position transition between indoor and outdoor environments. Relative range observations among pedestrians, camera observations, UWB range and Wi-Fi RSSI measurements were performed. All users transition from outdoor to indoor environments and thus, each pedestrian starts to lose GNSS signals successively. Once all pedestrians are indoors, GNSS observations are not available to any of them and therefore the users rely on relative UWB ranges, Wi-Fi measurements, and camera observations, for localizing all users cooperatively. For further details about the campaign the reader is referred to Kealy et al. (2019) and Retscher et al. (2020a).

3.1 Indoor UWB Localization Results

The UWB indoor experiment of the campaign aims at the investigation of the possibility of calibrating the UWB system in order to compensate for the effects of the static parts of the environment on UWB measurements, hence obtaining an improvement of the overall positioning accuracy (Retscher et al. 2020a). 14 Pozyx UWB anchors were fixed on the walls in the hallway and calibration and validation range measurement data sets were collected on 35 checkpoints along the corridor. 27 of these checkpoints were observed both during the calibration and validation data collection, whereas the remaining 8 were used only for validation. The users were moving in the test site in stop-and-go mode, whereas for the calibration data collection only a few persons (3 to 4, mostly involved in this experiment)

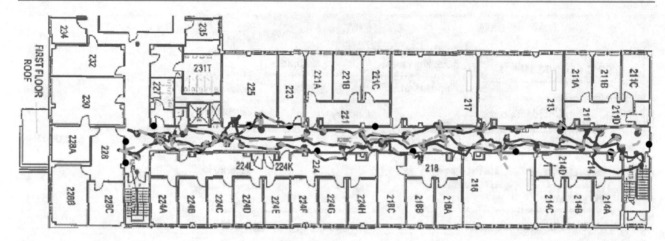

Fig. 2 Trajectory in the hallway estimated without calibration (solid blue line) and with the considered calibration model (green dashed line)

and in the validation up to 10 persons were moving around representing a more realistic and challenging scenario. For the calibration the median of the measured range errors was considered as an estimate of the measurement bias to be removed. Since the value of such a bias varies over all the area of interest, a calibration model with spatially varying additive bias was employed for each anchor by means of natural neighbour interpolation of the values computed on the 27 calibration points. Estimated trajectories obtained by a standard Extended Kalman filter are shown in Fig. 2 whereby either for the first trajectory no calibration (solid blue line) or for the second the aforementioned calibration model (dashed green line) are employed. A simple dynamic random walk model for the velocities of the device movement for integrating the ranges is assumed. Furthermore, the method took also advantage of a still-condition detection step similar as applied in Masiero et al. (2019). Figure 2 shows also anchor (black dots) and checkpoint (red dots) locations. As can be seen from the figure the improvement obtained with the proposed approach is quite significant and the considered approach can potentially be useful to reduce the effect of the systematic errors on the UWB measurements. Figure 3 presents the cumulative distribution function (CDF) of the 2D positioning errors for the results shown in Fig. 2. Distributions shown in Fig. 3 are obtained by taking into account of the 2D errors on the checkpoints shown as red dots in Fig. 2. Figure 3 confirms the improvement obtained with the considered calibration and tracking approach, e.g. the maximum 2D positioning error is almost 2 m smaller in the calibrated case with respect to the uncalibrated one.

3.2 Wi-Fi Localization Results

The aim of the Wi-Fi localization in the building was to achieve at least room-level or region-level granularity. For

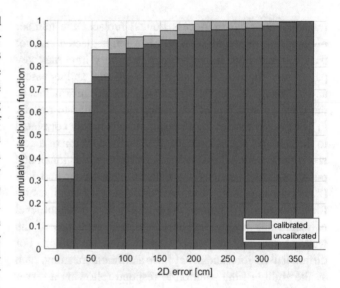

Fig. 3 CDF of the 2D errors of the position estimates of the 35 checkpoints

that purpose the test area in the building was segmented in cells including rooms and sections of 4 m in length in the corridor as well as entrances or exits. The localization method chosen was location fingerprinting. In the training phase of the fingerprinting, 200 RSSI scans of the visible Access Points (APs) at different locations were simultaneously collected by 10 different mobile devices to be able to locate a user in the positioning phase who has scanned again for the APs. In this phase, Bayesian inference is applied to calculate the probability that a user is at a certain location given a specified observation. Then the most likely location of the mobile device can be estimated. Thereby the accuracy of the statistical distribution model directly affects the final performance of the probabilistic fingerprint positioning (see e.g. Xia et al. 2017). The authors in Li et al. (2018) proposed a statistical approach to localize the mobile user to room level accuracy based on the Multivariate Gaussian Mixture Model

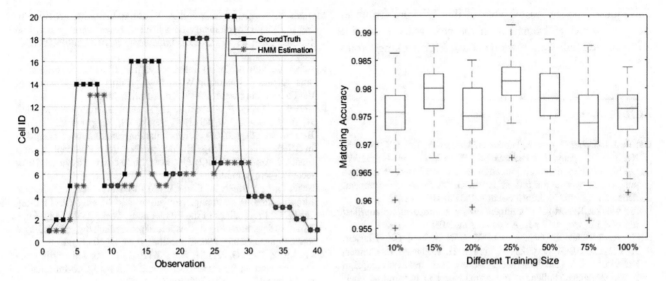

Fig. 4 Example of a kinematic walking trajectory (left) and matching probability rates for different training data sizes (right)

(MVGMM). A Hidden Markov Model (HMM) is applied to track the mobile user, where the hidden states comprise the possible room locations and the RSSI measurements are taken as observations. Due to the segmentation of the test area in different cells the transition matrix in the HMM is defined in such a way that only adjacent cells have non-zero transition probability while the transition probability between isolated cells is zero. In total, 11 kinematic walking trajectories were carried out with the different smartphones. Figure 4 (left) shows an example of an obtained trajectory of one smartphone user who moved in the study area between different defined cells. Figure 4 (right) presents the corresponding matching probabilities with different training sizes. The trajectories along the reference points could be obtained with matching success rates of up to 97%. It can be seen that the proposed method is nearly insensitive to the size of the training samples, even presenting more robust localization accuracy to lower sample sizes. This result is similar to the work in Zhou (2006) where the authors found that, given dense training samples for the area may introduce more noise to distinguish from other areas. It can be finally summarized that the proposed system and algorithm demonstrated a reliable room location awareness system in a real public environment.

4   Concluding Remarks and Outlook

In the presented benchmarking measurement campaign, the main focus was on CL of different sensor platforms, i.e., vehicles, bicyclists and pedestrians, in GNSS-denied/challenged in-/outdoor and transitional environments.

For this paper, pedestrian users wearing a specially designed sensor helmet navigated jointly in a neighborhood in an indoor positioning application. An overview of the field experimental schemes, set-ups, characteristics and sensor specifications along with the main results for the positioning are presented and further details may be found in Kealy et al. (2019). In these analyses trajectories of pedestrians walking around in an indoor office environment could be obtained on the decimeter-level using UWB and with Wi-Fi fingerprinting matching success rates of around 97% were achieved for assigning the user to the correct cell, i.e., either a room or section of the hallway. Further data processing and analyses of a CL solution is currently in progress and the results for UWB navigation presented in Gabela et al. (2019) indicate significant performance improvements of users navigating within a neighborhood. Positioning accuracies on the decimetre level are achieved for two moving users even at the end of the hallway where the geometry of the range measurement to the anchors is not the best. Ranges between the users constrain and improve the solution in this respect. Future work is especially concentrated on analyses of the localization accuracies and performance in the transitional and indoor environments. Apart from absolute localization of the users, dead reckoning with the inertial sensors is a further key element of future investigations. Especially the use of the smartphone sensors in combination with Wi-Fi and cameras is considered as a smartphone was mounted on the helmet which recorded at the same time Wi-Fi RSSI's, videos and the measurements of the inertial sensors as well as magnetometer and barometer. The extensive data set of the campaign are available for researchers from the joint IAG and FIG working group on request. The successful work of this Working Group in the past period will continue in the next years as a joint effort of IAG Sub-Commission 4.1 and FIG Working Group 5.5.

References

Gabela J, Retscher G, Goel S, Perakis H, Masiero A, Toth CK, Gikas V, Kealy A, Koppanyi Z, Błaszczak-Bąk W, Li Y, Grejner-Brzezinska DA (2019) Experimental evaluation of a UWB-based cooperative positioning system for pedestrians in GNSS-denied environment. Sensors 19(23):5274. https://doi.org/10.3390/s19235274

GNSS Market Report (2017). https://www.gsa.europa.eu/system/files/reports/gnss_mr_2017.pdf. Accessed Aug 2019

Kealy A, Retscher G, Gabela J, Li Y, Goel S, Toth CK, Masiero A, Błaszczak-Bąk W, Gikas V, Perakis H, Koppanyi Z, Grejner-Brzezinska DA (2019) A benchmarking measurement campaign in GNSS-denied/challenged indoor/outdoor and transitional environments, FIG Article of the Month, July. http://fig.net/resources/monthly_articles/2019/kealy_etal_july_2019.asp

Li Y, Williams S, Moran B, Kealy A, Retscher G (2018) High-dimensional probabilistic fingerprinting in wireless sensor networks based on a multivariate Gaussian mixture model. Sensors 18(8):2602. https://doi.org/10.3390/S18082602

Masiero A, Fissore F, Guarnieri A, Pirotti F, Vettore A (2019) Aiding indoor photogrammetry with UWB sensors. Photogramm Eng Remote Sens 85(5):369–378. https://doi.org/10.14358/PERS.85.5.369

Retscher G, Leb A (2019) Influence of the RSSI scan duration of smartphones in kinematic Wi-Fi fingerprinting. In: Proceedings of the FIG working week, 22–26 April, Hanoi, 15 pp

Retscher G, Kealy A, Gabela J, Li Y, Goel S, Toth CK, Masiero A, Błaszczak-Bąk W, Gikas V, Perakis H, Koppanyi Z, Grejner-Brzezinska DA (2020a) A benchmarking measurement campaign in GNSS-denied/challenged indoor/outdoor and transitional environments. J Appl Geod 14(2) (accepted). https://www.fig.net/fig2020/technical_program.htm

Retscher G, Li Y, Kealy A, Gikas V (2020b) The need and challenges for ubiquitous Positioning, Navigation and Timing (PNT) using Wi-Fi. In: Proceedings of the FIG working week, 10–14 May, Amsterdam, 18 pp. https://fig.net/resources/proceedings/fig_proceedings/fig2020/papers/ts05g/TS05G_retscher_li_et_al_10335.pdf

Xia S, Liu Y, Yuan G, Zhu M, Wang Z (2017) Indoor fingerprint positioning based on Wi-Fi: an overview. ISPRS Int J Geo-Information 6:135. https://doi.org/10.3390/ijgi6050135

Zhou R (2006) Wireless Indoor Tracking System (WITS). Aktuelle Trends in der Software Forschung, doIT Software-Forschungstag, dpunkt Verlag, Heidelberg, pp 163–177

A Method to Correct the Raw Doppler Observations for GNSS Velocity Determination

Kaifei He, Tianhe Xu, Christoph Förste, Zhenjie Wang, Qiang Zhao, and Yongseng Wei

Abstract

In the application of GNSS in the velocity determination, it is often the case that some GNSS receivers give an opposite sign for the raw Doppler observations which do not correspond to the real Doppler shift. This is caused by different methods of the GNSS signal processing in different GNSS receivers. If the velocities of kinematic platforms are calculated by using raw Doppler observations from the GNSS receiver directly, the directions of the estimated velocities may be reversed, and the value of the velocity is wrong with respect to the actual movement. This would lead to incorrect results, and unacceptable for research and applications. To overcome this problem, a new method of sign correction for raw Doppler observations is proposed in this study. This algorithm constructs a correction function based on the GNSS carrier-phase-derived Doppler observations. To test this approach, GNSS data of GEOHALO airborne gravimetric missions have been used. The results show that the proposed method, which is straightforward and practical, can produce the correct velocity for a kinematic platform in any case, independent of the internal hardware structure and the specific way of the signal processing of the GNSS receivers in question.

Keywords

Doppler observation · GNSS · Kinematic platform · Sign correction · Velocity determination

1 Introduction

GNSS is a cost-effective means to obtain reliable and high-precision velocities by exploiting the receiver raw Doppler (RD) and carrier-phase-derived Doppler (CD) measurements (Szarmes et al. 1997). The RD method is the most widely used technique and usually has a cm/s accuracy (Wang and Xu 2011). The CD method consists in differencing successive carrier phases observations, enables accuracies at the mm/s level (Freda et al. 2015). Since the CD is computed over a longer time span than the raw one, the random noise is averaged and suppressed. Therefore, very smooth velocity estimates in low dynamic environments can be obtained from CD measurements if there are no undetected cycle-slips (Serrano et al. 2004b). However, the functional model of the RD method is stricter than the CD method, and its velocity results are more reliable when a sudden change of the vehicle status occurs, such as braking, turning, and accelerating (Wang and Xu 2011). Thus, the RD method has been investigated here in this study.

In the applications of velocity determination using the RD observations, the sign of Doppler observations from some GNSS receivers is opposite to the real Doppler shift. If the velocity of the kinematic platform is calculated by using the RD observations directly from the GNSS receiver, the direction of the velocity of the kinematic platform will be

K. He (✉) · Z. Wang · Q. Zhao · Y. Wei
College of Oceanography and Space Informatics, China University of Petroleum (East China), Qingdao, China
e-mail: kfhe@upc.edu.cn

T. Xu
Institute of Space Science, Shandong University, Weihai, China

C. Förste
GFZ – German Research Centre for Geosciences, Potsdam, Germany

J. T. Freymueller, L. Sanchez (eds.), *Beyond 100: The Next Century in Geodesy*,
International Association of Geodesy Symposia 152, https://doi.org/10.1007/1345_2020_119

reverse, and the value of the velocity will be false. Thus, a sign correction method of the RD observations for GNSS velocity determination was studied in this article.

2 Velocity Determination Using GNSS Raw Doppler Observations

The GNSS RD observations, the Doppler shift $D_{r,j}^s$ between the receiver r and the GNSS satellite s at the frequency channel j can be given as (Hofmann-Wellenhof et al. 2008)

$$D_{r,j}^s = f_j^s - f_{r,j}^s = \frac{V_{\rho_r^s}}{c} f_j^s = \frac{V_{\rho_r^s}}{\lambda_j^s} \quad (1)$$

where f_j^s denotes the emitted frequency j of satellite s, $f_{r,j}^s$ is the received frequency j from satellite s, $V_{\rho_r^s}$ is the radial velocity along the range between the receiver r and the satellite s, c means the speed of electromagnetic waves in vacuum and λ_j^s denotes the wavelength. $D_{r,j}^s$ has a negative sign when the receiver and the transmitter move away from each other and a positive sign when they approach each other. Equation (1) for the observed Doppler shift scaled to range rate can be written as

$$V_{\rho_r^s} = \lambda_j^s \cdot D_{r,j}^s = \dot{\rho}_r^s + c \cdot \left(d\dot{i}_r - d\dot{i}^s \right) + \dot{T}_r^s - \dot{I}_{r,j}^s + \varepsilon_{r,j}^s \quad (2)$$

where the derivatives with respect to time are indicated by a dot, $\dot{\rho}_r^s$ means for the geometric range rate between the receiver r and the satellite s, the signs $d\dot{i}^s$ and $d\dot{i}_r$ are the satellite clock drift and receiver clock drift, respectively, $\dot{T}$ and $\dot{I}$ denote the tropospheric and the ionospheric delay rate, respectively, and $\varepsilon_{r,j}^s$ denotes all non-modeled error sources (e.g. multipath error) and the effect of the observational noise. The precise velocity estimation of the kinematic station can be directly obtained by the classic Least-Squares adjustment (Koch 1999), if Doppler observations from more than four GNSS satellites have been measured.

Normally, it is no problem to obtain the velocity information from the GNSS RD observations, but, in the case of some GNSS receivers (such as in the chosen experiments in this paper), the signs of the RD observations are opposite to the real Doppler shift. This phenomenon is caused by different methods of GNSS signal processing in different GNSS receivers. In such situations, the direction of the estimated velocity will be reversed and the value of velocity will be false compared with the actual movement if the velocity of a kinematic platform is calculated by directly using the RD observations from the GNSS receiver. To solve this problem, an algorithm has to be developed to obtain the correct velocity from GNSS RD observations.

3 A New Method of Sign Correction for GNSS Raw Doppler Observations

In order to solve the above-named problem, a correction function is constructed based on the CD observations. Then, the GNSS RD observations will be corrected using the correction function.

3.1 Carrier-Phase-Derived Doppler Observations

The CD observations can be obtained by differencing carrier phase observations in the time domain, normalizing them with the time interval of the differenced observations or by fitting a curve to successive phase measurements, using polynomials of various orders (Serrano et al. 2004a). At present, the first order central difference is one of the most popular methods for obtaining the virtual Doppler observations (Wang and Xu 2011). Based on the fundamental GNSS carrier-phase observation, the CD observation is given by

$$\dot{\varphi}_t = \frac{1}{2} \left(\frac{\varphi_{t+\delta t} - \varphi_t}{\delta t} + \frac{\varphi_t - \varphi_{t-\delta t}}{\delta t} \right) = \frac{\varphi_{t+\delta t} - \varphi_{t-\delta t}}{2\delta t} \quad (3)$$

where $\dot{\varphi}_t$ (namely the CD observation) denotes the variation rate of the raw carrier phase observations φ_t. Here, the carrier phase observations φ_t should don't have the cycle slip. If the cycle slip exist, these observations φ_t should be omit. t means the observation time and δt denotes the data sample interval.

The resulting observation equation for velocity determination can be expressed as

$$V_{\rho_r^s} = \lambda_j \dot{\varphi}_{r,j}^s = \dot{\rho}_r^s + c \cdot \left(d\dot{i}_r - d\dot{i}^s \right) + \dot{T}_r^s - \dot{I}_{r,j}^s + \varepsilon_{r,j}^s \quad (4)$$

Here, the CD observation $\dot{\varphi}_{r,j}^s$ of Eq. (4) has a similar function as the RD $D_{r,j}^s$ of Eq. (2). Thus, this relationship can be used to solve the problem as already stated in our previous letter.

3.2 Construction of the Sign Correction Function

The RD observations should, of course, have the same sign as the CD observations. Therefore, a sign correction function $f(j)$ is constructed as

$$f(j) = \begin{cases} 1, & if \left(D_{r,j}^s \cdot \dot{\varphi}_t > 0 \right) \\ -1, & if \left(D_{r,j}^s \cdot \dot{\varphi}_t < 0 \right) \end{cases} \quad (5)$$

Here, if the RD observation $D_{r,j}^s$ has the same sign as the CD observation $\dot{\varphi}_t$, the value of the sign correction function $f(j)$ will be $+1$. Otherwise, the value of the sign correction function $f(j)$ will be -1 when the RD observation $D_{r,j}^s$ has a different sign than the CD shift $\dot{\varphi}_t$. In this correction function, the CD observations were calculated by Eq. (3), but do not need to be calculated for all epochs, and just several epochs at the beginning are enough, such as the first 10 epochs (here, should be omitted the epochs that cycle slip occurrence).

3.3 Correction Method for the GNSS Raw Doppler Observations

After constructing the sign correction function, the GNSS RD observations can be corrected by

$$D_{r,j,new}^s = f(j) \cdot D_{r,j}^s \qquad (6)$$

where the GNSS RD observations $D_{r,j}^s$ have been modified by the sign correction function $f(j)$. If the RD observations $D_{r,j}^s$ has another sign than the CD observations $\dot{\varphi}_t$, their sign will be changed by applying this new method using Eq. (6). Then, the corrected Doppler observation $D_{r,j,new}^s$ can be used for the GNSS velocity determination. To illustrate the new method of sign correction, one example is given in the following.

4 Experiments and Analysis

To investigate this new method, the GNSS data of the airborne gravimetry campaign GEOHALO mission over Italy 2012 on June 6, 2012 (He et al. 2016) were chosen for testing. The selected kinematic station was AIR5 at the front part of the HALO aircraft. The station REF6, installed by GFZ next to the runway of the German Aerospace Centre (DLR) airport in Oberpfaffenhofen, Germany, was selected as the reference station. The GNSS receiver types of AIR5 and REF6 were both JAVAD TRE_G3T DELTA. The trajectory of the HALO aircraft on June 6, 2012 and the position of the selected reference station REF6 are shown in Fig. 1. The selected data contain GLONASS and GPS observations with a sampling rate of 1 Hz.

The HALO_GNSS software (He 2015) was used for the GNSS data processing. The calculated trajectory of the HALO aircraft is shown in Fig. 2 as latitude, longitude and height components, respectively. The two significant humps on the height curve correspond to crossing the Alps. In order to investigate the capability of the proposed new approach, two experimental figures were designed as follows.

Figure 3 GNSS velocity determination using the RD observations directly without any correction. The velocity

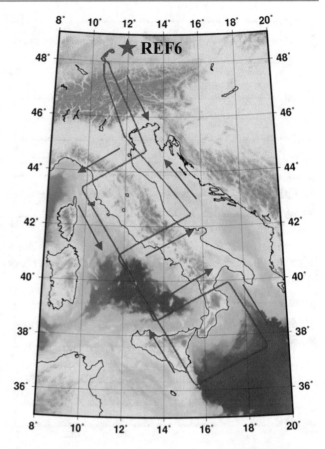

Fig. 1 HALO aircraft flight trajectory on June 6, 2012 and the location of the selected reference station REF6

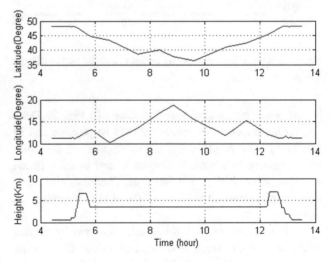

Fig. 2 Components of the flight trajectory of the HALO aircraft on June 6, 2012

results are shown in Fig. 3 as north, east and up components, respectively. The incorrect velocity results can be found in Fig. 3 compared with the trajectory components in Fig. 2. For instance, the latitude component of the trajectory of the HALO aircraft at nearly GPST 10:00 was increasing, and

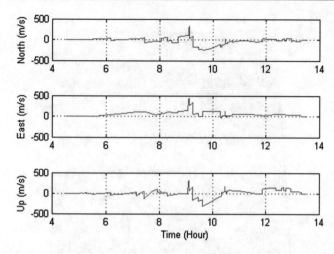

Fig. 3 Velocity components of HALO aircraft calculated from GNSS RD observations

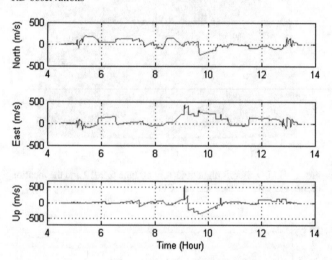

Fig. 4 Difference between velocity results calculated by the RD and CD

Table 1 The statistical results of difference between velocity results calculated by the RD and RD (Unit: m/s)

Figure		Directions	Min	Max	Mean	RMS
1	RD vs. CD	North	−246.54	198.11	3.20	93.38
		East	−148.17	432.84	61.19	120.76
		Up	−345.91	548.26	−12.63	109.22
2	Corrected RD vs. CD	North	−0.15	0.36	0.00	0.01
		East	−0.21	0.33	0.00	0.01
		Up	−0.61	0.43	0.00	0.03

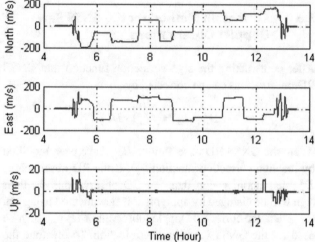

Fig. 5 Velocity components of the HALO aircraft calculated by GNSS RD observations corrected by the proposed approach

its longitude component was decreasing in Fig. 2. Thus, the north component of velocity should be positive and its east component negative, but the corresponding velocities were negative in the north component and positive in the east component in Fig. 3 at that time. Therefore, the direction of the calculated velocity of the HALO aircraft were reversed.

To analyze the value of this results, the velocity results, calculated by CD were used as "true value" to compare with the results of Fig. 3. The difference of the velocity results calculated by CD and RD are plotted in Fig. 4 and its statistic results given in Table 1. It is shown that the value of velocity results of Fig. 3 were false with respect to the actual movement if the velocity is calculated by directly using the RD observations from the GNSS receiver.

Figure 5 GNSS velocity determination using the RD observation corrected by the proposed approach. The veloc-

ity results are shown in Fig. 5 as north, east and up components, respectively. Compared with Fig. 3, the correct velocity results can be obtained by using the proposed approach. For instance, the corresponding north component of velocity has a positive value and its east component was negative at nearly GPST 10:00 in Fig. 5. Therefore, this example also shows that the proposed approach can correct the error that appeared in Fig. 3.

Compare with mentioned true value, the difference between them is plotted in Fig. 6 and its statistic results given in Table 1 as well. It is shown that the value of velocity results of Fig. 5 correspond with the actual movement if the velocity is calculated by using the corrected RD observations from the GNSS receiver. The reason for the difference between the velocity results calculated by corrected RD and the CD will be discussed in follows.

Normally, it is no problem to obtain the correct velocity information from GNSS RD observations for kinematic platform, but sometimes, such as in the chosen experiments, a problem occurs in the GNSS velocity determination when using the RD observations directly, see Fig. 3. When the proposed method was applied, the results of GNSS velocity determination by using the RD were corrected in such a way

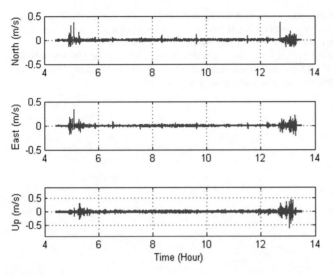

Fig. 6 The difference between velocity results calculated by the corrected RD and CD

that the direction and the magnitude of the estimated velocity corresponded to the actual movement.

The difference between velocity results calculated by corrected RD and the CD, shown as in Fig. 6, have two reasons. The one is the noise of observations that the RD measurements are usually noisier than the CD measurement. The other one is the movement of kinematic platform. The CD observations are average variation rates during a time interval $2\delta t$, see Eq. (3), which are more disturbed than the RD observations $D_{r,j}^{s}$ at the epoch time t, see Eq. (1). The sudden changes of the state of the HALO aircraft, see Figs. 2 and 5, correspond to the large spikes in Fig. 6. Therefore, the GNSS RD observations are more suitable to calculate the velocity results for highly dynamic platform.

5 Summary

This study focuses on GNSS velocity determination by using RD observations. There could occur a problem when using these observations directly. The direction of the velocity might be reversed and its magnitude false compared with the actual movement since the sign of the RD observations is opposite to the real Doppler shift. The reason for this is different signal processing methods in different types of GNSS receivers. It is hard for users to distinguish between the intrinsic methods of signal processing for every different type of GNSS receiver. Therefore, a new method was

proposed to correct the sign of the GNSS RD observations according to a proposed special sign correction function. To test this approach, GNSS data of GEOHALO airborne gravimetric campaigns have been used. The results show that the proposed method can yield the correct velocity of a kinematic platform in any case, without taking into account the internal structure and means of signal processing of the particular GNSS receiver.

Acknowledgements We express our appreciation to our colleagues of GFZ for their cooperation, discussions and providing GNSS data. This work is supported by the National Natural Science Foundation of China (41604027, 41574013), the financial support by Qingdao National Laboratory for Marine Science and Technology (QNLM2016ORP0401), the Key R&D Program of China (2016YFB0501700, 2016YFB0501705), the Shandong Provincial Natural Science Foundation (ZR2016DQ01, ZR2019MD005, ZR2016DM15), the Fundamental Research Funds for the Central Universities of China (18CX02054A) and Source Innovation Plan of Qingdao, China (17-1-1-100-jch).

References

Freda P, Angrisano A, Gaglione S, Troisi S (2015) Time-differenced carrier phases technique for precise GNSS velocity estimation. GPS Solutions 19:335–341

He K (2015) GNSS kinematic position and velocity determination for airborne gravimetry. PhD thesis (scientific technical report 15/04), GFZ - German Research Centre for Geosciences

He K, Xu G, Xu T, Flechtner F (2016) GNSS navigation and positioning for the GEOHALO experiment in Italy. GPS Solutions 20:215–224. https://doi.org/10.1007/s10291-014-0430-4

Hofmann-Wellenhof B, Lichtenegger H, Wasle E (2008) GNSS–global navigation satellite systems: GPS, GLONASS, Galileo, and more. Springer, Wien

Koch KR (1999) Parameter estimation and hypothesis testing in linear models. Springer, Wien

Serrano L, Kim D, Langley RB (2004a) A single GPS receiver as a real-time, accurate velocity and acceleration sensor. Paper presented at the 17th International Technical Meeting of the Satellite Division of the Institute of navigation, Long Beach, CA, USA, September 21–24

Serrano L, Kim D, Langley RB, Itani K, Ueno M (2004b) A gps velocity sensor: how accurate can it be?–a first look. Paper presented at the ION NTM, San Diego, CA, USA, January 26–28

Szarmes M, Ryan S, Lachapelle G, Fenton P (1997) DGPS high accuracy aircraft velocity determination using Doppler measurements. Paper presented at the International Symposium on Kinematic Systems (KIS), Banff, AB, Canada, June 3–6

Wang Q, Xu T (2011) Combining GPS carrier phase and Doppler observations for precise velocity determination. Sci China Phys Mech Astron 54:1022–1028

Assessment of a GNSS/INS/Wi-Fi Tight-Integration Method Using Support Vector Machine and Extended Kalman Filter

Marco Mendonça and Marcelo C. Santos

Abstract

Wi-Fi derived positions have been used in the past few years as a complementary source of positioning information for GNSS and Inertial Systems (INS). Ubiquitous positioning that transitions from indoors to outdoors and vice-versa is currently a hot topic of research. In this context, this study aims to analyze the potential of directional antennas sequentially tracking Wi-Fi signals on the 11 channels around the 2.4 GHz frequency in order to serve as an integrated signal for GNSS and INS positioning. Considering, as an example, a single point positioning (SPP) strategy coupled with an INS, the use of directional antennas can be beneficial in order to provide absolute directions of travel by the means of a Support Vector Machine (SVM) lane matching. In order to test the given hypothesis, real-world experiments were performed in areas with and without obstruction in an urban environment. Using a post-processed, smoothed in both forward and backward modes, and finally edited post-processed kinematic (RTK) solution as a reference, the solution integrating SPP GNSS, INS and Wi-Fi was assessed in terms of accuracy. Preliminary results show that such a combination of the directional antennas along with GNSS and INS and their respective SVM and EKF filters, can provide sub-meter accuracy at all times without the need of precise orbits or differential corrections, increasing solution availability, reliability and accuracy on a scalable and cost-effective way.

Keywords

GNSS · INS · Sensor integration · SVM · Vehicle navigation · Wi-Fi

1 Introduction

Sensor integration techniques are a contemporary topic of research, since mobile platforms can now achieve the computational power required for such tasks. The use of Global Navigation Satellite Systems (GNSS) as a source of positioning, albeit widespread in applications, has limitations particularly noticeable in urban environments. The main damaging effects on GNSS positioning in urban environments are signal obstructions and reflections, causing problems with both signal quality (usually yielding low signal-to-noise ratios), and low number of visible satellites. Several studies have been performed in order to quantify, analyse, and overcome such limitations using different techniques, such as solutions using new GNSS signals (Hsu et al. 2015), novel mathematical models to constrain the accumulating INS errors (Grejner-Brzezinska et al. 2001), sensor integration techniques, and signal-of-opportunity concepts (Groves 2011). With the growing demand for accurate and reliable urban positioning fueled by the advent and popularization of autonomous vehicles, improvements in this area are not only of academic value, but also of immediate practical applications. In this context, the cost and processing power requirements of the solutions are of paramount importance. Accurate and ubiquitous positioning equipment, such as

M. Mendonça (✉) · M. C. Santos
University of New Brunswick, Department of Geodesy and Geomatics Engineering, Fredericton, NB, Canada
e-mail: marco.mendonca@unb.ca; msantos@unb.ca

© The Author(s) 2020
J. T. Freymueller, L. Sanchez (eds.), *Beyond 100: The Next Century in Geodesy*,
International Association of Geodesy Symposia 152, https://doi.org/10.1007/1345_2020_120

GNSS/INS integrated NovAtel SPAN®, may have comparable costs to a semi-autonomous vehicle, such as the Tesla Model 3, therefore, not being capable of reaching mass-market applications.

With the aforementioned situation in mind, and considering the challenges of positioning in urban areas, the integration of cost-effective sensors and improvements in mathematical models are viable and tested alternatives to overcome such challenges (Grejner-Brzezinska et al. 2001; Groves 2008). In this study, a combination of Wi-Fi signals recorded by directional antennas, an INS and a GNSS (GPS + GLONASS) receiver is studied in order to asses how information from directional Wi-Fi antennas can be integrated in a GNSS/INS solution and what is the benefit of it. The remainder of this paper is divided among the following sections: a brief review of traditional GNSS/INS integration techniques, an overview of Wi-Fi positioning techniques, the development of an integration technique between the three systems, experiment design, and, finally, results and conclusions.

2   GNSS/INS Integration Techniques

Amongst several possible ING/GNSS integration techniques (Groves 2008), in this paper, the loosely-coupled integration is explored and used as basis for the integration with Wi-Fi derived information. Figure 1 shows an overview of how the loosely-coupled integration is performed. Block 1 represents a traditional GNSS positioning filter. This filter can output positions, velocity and timing (PVT) from any positioning method, such as single point positioning (SPP), precise point positioning (PPP) or real-time kinematic (RTK). Block 2

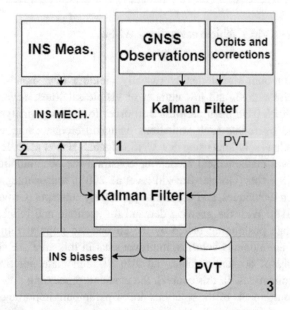

Fig. 1 Overview of a loosely-coupled integration scheme

represents an INS mechanization method, where positions are integrated to a current epoch using acceleration and angular velocity measurements. Without a long-term source of stability and constraint and due to the integrative nature of the INS mechanization procedure, errors rapidly accumulate, making the propagated coordinate unusable due to a high and exponentially growing bias. Referring again to Fig. 1, a closed-loop solution is estimated within Block 2. Finally, on Fig. 1, Block 3 represents the integration procedure of GNSS-derived coordinates and INS-derived displacements. A Kalman filter with the following state vector was then implemented on Block 3:

$$x_t = [r_3, v_3, \Psi_3, b_a, b_g], \tag{1}$$

where the following vectors are represented: r_3 for position, v_3 for velocity, Ψ for vehicle attitude, and b_a and b_g for accelerometer and gyroscope biases, respectively. The measurement vector z_t is given as:

$$z_t^{LC} = [\Delta r_x, \Delta r_y, \Delta r_z, \\ \Delta v_x, \Delta v_y, \Delta v_z] \tag{2}$$

where Δr is the difference between the estimated coordinates from GNSS and the propagated coordinates from INS at the same epoch on each axis, and the equivalent for velocity in Δv.

3   Outdoor Wi-Fi Positioning

In general lines, an outdoor Wi-Fi positioning system is comprised of two phases: training and positioning. Figure 2 shows a Wi-Fi positioning system as developed by Lu et al. (2010). In this system, which served as a basis in this study, a survey of a predefined route is performed. In this route, coordinates of the reference points (RP) are estimated and the local Wi-Fi networks (a mixture of public and private access points) received signal strength (RSS) are scanned and stored. The selection of the reference points is based on the latency of the Wi-Fi scanning. In this study, on average, one point was scanned at every 1.2 s in multiple passes over the same lane. With the data acquired during the training phase, a one-class support vector machine (SVM) unsupervised classifier is applied to assess if the clusters of RSS measurements represent properly the different *lanelets*. Once this capability is identified, a standard SVM classifier is then trained in order to optimally identify sections with similar RSS measurements. Finally, each one of those sections is linked to a *lanelet* (Bender et al. 2014), by an optimal non-linear SVM classifier (Steinwart and Christmann 2008). The result of this system is a trained model that is able

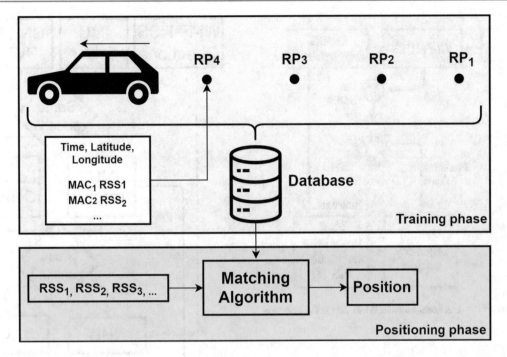

Fig. 2 Overview of a Wi-Fi/GNSS integration method (Lu et al. 2010)

to identify, given a set of measurements from the Wi-Fi antennas $z_t = [MAC_1, RSS_1; \ldots MAC_n, RSS_n]$, to which *lanelet* it most likely belongs. Since the number of RSS observations varies from area to area, only the 20 strongest measurements are used as an input for the SVM classifier. In case less observations are present, the dataset is padded with zeros, up to a lower limit of five observations. These values were empirically determined during the one-class SVM clustering. Figure 3 shows a section of the surveyed area. In this example, each numbered *lanelet* has three attributes: azimuth, maximum posted speed and average GNSS Geometry Dilution of Precision (GDOP), roughly representing how well the GNSS solution is expected to be in this region. Each *lanelet* has the empirical limit of two blocks of length, or if one of the three mentioned properties is significantly different at a certain point forward. It is important to notice that in this system, the absolute location derived from the Wi-Fi RSS measurements is not important, as long as the correct *lanelet* is identified and its attributes are applied in the integration filter. This approach makes it possible to extract useful information from Wi-Fi signals that can be directly used in vehicle control systems.

Fig. 3 Example of *lanelets* on a section of the surveyed area

4 Integration Between GNSS, INS and Wi-Fi

Given the rapid diverging characteristic of INS position propagation, in the absence of GNSS for a considerable period of time – situation common in urban areas – any

system relying on this information may suffer from unreliable position estimates. On an autonomous vehicle scenario, this situation may cause a disengagement of the auto-pilot or possibly accidents.

As an additional information, a radio map (equivalent to the database in Fig. 2) was built based on the work by Lu

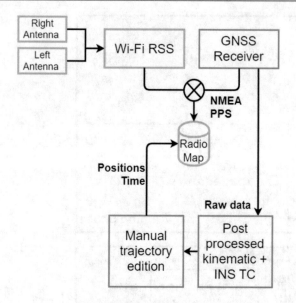

Fig. 4 Overview of the directional antenna Wi-Fi data collection system

et al. (2010) with an added layer of post-processing and editing for improved accuracy. This database is derived from the schematic on Fig. 4. Once the SVM model estimates a probable match for the measurements z_t, information is then extracted from that region's *lanelet* and integrated into the GNSS/INS loosely-coupled filter during the application of the non-holonomic constraint (NHC) (Groves 2008). The NHC is applied after the main integration filter and is responsible for constraining the vehicle movement on forwards. The cross-track and up velocities are constrained as zero, and any movement in those directions is then treated as a measurement residual during the filtering. The NHC is applied by the means of a Kalman filter with the measurement vector as:

$$z_t^{NHC} = [v_c \, v_{up}], \qquad (3)$$

where v_c and v_{up} are the cross-track and up velocities, respectively, estimated during the loosely-coupled integration, and partnered with the following design matrix:

$$H = [0_{2,3} \begin{pmatrix} 1 & 0 & 0 \\ 0 & 1 & 0 \end{pmatrix} C_b^n \, 0_{2,3} \, 0_{2,3} \, 0_{2,3}], \qquad (4)$$

where C_b^n is the rotation matrix between the body-frame (same reference as z_t) and the navigation (ENU) frame. It is possible to see on Eq. 5 that the body-to-navigation rotation matrix C_b^n is a function of the travel direction angle *yaw* along with the vehicle's pitch and yaw. Retrieving the yaw angle from the Wi-Fi radio map, a vehicle motion constraint

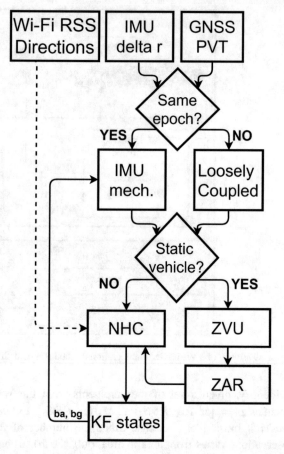

Fig. 5 Overview of the GNSS/INS/Wi-Fi integration algorithm

(Groves 2008) is then created and C_b^n becomes (Rogers 2007):

$$C_b^n = \begin{vmatrix} c\psi c\theta & c\psi s\theta s\phi - s\psi c\phi & c\psi s\theta s\phi + s\psi c\phi \\ -s\Psi c\theta & -s\psi s\theta s\phi - c\psi c\phi & c\psi s\phi - s\psi s\theta c\phi \\ s\theta & -c\theta s\phi & -c\theta c\psi \end{vmatrix}, \qquad (5)$$

where ϕ, θ, and ψ are the roll, pitch and yaw, or azimuth, angles respectively, and c and s are cosine and sine functions, also respectively.

By applying the roll and pitch estimates from the filter before the NHC and updating yaw, the system biases, along with all other estimates, are all updated with the external information. Figure 5 shows an overview of the integration methodology. The updated parameters after the NHC are then used as input for the INS mechanization, and updated again during the next INS/GNSS integration. Other techniques to integrate external yaw information to the filter have been explored in literature (Falco et al. 2013;

Angrisano 2010), and will be investigated in future versions of this study.

5 Experiment and Results

In order to assess the use of the Wi-Fi derived information, an experiment was performed in Fredericton, NB, Canada. The experiment consisted of a 73-min drive in several areas with different satellite coverage, obstruction, and multipath characteristics. Figure 6 shows the equipment of top of the vehicle. Inside, right above the vehicle's rear axis, a KVH TG60000 tactical inertial system was deployed. Since the RSS measurements are observed by directional antennas and input in the SVM training phase, it is clear that antennas pointing in opposite directions will have different measurements for the same location. This represents an important assumption of this study, and more segmented information for the hyper-plane model to better cluster the input information.

For the SVM training algorithm, data from several previous surveys in the area were utilized. With a rate of 90% of training data for 10% of test, the accuracy of the classifier was 82% (N = 7,422) on the correct *lanelet*. On a more detailed analysis of the misclassifications (about 130 points out of 7,422), it was found that they happened in situations where the vehicle was stopped at street crossings, and the classifier could not differentiate between the end of three or more *lanelets*. The misclassifications in this stance do not represent a risk for the system integrity, since sudden "jumps" from one *lanelet* to another can be easily ruled out by the INS measurements.

In this section, for brevity's sake, one particularly harsh section of the full experiment was selected to assess the method.

Figure 7 shows the analyzed route with the results from an edited post-processed kinematic (PPK) used as the reference. Finally, by forcing the integration with the proper yaw angle of the road, the solution is smoothed and kept in the driving lanes, as Fig. 10 shows. Figure 8 shows the results from a

Fig. 6 Hardware of the initial test campaign. Septentrio APS3G multi-GNSS receiver on the center, and two TPLINK directional Wi-Fi (2.4 GHz band) antennas pointing sideways

Fig. 7 Downtown Fredericton area with edited PPK reference shown in yellow

Fig. 8 SPP results in the test area

Fig. 9 SPP and INS loosely-coupled integration results

Table 1 Horizontal component RMS of the techniques evaluated versus post-processed RTK

Technique	RMS_{HORZ}
SPP	2.914 m
SPP+INS	1.219 m
SPP+INS+Wi-Fi/SVM	0.970 m

single-point positioning (SPP) in the same area as Fig. 7. The processing strategy utilized ion-free observables, a 15° elevation mask, and *Saastamoinen* as tropospheric model. It is possible to see that in two particular areas on the bottom street, the number of satellites drops and the multipath effect is more pronounced, generating jumps in the trajectory. Figure 9 shows the results from a standard loosely-coupled integration in the same area. Even with criteria in place to avoid integration when the GNSS data is not reliable (number of satellites greater than 4 and reported horizontal standard deviation lower than 5 m), in the areas where the absolute positions are not reliable, the integration itself returns values out of the driving lane, and out of the streets on some occasions.

Table 1 summarizes the RMS values using the PPK solution as reference. The RMS of the GNSS+INS+Wi-Fi/SVM integrated solution is 20% improved from the GNSS+INS only, and 66% better than the SPP solution only.

6   Conclusions

From Table 1 and Fig. 10, it is possible to see the already explored in literature effect of an external directional constraint on navigation. This paper explores the novel possibility of integrating an SVM classification on Wi-Fi RSS data to generate such directional constraints to be integrated in the filter. Future version of this study will explore other methods of integrating yaw measurements, and explore more fine system calibration techniques to improve the already promising results achieved.

Fig. 10 SPP, INS and external yaw constraint on a loosely-coupled filter

References

Angrisano A (2010) GNSS/INS integration methods. Ph.D. thesis, Universita' Degli Studi Di Napoli

Bender P, Ziegler J, Stiller C (2010) Lanelets: efficient map representation for autonomous driving. In: IEEE intelligent vehicles symposium, proceedings (Iv), pp 420–425. https://doi.org/10.1109/IVS.2014.6856487

Falco G, Campo-Cossío M, Puras A (2013) MULTI-GNSS receivers/IMU system aimed at the design of a heading-constrained tightly-coupled algorithm. In: 2013 International conference on localization and GNSS, ICL-GNSS 2013, June. https://doi.org/10.1109/ICL-GNSS.2013.6577263

Grejner-Brzezinska DA, Yi Y, Toth CK (2001) Bridging GPS gaps in urban canyons: the benefits of ZUPTs. Navigation 48(4):216–226. https://doi.org/10.1002/j.2161-4296.2001.tb00246.x

Groves PD (2008) Principles of GNSS, inertial, and multisensor integrated navigation systems, vol. 2. Artech House, London

Groves, P.D.: Shadow matching: a new GNSS positioning technique for urban canyons. J Navig 64(3):417–430 (2011). https://doi.org/10.1017/S0373463311000087

Hsu L-T, Gu Y, Chen F, Wada Y, Kamijo S (2015) Assessment of QZSS L1-SAIF for 3D map-based pedestrian positioning method in an urban environment. In: Proceedings of the 2015 international technical meeting of the Institute of Navigation, Dana Point, January 2015, pp 331–342

Lu H, Zhang S, Dong Y, Lin X (2010) A Wi-Fi/GPS integrated system for urban vehicle positioning. In: IEEE conference on intelligent transportation systems, proceedings, ITSC, pp 1663–1668. https://doi.org/10.1109/ITSC.2010.5625268

Rogers RM (2007) Applied mathematics in integrated navigation systems, 3rd edn. https://doi.org/10.2514/4.861598

Steinwart, I., Christmann, A.: Support Vector Machines. Information Science and Statistics. Springer New York, New York, NY (2008). https://doi.org/10.1007/978-0-387-77242-4

Enhancing Navigation in Difficult Environments with Low-Cost, Dual-Frequency GNSS PPP and MEMS IMU

Sudha Vana and Sunil Bisnath

Abstract

The Global Navigation Satellite System (GNSS) Precise Point Positioning (PPP) technology benefits from not needing local ground infrastructure such as reference stations and accuracy attained is at the decimetre-level, which approaches real-time kinematic (RTK) performance. However, due to its long position solution initialization period and complete dependence on the receiver measurements, PPP finds limited utility. The emergence of low-cost, micro-electro-mechanical sensor (MEMS) inertial measurement units (IMUs) has prompted research in integrated navigation solutions with the PPP processing technique. This sensor fusion aids to achieve continuous positioning and navigation solution availability when there are insufficient numbers of navigation satellites visible. In the past, research has been conducted to integrate high-end (geodetic) GNSS receivers with PPP processing and MEMS IMUs, or low-cost, single-frequency GNSS receivers with point positioning processing and MEMS IMUs. The objective of this research is to investigate and analyze position solution availability and continuity by integrating low-cost, dual-frequency GNSS receivers using PPP processing with the latest low-cost, MEMS IMUs to offer a complete, low-cost navigation solution that will enable continuously available positioning and navigation solutions, even in obstructed environments. The horizontal accuracy of the developed low-cost, dual-frequency GNSS PPP with MEMS IMU integrated algorithm is approximately 20 cm. During half a minute of simulated GNSS signal outage, the integrated solution offers 40 cm horizontal accuracy. A low-cost, dual-frequency GNSS receiver PPP solution integrated with a MEMS IMU forms a unique combination of a total low-cost solution, that will open a significant new market window for modern-day applications such as autonomous vehicles, drones and augmented reality.

Keywords

Global Navigation Satellite System (GNSS) · Global positioning system (GPS) · Inertial measurement unit (IMU) · Low-cost navigation · Precise Point Positioning (PPP) · PPP + IMU

1 Introduction

Obtaining continuous and accurate navigation solutions in any environment is a challenge because GNSS signals are obstructed in environments such as downtowns, tunnels or areas covered with foliage. Integrating the GNSS sensor with another self-contained navigation sensor such as an Inertial Measurement Unit (IMU) becomes necessary in such cases.

Presented at IUGG 2019 General Assembly.

S. Vana (✉) · S. Bisnath
Department of Earth and Space Science and Engineering, York University, Toronto, ON, Canada
e-mail: sudhav@yorku.ca

J. T. Freymueller, L. Sanchez (eds.), *Beyond 100: The Next Century in Geodesy*,
International Association of Geodesy Symposia 152, https://doi.org/10.1007/1345_2020_118

The advent of IMUs based on micro-electro-mechanical sensors (MEMS) has brought a whole new market of low-cost IMU sensors. MEMS IMU sensors are cheaper by price but also come with some significant in-built errors such as bias, noise and scale factor (Abd Rabbou and El-Rabbany 2015). Shin et al. (2005), Abdel-Hamid et al. (2006), Scherzinger (2000) and many other researchers have all performed integrated navigation of GNSS and MEMS IMU by applying Differential GPS (DGPS) or Real-Time Kinematic (RTK) techniques to improve continuity and accuracy of the navigation solution in the event of GNSS signal outage. Precise Point Positioning (PPP) is an augmentation technique that does not require a local reference station unlike the RTK technique (Zumberge et al. 1997). Precise orbit and clock information is broadcast via satellites or the Internet to the user. The performance accuracy achieved is decimetre- to centimetre-level. The initial convergence time of the PPP technique is minutes to the decimetre-level, which is one of its major drawbacks. PPP is a widely used technique for applications such as marine mapping, crustal deformation, airborne mapping, precision agriculture and construction applications (Seepersad 2012; Aggrey 2015; Aggrey and Bisnath 2017). PPP can be further augmented to reduce convergence period by applying satellite phase biases to obtain integer ambiguities and a priori atmospheric refraction information (Lannes and Prieur 2013; Teunissen and Khodabandeh 2015). This enables a stand-alone user-receiver to achieve RTK-like performance with a shorter convergence period, while limiting dependency on external infrastructure.

In the recent past, researchers have started applying the PPP technique to perform GNSS and MEMS IMU integration which has offered promising outcomes. Abd Rabbou and El-Rabbany (2015) experimented with GPS-PPP integration using a high-end GPS sensor and a MEMS IMU. The study showed decimetre-level accuracy with no GPS signal outages and during a 60 s of signal outage, sub-metre-level accuracies were demonstrated. Liu et al. (2018) conducted a study on integrating a low-cost, single-frequency (SF) GNSS with a MEMS IMU and were able to achieve centimetre- to decimetre-level accuracy with no GNSS signal loss. During a GNSS signal loss of 3 s, the solution performed at the metre level.

The motivation of this research is to assess and analyze the performance of the recently emerging low-cost, dual-frequency (DF) GNSS receivers in the market integrated with a relatively low-cost MEMS IMU. The research questions addressed are: (1) What is the accuracy performance of a low-cost DF GNSS PPP receiver integrated with a MEMS IMU? And (2) What is the accuracy performance of a low-cost DF GNSS PPP receiver integrated with a MEMS IMU during a 30 s GNSS signal outage?

Modern-day applications such as low-cost vehicle autonomy, augmented reality, and pedestrian dead reckoning demand decimetre-level accuracy with low-cost sensors. Therefore, integrated, low-cost, DF GNSS PPP with MEMS IMU has the potential to offer accurate, continuous and precise navigation solution for the next generation of applications which is the objective of this research.

2 Inertial Navigation System

The raw measurements from an IMU, specific force and turn rates, are converted into position, velocity, and attitude using the IMU mechanization process. Known position, velocity and attitude are also inputs to the mechanization. The equation for accelerometer and gyro measurements with errors is as given below in Eqs. (1) and (2) (Farrell 2008).

$$\widetilde{f}^a = (I - \delta SF_a)(f^a - \delta b_a - \delta nl_a - v_a) \tag{1}$$

$$\widetilde{\omega}^g_{ip} = (I - \delta SF_g)(\omega^g_{ip} - \delta b_g - \delta k_g - v_g) \tag{2}$$

$\widetilde{f}^a$ is the actual accelerometer measurement that accounts for all measurement errors, δSF_a is the accelerometer scale factor error, δb_a is the accelerometer bias, δnl_a is the accelerometer nonlinearity error, v_a is the measurement noise. $\widetilde{\omega}^g_{ip}$ is the actual gyro measurement accounting for all measurement errors, δSF_g is gyro scale factor, δb_g is gyro bias, δk_g represents gyro g-sensitivity, v_g is measurement noise. f^a and ω^g_{ip} are obtained in body-frame. After error compensations are made position and velocity can be calculated using IMU mechanization equations.

There are four main steps to the mechanization process: (1) Attitude update using the turn rates from gyroscopes; (2) reference frame conversion of specific force from body to the intended reference frame; (3) velocity update; and (4) position update.

Figure 1 depicts the IMU mechanization process. The details of mechanization in all the reference frames is explained in Groves (2013). For this research, the Earth-Centred-Earth-Fixed (ECEF) frame of mechanization is used because the range measurements from the GNSS satellites can be used directly in the estimation process. In the Fig. 1, specific force f^b from accelerometer and the angular rate ω^b_{ib} from gyroscope are output in body frame. They are converted into the ECEF frame using direction cosine matrices or quaternions. After gravity compensation and Coriolis correction, along with the initial position, velocity and attitude estimates, time integration gives the current position, velocity and attitude.

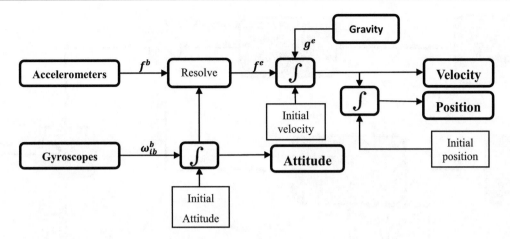

Fig. 1 Block diagram of IMU mechanization process (after Titterton et al. 2004)

Inputs to IMU mechanization are specific force f^b and turn rates ω_{ib}^b. Mechanization process including equations are described in detail in (Farrell 2008).

3 GNSS PPP/INS Tightly Coupled Kalman Filter

In this research, a tightly-coupled Extended Kalman Filter (EKF) is used to fuse the GNSS and IMU measurements. In a tightly-coupled integration architecture, raw measurements from the sensors are used, which enables continuous navigation during a GNSS signal outage. The typical error budget for GNSS PPP is listed in Table 1.

The inputs to the complementary Kalman filter are (1) code and phase measurements from a low-cost DF GNSS receiver corrected for atmosphere, relativistic errors and clock and orbit errors using the precise PPP corrections, and (2) predicted code and phase measurements that are formed using the IMU position and velocity with the satellite position and velocity. For this research work, the ionosphere-free (IF) model is used to avoid estimation of the ionosphere, which simplifies the number of states to be estimated.

The ambiguities estimated are float only. The mathematical model for IF PPP can be written as (Parkinson and Spilker 1996):

$$P = \rho + c\,(dt_r - dt^s) + T + c\left(B_p^r - B_p^s\right) + e_P \quad (3)$$

$$\varphi = \rho + c\,(dt_r - dt^s) + T + c\left(B_\varphi^r - B_\varphi^s\right) + N\lambda + e_\varphi \quad (4)$$

In Eqs. (3) and (4), dt_r and dt^s are the receiver clock error and satellite clock errors respectively, T is the tropospheric delay, B_p^r and B_p^s are the code bias for receiver and satellite, B_φ^r and B_φ^s are the phase bias for receiver and satellite, e_P and e_φ are the unmodelled errors in pseudorange and carrier phase measurements, and $N\lambda$ is the ambiguity term between the receiver and satellite on phase measurements.

Figure 2 provides the representation of the EKF integration of the GNSS-PPP and IMU.

In Fig. 2, f_b, w_b are the IMU specific force and turn rate measurements. These measurements are converted into position P_{IMU}, velocity V_{IMU} and attitude A_{IMU} from a known position, velocity and attitude by applying IMU mechanization process. Predicted ρ_{IMU}, φ_{IMU} are constructed by using the satellite position and velocity, which are corrected by applying the precise orbit and clock corrections. DF code and phase measurements ρ_{GNSS}, φ_{GNSS} are corrected for typical errors such as the errors mentioned in Table 1. The estimated output from the EKF are the error in IMU position δr^n, velocity δv^n attitude $\delta \varepsilon^n$ and biases b_g and b_a. P_{IMU}^e, V_{IMU}^e and A_{IMU}^e give the final IMU position, velocity and attitude.

The state vector consists of the navigation states, IMU states, and the GNSS only states. Navigation states include position error, velocity error and attitude error. While the inertial states consist of accelerometer and gyroscope biases. The GNSS states estimated are: GNSS receiver clock, as well

Table 1 PPP Error budget (Choy 2018)

Error source	Error (m)
Ionosphere delay	10–20
Troposphere delay	1–10
Relativistic	10
Multipath	1.0
Receiver measurement noise	0.1–0.7
SV orbit/clock	~0.01–0.1
Satellite phase centre variation	0.05–1
Solid earth tide	0.2
Ocean loading	0.05
Phase wind-up (ionosphere-free)	0.1
Receiver phase centre variation	0.001–0.01

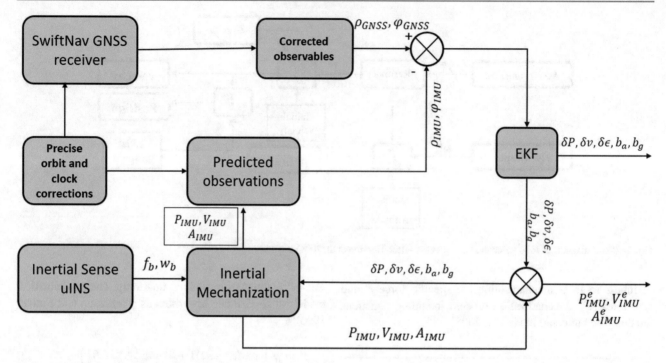

Fig. 2 Block diagram depicting PPP-INS tightly coupled integration

as drift and the float ambiguity terms. The state vector is represented mathematically in Eq. (5) (Jekeli 2012; Groves 2013, p. 201; Abd Rabbou and El-Rabbany 2015).

$$\delta x = \begin{bmatrix} \delta P & \delta v & \delta \varepsilon & \delta t_c & \dot{\delta t_c} & d_{tropo} & b_a & b_g & N_1 & N_2 & \dots & N_n \end{bmatrix}$$

(5)

In the Eq. (5), δP is the 3D position error, δv is the 3D velocity error, $\delta \varepsilon$ is the attitude error, δt_c GNSS receiver clock error, $\dot{\delta t_c}$ is GNSS receiver clock drift error, d_{tropo} is the troposphere wet delay, b_a and b_g are accelerometer and gyroscope bias and N_i float ambiguity of satellite i.

In continuous time, the transition matrix is given by Groves (2013).

$$F(t) = \begin{bmatrix} 0_{3x3} & I_{3x3} & 0_{3x3} & 0_{3x3} & 0_{3x3} \\ F_{23}^e & -2\Omega_{ie}^e & F_{21}^e & C_b^e & 0_{3x3} \\ 0_{3x3} & 0_{3x3} & -\Omega_{ie}^e & 0_{3x3} & C_b^e \\ 0_{3x3} & 0_{3x3} & 0_{3x3} & 0_{3x3} & 0_{3x3} \\ 0_{3x3} & 0_{3x3} & 0_{3x3} & 0_{3x3} & 0_{3x3} \end{bmatrix}$$

$$F_{21}^e = \begin{bmatrix} -\left(C_b^e f_{ib}^b \right) X \end{bmatrix} \qquad F_{23}^e = -\frac{2\widehat{\gamma_{ib}^e}}{r_{eS}^e(\widehat{L_b})} \frac{\hat{r}_{eb}^{e\,T}}{|\hat{r}_{eb}^e|}$$

Terms F_{21}^e and F_{23}^e are explained in detail in Groves (2013).

The measurement vector is the difference between corrected GNSS measurements, pseudorange, carrier phase and predicted IMU measurements. The measurement vector is given in Eq. (6)

$$z = \begin{bmatrix} \rho^i{}_{GNSS} - \rho^i{}_{INS} \\ \varphi^i{}_{GNSS} - \varphi^i{}_{INS} \\ \vdots \end{bmatrix}$$

(6)

In Eq. (6), ρ^i is the pseudorange of the satellite i and φ^i is the carrier phase measurement corresponding to satellite i. The design matrix will encompass the partial derivatives to the state terms related to GNSS. The other terms become zero.

$$H = \begin{bmatrix} \frac{\partial \rho_i}{\partial x} & \frac{\partial \rho_i}{\partial y} & \frac{\partial \rho_i}{\partial z} & 0_{3x3} & 0_{3x3} & 1 \ 0 \ 0 \ sne_i \ 0 \ 0 \ 0 \ 0 \ 0 \ \dots \\ \frac{\partial \Phi_i}{\partial x} & \frac{\partial \Phi_i}{\partial y} & \frac{\partial \Phi_i}{\partial z} & 0_{3x3} & 0_{3x3} & 1 \ 0 \ 0 \ sne_i \ 0 \ 0 \ 0 \ 0 \ \lambda \ \dots \\ \vdots & \vdots & \vdots & \vdots & \vdots & \vdots \ \vdots \ \vdots \ \vdots \ \vdots \ \vdots \ \vdots \ \vdots \ \vdots \ \vdots \ \dots \end{bmatrix}$$

(7)

In Eq. (7), $sne_i = \frac{1}{\sin(elevation)}$ is the mapping function for the troposphere wet delay component. λ is a partial derivative entry for the ambiguity terms $N\lambda$ which is wavelength corresponding to ambiguity N.

4 Field Test and Results

To test and evaluate the tightly-coupled EKF, kinematic data were collected at the York University main campus in Toronto, Canada. A low-cost, dual-frequency receiver,

Piksi-multi by Swift Navigation, and low-cost MEMS IMU, Inertial sense μINS, were used. The Piksi is a multi-constellation, dual-frequency receiver that can offer 0.75 m accuracy in horizontal (CEP 50 in SBAS mode) (SwiftNav 2012). The Piksi is also capable of offering RTK like cm-level solutions with fast convergence with horizontal accuracy of 1 cm. The inertial Sense μINS consists of a SF uBlox GNSS receiver, magnetometer, barometer, and IMU. The MEMS IMU onboard μINS is comparable to a tactical grade IMU based on the specifications (Inertial Sense 2013). The specifications of the InertialSense uINS is detailed in Table 2. As per the categorization of IMUs given in (Vector Nav Library 2008), uINS can be categorized as a tactical-grade IMU from the specifications given in Table 2. The antenna used in the experiment is a geodetic grade antenna by SwiftNav. Since, a geodetic grade antenna was used in the setup, the quality of the measurements were better than the ones acquired using a low-cost antenna such as a patch antenna.

Both the GNSS and IMU sensors were placed beside each other in the car trunk. The geodetic grade antenna was installed on the car roof. The data logged consisted of a multi-constellation carrier phase and pseudorange informa-

Table 2 Specifications of uINS MEMS IMU

Sensor	IMU-Gyros	IMU-Accels
Operating range	$\pm2,000°/s$	±16 g
Bias repeatability	$<0.2°/s$	<5 mg
In-run bias stability	$<10°/h$	$<40\,\mu g$
Random walk	$0.15°/\sqrt{h}$	0.07 m/s/$\sqrt{h}$
Non-linearity	$<0.1\%$ FS	$<0.5\%$ FS
Noise density	$0.01°/s/\sqrt{Hz}$	$300\,\mu g/\sqrt{Hz}$

tion. Thus, collected raw observables were then processed using the York PPP + IMU algorithm for validation.

Figure 3 represents the track of data collected at a parking lot near the York University Campus in Toronto, Canada. The data were collected on October 12, 2019, DOY 168 for a period of 24 min.

Logging data rate of GNSS observables was set to 5 Hz and the IMU data rate was set to 100 Hz. Novatel's Waypoint software was used to post-process the measurements in RTK mode for the same data used as the reference. The processing parameters used for the data are summarized in Table 3.

Figure 4 represents a plot of GNSS satellites available during the span of data collection and corresponding position dilution of precision (PDOP). The average number of satel-

Fig. 3 Track of data collected around York University, Toronto, Canada

Table 3 Processing parameters used for PPP + IMU TC algorithm

Constellations processed	GPS, GLONASS, Galileo, BeiDou
Elevation mask	10°
Processing type	Iono-free combination
Satellite orbits and clocks	CNES products (http://www.ppp-wizard. net/products/REAL_TIME/)
Observation processing mode	Dual-frequency, kinematic processing
Data format	RINEX 3.x

lites is 15 and the mean PDOP is 1.7. It is clear from Fig. 4 that the number of BeiDou satellites is much less compared to other constellations as the data were collected in North American region.

Figure 5 is a plot of horizontal and vertical error of the GNSS PPP and IMU solution when compared to the Novatel's Waypoint reference solution. The highlighted black box in the Fig. 3 is area where there are many trees and a signal outage took place. This can be seen in Fig. 5 as a jump in the position solution at 1,200 s. The rms error of the solution is 23 cm in the horizontal direction and 33 cm

in the vertical direction. The rms was calculated after the solution reaches convergence time which is 400 s and before signal outage due to trees happens at 1,200 s. The decimetre-level performance of the algorithm makes it appropriate and suitable for the applications that require decimetre-level accuracy in positioning for a lower price.

Given the number of states that are estimated for the purpose of navigation from Eq. (5), at least five satellite raw observables are necessary to compute the user position. The evaluation of the performance of EKF during GNSS outage was done by simulating a GNSS signal outage for 30 s in the track. During the 30 s of simulated GNSS signal outage, the algorithm was tested with only four GNSS satellites available. Figure 6 is the horizontal error when compared to the reference solution during the GNSS signal outage of 30 s. The blue solution is an error comparison with no outage while red-coloured error plot corresponds to the PPP + IMU performance during a 30 s outage. The simulated outage period is highlighted in black between 440 and 470 s in Fig. 6.

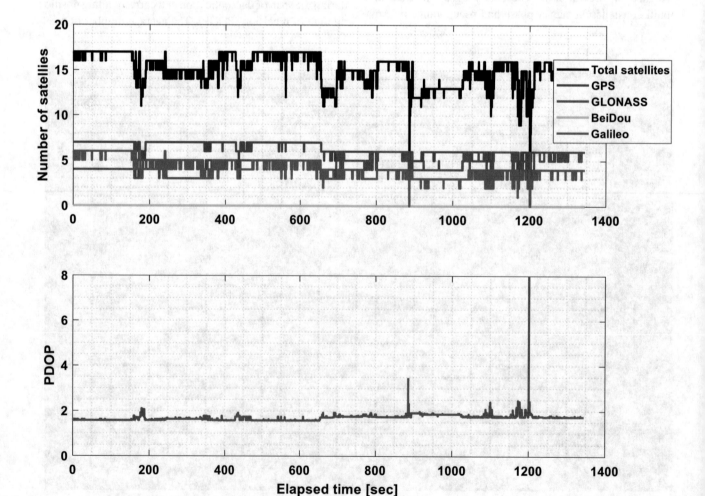

Fig. 4 Plot of available satellite and DOP vs elapsed time

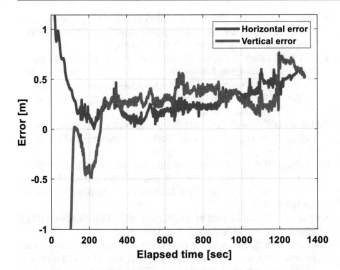

Fig. 5 Plot of horizontal and vertical error compared to the reference solution

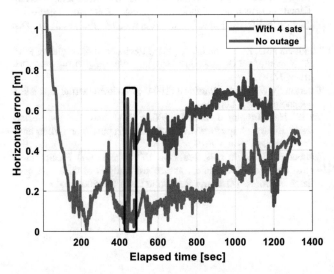

Fig. 6 Plot of horizontal error with respect to RTK Lib solution when 30 s outage was simulated

During the 30 s outage period, the solution performs with a 2D rms of 0.4 m horizontally and 1.2 m vertically. The rms of the solution starting from the outage to end of data is 0.55 m horizontally and 1.1 m vertically. It can be noticed

from Fig. 6 that the solution may not necessarily behave as it works when there is no GNSS signal outage after the outage, because every epoch of estimation process uses state estimate and covariance information from the previous epoch. The state vector and covariance information will vary based on the DOP and satellite information used in previous epoch. (Liu et al. 2018) indicated that using an SF GNSS PPP with MEMS IMU performs at a rms of 1 m with a 3 s GNSS signal outage and the accuracy was less than 10 m with a half minute of GNSS signal outage when there were only two satellites operating. The GNSS sensor used by (Liu et al. 2018) was Ublox EVK-M8U which has a SF GNSS chip as well a MEMS IMU in the package and global ionosphere map (GIM) products were used to reduce the ionosphere delays. A tightly-coupled algorithm using a low-cost, DF GNSS PPP with MEMS IMU performs at a less than the metre level rms error with a 30 s GNSS signal outage, which is 10 times better accuracy than a SF GNSS PPP with MEMS IMU solution.

5 Conclusions and Future Work

In this paper, the performance of a tightly-coupled EKF with low-cost DF GNSS PPP and MEMS IMU performance with and without outage were investigated. The algorithm performs at the decimetre-level of accuracy when there are no signal outages and it performs at the decimetre- to metre-level accuracy during a 30 s outage with four satellites available. The performance of DF GNSS PPP and MEMS IMU integrated system during outage proves to be 10 times better than SF GNSS PPP with MEMS IMU. The accuracy level of the algorithm seems promising for the next generation of applications that demand higher accuracy with lower price sensors. Table 4 gives a brief summary of accuracy of the low-cost DF GNSS PPP + IMU integrated solution with and without GNSS signal outage. As part of future work, resolving ambiguities for the low-cost, DF GNSS + IMU will results in less than decimetre level accuracy and perform at an rms error of decimetre level during a half minute of GNSS signal partial absence.

Table 4 Summary of horizontal and vertical rms with and without signal outage

Error [m]	During signal outage		Starting from outage		Overall (after convergence)	
	Horizontal	Vertical	Horizontal	Vertical	Horizontal	Vertical
Without GNSS signal outage	0.07	0.3	0.28	0.4	0.2	0.4
With 30 s GNSS signal outage (only 4 satellites available)	0.4	1.2	0.55	1.1	0.48	0.9

Acknowledgements The authors would like to thank Natural Sciences and Engineering Research (NSERC) and York University for financial support and German Research Centre for Geosciences (GFZ), International GNSS Services (IGS) and National Centre for Space Studies (CNES) for data provided.

References

Abd Rabbou M, El-Rabbany A (2015) Tightly coupled integration of GPS precise point positioning and MEMS-based inertial systems. GPS Solutions 19:601–609. https://doi.org/10.1007/s10291-014-0415-3

Abdel-Hamid W, Abdelazim T, El-Sheimy N, Lachapelle G (2006) Improvement of MEMS-IMU/GPS performance using fuzzy modeling. GPS Solutions 10:1–11

Aggrey JE (2015) Multi-GNSS precise point positioning software architecture and analysis of GLONASS pseudorange biases. MSc Thesis, York University

Aggrey J, Bisnath S (2017) Analysis of multi-GNSS PPP initialization using dual- and triple-frequency data. In: Proceedings of the 2017 international technical meeting of the institute of navigation. Monterey, California, pp 445–458

Choy S (2018) GNSS precise point positioning. https://www.unoosa.org/documents/pdf/icg/2018/ait-gnss/16_PPP.pdf. Accessed 5 Mar 2020

Farrell J (2008) Aided navigation: GPS with high rate sensors. McGraw-Hill, Inc, New York

Groves PD (2013) Principles of GNSS, inertial, and multisensor integrated navigation systems. Artech House, Norwood

Inertial Sense (2013) Inertial Sense. In: Inertial Sense. https://inertialsense.com/products/. Accessed 27 Sep 2019

Jekeli C (2012) Inertial navigation systems with geodetic applications. Walter de Gruyter, Berlin

Lannes A, Prieur J-L (2013) Calibration of the clock-phase biases of GNSS networks: the closure-ambiguity approach. J Geod 87:709–731. https://doi.org/10.1007/s00190-013-0641-4

Liu Y, Liu F, Gao Y, Zhao L (2018) Implementation and analysis of tightly coupled Global Navigation Satellite System Precise Point Positioning/Inertial Navigation System (GNSS PPP/INS) with insufficient satellites for land vehicle navigation. Sensors 18:4305. https://doi.org/10.3390/s18124305

Parkinson BW, Spilker JJ (1996) The global positioning system: theory and applications. American Institute of Aeronautics and Astronautics, Washington, DC

Scherzinger BM (2000) Precise robust positioning with inertial/GPS RTK. In: Proceedings of the 13th International Technical Meeting fo the Satellite Division of the Institute of Navigation (ION GPS). pp 115–162

Seepersad G (2012) Reduction of initial convergence period in GPS PPP data processing. MSc Thesis, York University

Shin E-H, Niu X, El-Sheimy N (2005) Performance comparison of the extended and the unscented Kalman filter for integrated GPS and MEMS-based inertial systems. ION NTM 2005, pp 961–969

SwiftNav (2012) Swift Navigation Precise Positioning Solutions|Piksi Multi, Duro, Duro Inertial RTK GNSS Receivers, Skylark Cloud Corrections Service, Starling Software Positioning Engine|SwiftNav. https://www.swiftnav.com/. Accessed 31 Dec 2019

Teunissen PJG, Khodabandeh A (2015) Review and principles of PPP-RTK methods. J Geod 89:217–240. https://doi.org/10.1007/s00190-014-0771-3

Titterton D, Weston JL, Weston J (2004) Strapdown inertial navigation technology. IET, London

Vector Nav Library I and I (2008) IMU and INS – VectorNav library. https://www.vectornav.com/support/library/imu-and-ins. Accessed 29 Jun 2019

Zumberge JF, Heflin MB, Jefferson DC et al (1997) Precise point positioning for the efficient and robust analysis of GPS data from large networks. J Geophys Res: Solid Earth 102:5005–5017

Part V

Monitoring and Understanding the Dynamic Earth with Geodetic Observations

Water Depletion and Land Subsidence in Iran Using Gravity, GNSS, InSAR and Precise Levelling Data

Jacques Hinderer, Abdoreza Saadat, Hamideh Cheraghi,
Jean-Daniel Bernard, Yahya Djamour, Masoomeh Amighpey,
Seyavash Arabi, Hamidreza Nankali, and Farokh Tavakoli

Abstract

Population growth, coupled with the expansion of exploitation of groundwater resources for agricultural and industrial purposes, has led Iran to face the necessity of proper use and sustainable management of existing water resources. In this study we will use the existing valuable geodetic data (gravity, GNSS, precise levelling, InSAR) in Iran to better understand the surface deformation and gravity variations caused by underground water depletion attributed to drastic pumping. Based on repetition of first order precise leveling network of Iran, about 44 subsidence areas are identified and continuous data collected by the Iranian permanent GNSS and geodynamic network (IPGN), as well as InSAR data, indicate strong elevation changes in some parts of the country. GRACE satellite gravity solutions over Iran also show a general gravity decrease between 2002 and 2016. New absolute gravity campaigns were performed in Iran in 2017 and 2018 in the frame of the TRIGGER French-Iranian program. Several new absolute gravity stations were established and former stations, first measured between 2000 and 2007, were repeated showing that the gravity values of many stations have changed in time. Most of these changes indicate a gravity decrease mostly linked to mass deficit due to water depletion. On the contrary some stations show a large gravity increase that can be merely explained by land subsidence itself linked to water depletion by poroelastic effects.

Keywords

Absolute gravity · GNSS · InSAR · Land subsidence · Precise levelling · Water depletion

1 Introduction

Land subsidence is one of the major hazards in Iran. It usually occurs due to both man-made and natural causes (Motagh et al. 2008). Today, due to the over-extraction of underground water over large parts of the country, the range of subsidence has expanded to urban areas. This can lead to damage of buildings or other constructions and lead to large costs for compensation. To avoid this, subsidence areas should be identified at an early stage. Here, geodetic measurement such as gravity, precise levelling and GNSS data along with InSAR techniques can help us to detect the surface deformation with accurate evaluation of the magnitude, distribution and spatial pattern of land subsidence. In this paper, we investigate the use of geodetic data to monitor the ongoing land subsidence over the country. According to the report of Geodesy and Land Surveying Department of the National Cartographic Center (NCC) of Iran, the repetition of the first-order leveling network has led to detect more than 44 subsidence areas, with a total extension of at least 1,200 km^2 estimated (Amighpey and Arabi 2016), which indicates that vast areas of the country are affected by the phenomenon

J. Hinderer (✉) · J.-D. Bernard
EOST (Ecole et Observatoire des Sciences de la Terre)/Institut de Physique du Globe de Strasbourg, UMR7516 CNRS/Université de Strasbourg, Strasbourg Cedex, France
e-mail: jhinderer@unistra.fr

A. Saadat · H. Cheraghi · Y. Djamour · M. Amighpey · S. Arabi · H. Nankali · F. Tavakoli
National Cartographic Center (NCC) Meraj Avenue, Tehran, Iran

© The Author(s) 2020
J. T. Freymueller, L. Sanchez (eds.), *Beyond 100: The Next Century in Geodesy*,
International Association of Geodesy Symposia 152, https://doi.org/10.1007/1345_2020_125

of subsidence. Besides to the leveling technique we also considered the InSAR (Interferometric synthetic aperture radar) method with the help of Sentinel 1 data for the recent 2017–2019 period. Finally the analysis of the time-series of permanent GNSS stations further confirms the importance of subsidence in many parts of Iran.

Another way to investigate water depletion is to estimate the changes in time of the gravity field. This can be done from space thanks to the GRACE mission which provides a view of the gravity changes at large scales (hundreds of km) as well as at the Earth's surface where the measurements are of more local interest. Water depletion effects leading to a decrease in gravity are well known in Middle East (Longuevergne et al. 2013; Voss et al. 2013; Joodaki et al. 2014; Darama 2014; Forootan et al. 2017). The repetition of absolute gravity campaigns between 2000 and 2018 shows that the gravity values of many stations have changed in time due to the vertical movements and/or water depletion in Iran.

It is interesting to note that any ground gravimeter will see two opposite effects linked to water depletion:

- the Newtonian attraction of the water masses which is directly function of the water table height changes multiplied by porosity (less water less gravity) (42 μGal/m of water)
- the effect of uplift or subsidence of the ground because of the vertical gradient (−306 μGal/m for the Free Air Gradient) (subsidence leads to gravity increase and vice versa).

Since subsidence is usually linked to underground water depletion in any poroelastic model (Burbey 2001), the two effects are opposite in sign but, according to the site characteristics, one effect can be larger than the other leading to gravity increase or decrease.

Finally let us mention that, on the contrary to surface gravity, GRACE space gravity is not sensitive to ground displacement as discussed for instance in Crossley et al. (2012).

2   Geodetic and Gravity Data

National Cartographic Center (NCC) of Iran is responsible for all activities in the field of mapping and surveying all over the country. For this purpose, the national geodetic reference frame it provided by the implementation of fundamental networks. These base networks are divided into different types, such as Precise Leveling network, Gravity network and, finally, Permanent Geodynamic and GNSS network. Iran belongs to a part of the globe with high seismic hazard. Establishing geodynamic networks can help us to better understand the Earth's crust movements and behavior of the tectonic deformation in different active parts of Iran

(e.g. Djamour et al. 2010). The objectives of implementation of Iranian permanent geodynamic and GNSS network (IPGN) are mainly a better understanding of tectonic deformation and the estimation of potential for future earthquake hazards.

Measurements of the first order precise levelling network of Iran include more than 33,000 km, carried out from 1980 up to 1996 using optical levels. After that, these measurements were repeated in 2001 and were completed in 2009 using new digital levels. The levelling networks of Iran are categorized in three orders (all exceed 90,000 km) based on the scopes, specification and achieved precision.

We applied Sentinel-1 interferometric wide-swath images covering the 2017–2019 period with small baseline time-series approach to estimate subsidence rate in Iran. The Sentinel-1 mission from ESA (European Space Agency) consists of two polar-orbiting satellites, operating day and night performing C-band synthetic aperture radar imaging. The average velocities acquired from InSAR time series were projected onto vertical direction to compare with our geodetic results, assess the accuracy of them and validate them.

Absolute gravity campaigns have been done in Iran in the period 2000–2007 and some of these measurements could be repeated in 2017 and 2018 during more recent campaigns. Gravity measurements with absolute gravimeter FG5 type (Micro-g Solutions Inc.) were performed in Iran for three main applications: first to establish zero-order gravity network as fundamental points to extend gravity data in relative sense over the country and provide Iranian gravity calibration line, second to use gravity to have more insight in the determination of vertical motion in tectonic areas and finally to investigate the continental water changes in the arid region of Iranian central desert.

The zero-order gravity network of Iran, which includes 35 stations throughout the country, was designed in 1997. Most of its stations were measured by absolute gravimeter FG5-206, in collaboration with the University of Montpellier-II and Strasbourg during 2000 to 2007 in several field campaigns with 0.001–0.002 mGal level of uncertainty (Hinderer et al. 2003). After that, the first and second order gravity networks, including 670 and 1909 stations respectively, were also extended over the country by connecting to the zero-order gravity network in relative sense using relative gravimeters CG-3 M and CG5. In order to do regional gravity field modelling, a 5′ × 5′ dense gravity data including 20,437 points has been measured in Iran.

We finally also considered gravity data at monthly rate from the GRACE (Gravity Recovery and Climate Experiment) twin satellites to point out the large scale (typically 400 km × 400 km pixels) gravity changes over Iran from 2002 to 2016.

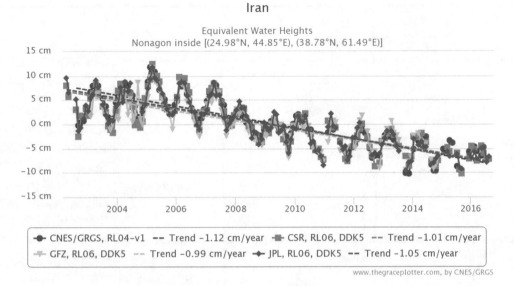

Fig. 1 GRACE observations of the water depletion in Iran for the 2002–2016 period

3   Results

We present now the main results first in gravity from space and ground (AG) and second from the three geodetic techniques available to us (GNSS, leveling, InSAR). Everytime land subsidence was evidenced by GNSS and leveling, InSAR data were also considered to provide the most accurate vertical ground motion to correct surface gravity.

3.1   Space Gravimetry (GRACE)

Figure 1 shows the gravity decrease in Iran from 2002 to 2016 as observed by GRACE satellite gravity observations. Different solutions from different processing centers (CNES/GRGS, CSR, GFZ, JPL) are presented that agree both on the annual signal and on the long term trend that amounts to a value close to −1 cm/year (in terms of equivalent water height).

3.2   Absolute Gravity

Several absolute gravity measurements done in the first epoch (2000–2007) could be repeated in 2017 and 2018 all over Iran. Figure 2 shows the results that merely can be divided into two different zones: gravity decrease in the northern part ranging from −0.8 μgal/year to −5.9 μgal/year and gravity increase in the southern part ranging from 0.8 μgal/year to 13 μgal/year.

3.3   Geodetic and InSAR Results

Figure 3 summarizes all the stations exhibiting subsidence with estimated vertical displacement rates (in mm/year) of permanent IPGN observations and repeated leveling measurements in Iran. These geodetic results have been used to locate potential land subsidence areas in Iran where InSAR data were investigated to detect and monitor the spatial and temporal pattern of subsidence (Motagh et al. 2017). We processed Sentinel-1 images where IPGN and leveling results showed land subsidence.

Figure 4 shows the numerous subsidence regions in Iran and the maximum rate of subsidence (in mm/year) in each region found from InSAR data.

4   Discussion

The average water depletion all over Iran found from GRACE data of the order of 1 cm/year (see Fig. 1) and confirms previous studies (Forootan et al. 2014; Afshar et al. 2016; Khaki et al. 2018). In terms of gravity, this amounts to a decrease of 0.42 μGal/year if elasticity of the Earth is neglected. However this large scale mean rate is different from the rate derived from surface gravity measurements since 1/satellite data are not sensitive to the surface vertical motion on the contrary to ground data (see Eq. 1 in Crossley et al. 2012) and 2/local ground water decline can be large (a few tens of cm/year) as shown by piezometric measurements in Iran (Haghighi and Motagh 2019).

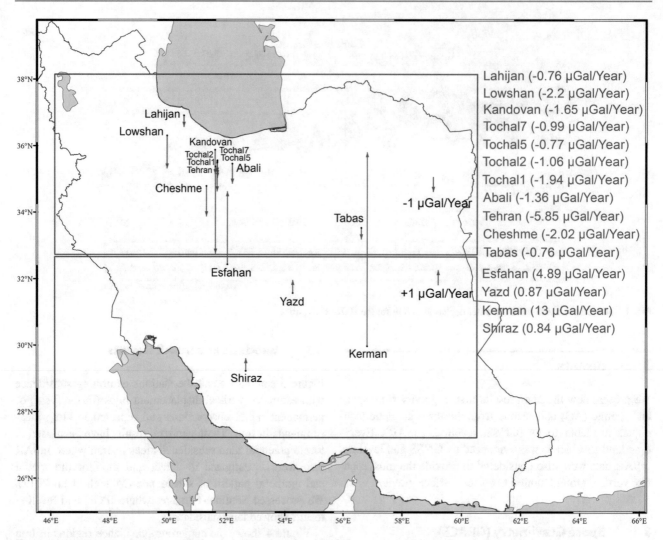

Lahijan (-0.76 µGal/Year)
Lowshan (-2.2 µGal/Year)
Kandovan (-1.65 µGal/Year)
Tochal7 (-0.99 µGal/Year)
Tochal5 (-0.77 µGal/Year)
Tochal2 (-1.06 µGal/Year)
Tochal1 (-1.94 µGal/Year)
Abali (-1.36 µGal/Year)
Tehran (-5.85 µGal/Year)
Cheshme (-2.02 µGal/Year)
Tabas (-0.76 µGal/Year)
Esfahan (4.89 µGal/Year)
Yazd (0.87 µGal/Year)
Kerman (13 µGal/Year)
Shiraz (0.84 µGal/Year)

Fig. 2 Absolute gravity changes (in µGal/year) resulting from the 2017–2018 repetition of the Iranian fundamental network

Despite the different time periods of the results, the rate of the vertical displacements estimated from geodetic tools are consistent with each other; for instance see the agreement between the GNSS stations (FHRJ, FDNG, NISH, GRGN, GGSH) and the closest respective leveling stations (CVDE, CUDC, BCBK, ASBA, AGHA). The comparison of Figs. 3 and 4 also shows the agreement between leveling stations and InSAR results in several regions like the Kerman one for instance (see below the specific values for Kerman station).

From our AG measurements we think that the gravity decreases seen in the northern part of Fig. 2 are mostly due to water depletion attraction with small contribution from ground subsidence. The largest gravity decrease is found for Tehran (−5.8 µGal/year) but −1 to −2 µGal/year effects are frequent. A typical example of the northern part is the Tehran station where the numerous repeated AG measurements since 2000 indicated a large decrease rate of several µGal/year (that seems even to accelerate in recent years) due to underground water depletion, well documented by various piezometric records in the vicinity of the station, although there is no significant vertical displacement (less than 4 mm/year). The Tehran station does not fall into the main subsidence bowl in Western Tehran as presented by Haghighi and Motagh (2019) which might be the reason why the subsidence does not affect the gravity measurement.

On the contrary, the gravity increases seen in the southern part of Fig. 2 are mostly due to ground subsidence linked to water depletion with only a small contribution from attraction. The largest gravity effects are for Espahan (+5 µGal/year) and Kerman (+13 µGal/year). A typical example of the southern part is the Kerman station exhibiting large subsidence rates: – 3.9 cm/year from InSAR (2017–2019), – 4.3 cm/year from GPS (2011–2017) and – 4.0 cm/year from leveling (1987–2006) while gravity increases by 13 µGal/year. We refer to Setyawan et al. (2015) for a similar study in Indonesia combining gravity and height changes due to declining groundwater. Taking into account a subsidence rate close to 4 cm/year and a gravity increase

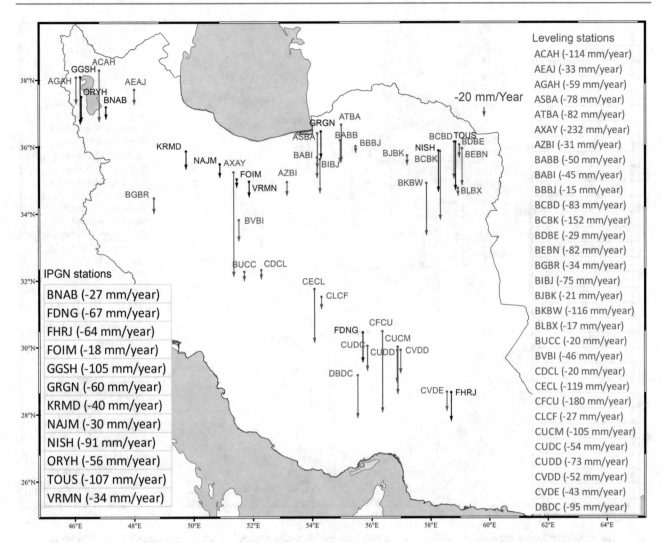

Fig. 3 Subsidence rate of IPGN and Leveling stations. The black arrows show the IPGN rates and the red ones indicate the estimated subsidence rate of leveling stations

of 13 μGal/year would lead to a gravity/height ratio of −3.2 μGal/cm which is very close to measured free air gradient (−3.1 μGal/cm); however there might also be an attraction effect (gravity decrease) due to water depletion which would enhance this ratio. This effect is yet unknown because of the lack of piezometric records in the vicinity.

5   Conclusion

Repetitive and continuous observations based on the Iranian Fundamental Geodetic networks indicate a high rate of subsidence in some parts of the country. These changes are more than a few decimeters per year in some areas. Most subsidence areas in Iran are located in plains and centers with known over-extraction of underground water. This reveals that there is a need for better management of groundwater resources. Geodetic observations (leveling,

GNSS, inSAR) yield very accurate data for monitoring the deformation of the Earth's crust. Absolute gravity as well as space gravimetry data (GRACE) reveal significant long-term gravity changes related to water depletion. By combining these data with other data from geological resources, valuable information can be obtained for decision making. Combining these data with radar interferometry techniques along with other geological resources can help authorities in crisis management to make better decisions. There is a need to repeat the zero-order and hydrological stations in Central Iran in the upcoming years (in the frame of TRIGGER French-Iranian program).

References

Afshar AA, Joodaki GR, Sharifi MA (2016) Evaluation of groundwater resources in Iran using GRACE gravity satellite data. J Geomatic Sci Technol 5(4):73–84. http://jgst.issge.ir/article-1-381-en.html

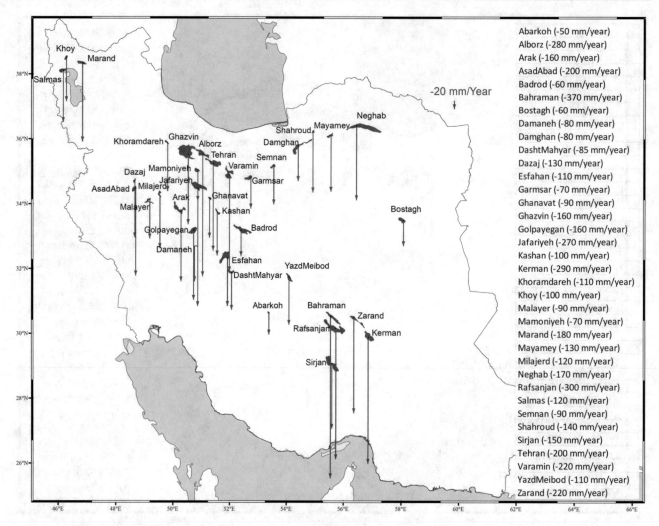

Fig. 4 The subsidence area of some plains in Iran (red polygons) from InSAR results obtained by processing Sentinel-1 data between 2017 and 2019. The arrows show the maximum vertical displacement rate in each region

Amighpey M, Arabi S (2016) Studying land subsidence in Yazd province, Iran, by integration of InSAR and levelling measurements. Remote Sensing App Soc Environ 4:1–8

Burbey TJ (2001) Stress-strain analyses for aquifer-system characterization. Ground Water 39(1):128–136

Crossley D, de Linage C, Boy J-P, Hinderer J, Famiglietti J (2012) A comparison of the gravity field over Central Europe from superconducting gravimeters, GRACE, and global hydrological models, using EOF analysis. Geophys J Int 189:877–897

Darama Y (2014) Comment on "Groundwater depletion in the Middle East from GRACE with implications for transboundary water management in the Tigris-Euphrates-Western Iran region" by Katalyn A. Voss et al. Water Resour Res 50:754–757. https://doi.org/10.1002/2013WR014084

Djamour Y, Vernant P, Bayer R, Nankali HR, Ritz J-F, Hinderer J, Hatam Y, Luck B, Le Moigne N, Sedighi M, Khorrami F (2010) GPS and gravity constraints on continental deformation in the Alborz mountain range, Iran. Geophys J Int 183:1287–1301. https://doi.org/10.1111/j.1365-246X.2010.04811.x

Forootan E, Rietbroek R, Kusche J, Sharifi MA, Awange J, Schmidt M, Omondi P, Famiglietti J (2014) Separation of large scale water storage pat- terns over Iran using GRACE, altimetry and hydrological data. J Remote Sens Environ 140:580–595. https://doi.org/10.1016/j.rse.2013.09.025

Forootan E, Safari A, Mostafaie A, Schumacher M, Delavar M, Awange J (2017) Large-scale total water storage and water flux changes over the arid and semi- arid parts of the middle east from GRACE and reanalysis products. Surv Geophys 38:591–615. https://doi.org/10.1007/s10712-016-9403-1

Haghighi MH, Motagh M (2019) Ground surface response to continuous compaction of aquifer system in Tehran, Iran: results from a long-term multi-sensor InSAR analysis. Remote Sens Environ 221(2019):534–550

Hinderer J, Sedighi M, Bayer R, Ghazavi K, Amalvict M, Luck B, Nilforoushan F, Masson F, Peyret M, Djamour Y, Kouhzare A (2003) The absolute gravity network in Iran: an opportunity to analyse gravity changes caused by present-day tectonic deformation. In: Proc. IMG-2002 (Instrumentation and metrology in Gravimetry), vol 22. Cahiers du Centre Européen de Géodynamique et de Séismologie, Luxembourg, pp 137–141

Joodaki G, Wahr J, Swenson S (2014) Estimating the human contribution to groundwater depletion in the middle east, from GRACE data, land surface models, and well observations. Water Resour Res 50:26792692. https://doi.org/10.1002/2013WR014633

Khaki M, Forootan E, Kuhn M, Awange J, van Dijk AIJM, Schumacher M, Sharifi MA (2018) Determining water storage depletion within Iran by assimilating GRACE data into the W3RA hydrological model. Adv Water Resour 114(2018):1–18

Longuevergne L, Wilson C, Scanlon BR, Cretaux JF (2013) GRACE water storage estimates for the Middle East and other regions with significant reservoir and lake storage. Hydrol Earth System Sci 17(12):4817–4830. https://doi.org/10.5194/hess-17-4817-2013

Motagh M, Walter TR, Sharifi MA, Fielding E, Schenk A, Anderssohn J, Zschau J (2008) Land subsidence in Iran caused by widespread water reservoir overexploitation. Geophys Res Lett 35:L16403. https://doi.org/10.1029/2008GL033814

Motagh M, Shamshiri R, Haghshenas Haghighi M, Wetzel H, Akbari B, Nahavandchi H, Roessner S, Arabi S (2017) Quantifying groundwater exploitation induced subsidence in the Rafsanjan plain, south-eastern Iran, using InSAR time-series and in situ measurements. Eng Geol 218:134–151

Setyawan A, Fukuda Y, Nishijima J, Kazama T (2015) Detecting land subsidence using gravity method in Jakarta and Bandung area, Indonesia. Procedia Environ Sci 23:17–26

Voss KA, Famiglietti JS, Lo M, de Linage C, Rodell M, Swenson SC (2013) Groundwater depletion in the Middle East from GRACE with implications for transboundary water management in the Tigris-Euphrates-Western Iran region. Water Resour Res 49:904–914. https://doi.org/10.1002/wrcr.20078

Past and Future Sea Level Changes and Land Uplift in the Baltic Sea Seen by Geodetic Observations

M. Nordman, A. Peltola, M. Bilker-Koivula, and S. Lahtinen

Abstract

We have studied the land uplift and relative sea level changes in the Baltic Sea in northern Europe. To observe the past changes and land uplift, we have used continuous GNSS time series, campaign-wise absolute gravity measurements and continuous tide gauge time series. To predict the future, we have used probabilistic future scenarios tuned for the Baltic Sea. The area we are interested in is Kvarken archipelago in Finland and High Coast in Sweden. These areas form a UNESCO World Heritage Site, where the land uplift process and how it demonstrates itself are the main values. We provide here the latest numbers of land uplift for the area, the current rates from geodetic observations, and probabilistic scenarios for future relative sea level rise. The maximum land uplift rates in Fennoscandia are in the Bothnian Bay of the Baltic Sea, where the maximum values are currently on the order of 10 mm/year with respect to the geoid. During the last 100 years, the land has risen from the sea by approximately 80 cm in this area. Estimates of future relative sea level change have considerable uncertainty, with values for the year 2100 ranging from 75 cm of sea level fall (land emergence) to 30 cm of sea-level rise.

Keywords

Baltic Sea · Geodetic time series · Land uplift · Sea level rise

1 Introduction

Fennoscandia is a geodynamically active region due to the relatively recent (20–10 thousand years before present) demise of the large ice sheets that covered this region at the last glacial maximum. The weight of the ice sheets pressed the Earth's crust down and now, in the process called post glacial rebound, the crust is slowly uplifting as the Earth relaxes to a state of isostatic equilibrium. Fennoscandia has the highest number of land uplift related observations in the world (e.g. Poutanen and Steffen 2014), and the phenomenon has also been studied for centuries (see e.g. Ekman 1999, 2009; Steffen and Wu 2011).

One sign of the uniqueness of the Fennoscandian uplift is the World Heritage Site status the Kvarken Archipelago (Finland) and High Coast (Sweden) have obtained (Fig. 1). As part of the Lystra project, funded by EU Interreg Botnia-Atlantica program, which main aim is to update the knowledge and information materials for visitors of the World Heritage Site, we have been looking at the geodetic data available for the area to understand the land uplift patterns and magnitudes better (Peltola 2019) and to compare them to sea level rise scenarios to understand how the coastline might move in the future (Huuskonen 2020).

One active operator in geodesy in the Nordic area is the Nordic Geodetic Commission (NKG). The NKG was established 1953 to enhance the co-operation between geodesists in the Nordic countries (Denmark, Finland, Iceland, Norway and Sweden), nowadays also Baltic countries (Estonia, Latvia and Lithuania) are active. There are three NKG activi-

M. Nordman · A. Peltola · M. Bilker-Koivula · S. Lahtinen
National Land Survey of Finland, Finnish Geospatial Research Institute, Kirkkonummi, Finland

M. Nordman (✉)
Aalto University, School of Engineering, Espoo, Finland
e-mail: maaria.nordman@aalto.fi

© The Author(s) 2020

J. T. Freymueller, L. Sanchez (eds.), *Beyond 100: The Next Century in Geodesy*,
International Association of Geodesy Symposia 152, https://doi.org/10.1007/1345_2020_124

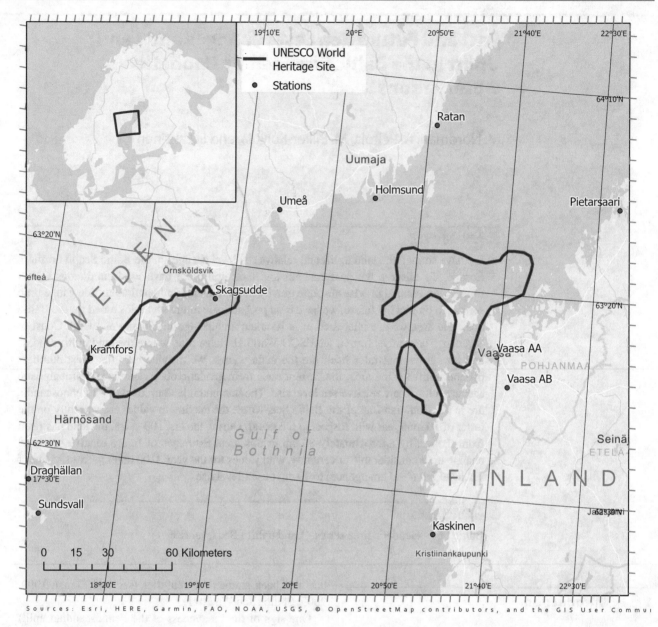

Fig. 1 The UNESCO World Heritage Site of Kvarken and High Coast in Northern Europe together with the geodetic observation points of the area

ties that are of interest for our study. Firstly, the NKG2016LU land uplift model (Vestøl et al. 2019) that was computed in co-operation in the Working Groups for Geodynamics and for Geoid and Height Systems. Secondly, the absolute gravity measurements that are discussed and published in co-operation in the Working Group of Geodynamics (e.g. Olsson et al. 2019). And thirdly, the NKG Analysis Centres, under the Working Group for the Reference Frames, which continuously process the GNSS data from the permanent GNSS stations (Lahtinen et al. 2018).

In this study we show comparisons of land uplift rates from selected geodetic techniques as well as probabilistic future scenarios for the Kvarken and High Coast areas. The

second chapter describes the data and methods chosen, the third chapter shows and discusses the results. The last chapter is left for conclusions.

2 Data and Methods

2.1 Geodetic Data

The GNSS uplift values used in the present study were the trends from the NKG Analysis Centre solution by Lahtinen et al. (2019). There were altogether eight stations in the study area, one in Finland and seven in Sweden. The longest time series were approximately 20 years in length (Vaasa, Umeå and Sundsvall). The other stations have approximately 10 years of time series. The uncertainties in the time series are less than 0.1 mm/year. GNSS gives absolute uplift rates in a global reference frame. These rates must be converted to uplift rates relative to the geoid, as measured by repeated levelling, to be able to compare them to the sea level height changes. We do this using the differences between the absolute and relative versions of the NKG2016LU land uplift model (NKG2016LU_abs and NKG2016LU_lev, see below). The difference is in the order of 0.6 mm/year for our study area. The elastic deformation from present day glacier melt has an effect on the land uplift rates in Fennoscandia (Simon et al. 2018). For now, the effect is not included as the signal needs to be adequately approximated. In future studies the effect should be considered.

The absolute gravity (AG) observations were obtained from the NKG gravity database, published in Olsson et al. (2019), with some additions for the last years (2016–2018, pers. comm. P. Olsson, pers. comm. M. Bilker-Koivula). The gravity data gives information on the land uplift independent from the other observables or models. There are altogether five absolute gravity stations in the area with long time spans between 8 (Kramfors) and 20 years (Skellefteå, Vaasa AA and AB). Here we use only data measured with the FG5-type instruments (Niebauer et al. 1995). At Skellefteå and the Vaasa stations also measurements exist made with the JILAg-5 instrument (Faller et al. 1983). However, there is a suspicion that the JILAg-5 measurements may have an offset with respect to the FG5 measurements, as such an offset has been found for other JILAg instruments (Timmen et al. 2008; Lambert et al. 2001; Pálinkáš et al. 2013). Also, Olsson et al. (2019) suggest that the JILAg-5 may have an offset and leave JILAg data out of the final solution. To avoid offset problems, we use only FG5 data in this study. The FG5 data have uncertainties of 2–3 μGals (where 1 μGal is 10^{-8} m/s^2).

Absolute gravity measurements are subject to changes in the environment such as variations in local hydrology. If information on the hydrological signal is available this can be used to reduce the variation of the gravity time series (see e.g. Ophaug et al. (2016), Lambert et al. (2006) and Mikolaj et al. (2015)). Modelling of the hydrological signal is however out of the scope of this paper and we assume that our long time series will cancel out any seasonal variations in gravity.

To obtain the land uplift from the gravity change values, a ratio of -0.163 μGal/mm was used (Olsson et al. 2015) with the uncertainty $\pm$ 0.016 μGal/mm (Ophaug et al. 2016). The ratio is theoretically predicted for a visco-elastic Earth model and confirmed by absolute gravity observations (Olsson et al. 2019). Like the GNSS uplift rates, the gravity derived uplift rates are absolute, as the absolute gravity observations give

the rate of change with respect to the center of mass. Thus, correction for geoid rise needs to be added.

The land uplift model that was used for the comparison is the semi-empirical NKG2016LU model (Vestøl et al. 2019). The model combines GNSS time series and levelling as observations with a GIA (glacial isostatic adjustment) model covering the areas where observations are sparse (e.g. on the sea). The model's uplift rates are available relative to ellipsoid (NKG2016LU_abs) or to geoid (NKG2016LU_lev). We chose the latter. The model comes with uncertainty values that are in general slightly less than 0.2 mm/year.

2.2 Sea Level Data

The sea level data was obtained from the PSMSL data base (Holgate et al. 2013). The data consist of yearly mean averages for all tide gauge stations near the study area. There are three tide gauges on the Finnish side and two on the Swedish side. The time periods for the stations vary somewhat, the earliest observations start in the early twentieth century and all, except one station, are still running. The problem with the tide gauges is that the eustatic sea level change, meaning the global sea level change because of mass and volume changes of the oceans, has changed over the years (e.g. Johansson et al. 2004; Dangendorf et al. 2019), thus using a single number for long time series would produce optimistic numbers for the land uplift rates. In the Baltic Sea the trend in tide gauge record has been estimated to be stable up to 1980, after which it has started to change (Johansson et al. 2003, 2014). We, therefore, decided to use time series only up to 1980 for the land uplift estimates, to diminish the effect of the changing global sea level. A uniform time period was chosen for all the stations after some research. In order to compare the tide gauge trends to geodetic trends, the eustatic sea level rise of 1.0 mm/year for twentieth century (Ekman and Mäkinen 1996) was removed.

2.3 Future Scenarios

In order to study how the World Heritage Site will change in the future, we compared our results of what has happened in the last 100 years to probabilistic future scenarios computed into 2100 by Pellikka et al. (2018). We took the 5%, 50% and 95% probability values for our comparison. The 50% is the median value expected at each site by 2100 and the 5% and 95% are low- and high-end scenarios. The three components affecting the sea level rise in the Baltic Sea are the land uplift, the global mean sea level rise and the meteorological component, i.e. the effect of westerly winds that might increase in the future. The scenarios in Pellikka et al. (2018) take into account these components on the Finnish coast. As we have stations also on the Swedish coast, we have paired the Swedish stations with Finnish stations that are located on approximately the same latitude from Pellikka et al. (2018), based on the assumption that the whole gulf would react similarly to future changes. To estimate the future scenarios we take the sea level scenario value from Table 3 in Pellikka et al. (2018), remove the land uplift they have used (listed in the same table) and add back the land uplift rate from our observations.

3   Results and Discussion

All the land uplift rates from different techniques are listed in Table 1. The uplift rates from geodetic data, i.e. GNSS and absolute gravity are shown in the first five columns of Table 1. All the trends are converted to trends relative to the geoid for comparison. In the sixth and seventh column are the changing sea levels ($\dot{S}$), which have a negative sign, as the sea level falls relative to ground. Different time periods give somewhat different results for the tide gauges. We tested different time periods for each tide gauge (not shown here), depending on when the tide gauges started operating, if there were data gaps or other issues. We concluded that the period up to 1979 shows the land uplift best and took, when possible, the same time period for all stations. The $\dot{H}_{w.mean}$ in the second last column was computed by weighing the observations by their variance, if there were more than one technique available, meaning that the GNSS time series with their small standard deviations give the biggest impact to the weighted mean and its standard deviation. The last column shows the rates from the NKG2016LU model for comparison.

3.1   Geodetic and Sea Level Data

As can be seen from Table 1, the GNSS and AG results agree quite well, especially within uncertainty. The Vaasa station has previously shown some anomalistic low GNSS-derived rates (Kierulf et al. 2014; Lidberg et al. 2009), which has now been addressed to changing surroundings of the station. For the Lahtinen et al. (2019) result the time series has been modified more drastically and fits now the overall picture better. The absolute gravity value of Vaasa AA is somewhat higher than expected. We suspect that this is due to changes in local hydrology over the years near the station. However, this is speculation and must be subject of further studies. The NKG2016LU values agree better with GNSS, which is not surprising as the model has GNSS values inside it.

The tide gauges have comparable values to geodetic measurements, in the order of 7–9 mm per year sea level fall. When comparing to previous studies (e.g. Ekman 1996; Pellikka et al. 2018) our values are in the same range.

3.2   Comparison of Geodetic and Sea Level Trends

The land uplift rates from different techniques show comparable results, as could be expected. The maximum rates are on the Swedish side, whereas the Finnish sites have slightly lower values. The absolute gravity shows somewhat higher values, which might relate to mass changes that are not induced by land uplift or that the ratio derived using visco-elastic Earth model is not optimal for the maximum uplift area. However, the underestimation of gravity rates estimated by Olsson et al. (2015) for the areas close to

Table 1 The land uplift rates from different techniques. $\dot{h}_{GPS}$ and $\dot{H}_{GNSS}$ are the absolute and relative vertical rates of the permanent GNSS stations. $\dot{g}_{AG}$ are the absolute gravity change rates from AG data and $\dot{h}_{AG}$ and $\dot{H}_{AG}$ are the absolute and relative vertical rates derived from the absolute gravity data. $\dot{S}_{1920-1979}$ are the sea level change for years 1920–1979 and for comparison we show also $\dot{S}_{1919-2018}$, change within the last 100 years (to be used in Table 2). $\dot{H}_{TG}$ are the relative land uplift rates from tide gauges. The bold numbers were used to compute the weighted mean $\dot{H}_{w.mean}$. $\dot{H}_{NKG}$ are the vertical rates from the NKG2016LU_lev land uplift model

Station	$\dot{h}_{GNS}$ mm/a	$\dot{H}_{GNSS}$ mm/a	$\dot{g}_{AG}$ μgal/a	$\dot{h}_{AG}$ mm/a	$\dot{H}_{AG}$ mm/a	$\dot{S}_{1920-79}$ mm/a	$\dot{S}_{1919-2018}$ mm/a	$\dot{H}_{TG}$ mm/a	$\dot{H}_{w.mean}$ mm/a	$\dot{H}_{NKG}$ mm/a
Pietarsaari						−8.5 ± 0.5	−7.1 ± 0.3	**9.5 ± 0.7**	9.5 ± 0.69	9.0 ± 0.2
Vaasa AA			−1.8 ± 0.2	10.8 ± 1.4	**10.2 ± 1.4**	−7.9 ± 0.5	−7.0 ± 0.2	**8.9 ± 0.7**	9.2 ± 0.59	8.8 ± 0.2
Vaasa AB	9.1 ± 0.06	**8.5 ± 0.06**	−1.6 ± 0.1	9.8 ± 1.3	**9.2 ± 1.3**				8.5 ± 0.06	8.6 ± 0.2
Kaskinen[a]						−7.3 ± 0.7	−6.3 ± 0.3	**8.3 ± 0.7**	8.3 ± 0.67	8.3 ± 0.2
Skellefteå[b]	10.3 ± 0.08	**9.6 ± 0.08**	−1.8 ± 0.1	11.0 ± 1.2	**10.4 ± 1.2**	−9.5 ± 0.5	−7.9 ± 0.3	**10.5 ± 0.7**	9.6 ± 0.08	9.5 ± 0.2
Bjuröklubb	10.0 ± 0.05	**9.4 ± 0.05**							9.4 ± 0.05	9.5 ± 0.2
Ratan	10.1 ± 0.10	**9.4 ± 0.10**	−1.7 ± 0.3	10.7 ± 2.3	**10.0 ± 2.3**	−8.5 ± 0.5	−7.7 ± 0.3	**9.5 ± 0.7**	9.4 ± 0.10	9.5 ± 0.2
Holmsund	10.0 ± 0.08	**9.4 ± 0.08**							9.4 ± 0.08	9.5 ± 0.2
Umeå	10.3 ± 0.01	**9.6 ± 0.01**							9.6 ± 0.01	9.6 ± 0.1
Kramfors	10.0 ± 0.06	**9.4 ± 0.06**	−1.7 ± 0.3	10.2 ± 2.0	**9.5 ± 2.0**				9.4 ± 0.06	9.2 ± 0.2
Sundsvall	9.5 ± 0.04	**8.9 ± 0.04**							8.9 ± 0.04	8.9 ± 0.2

[a]Kaskinen observations start in 1926
[b]Skellefteå (GPS and AG) was combined with Furuögrund tide gauge

the maximum uplift, are around 0.02 µGal/year, which is of an order smaller than the deviations we see between the absolute gravity rates and the gravity changes derived from the NKG2016LU_gdot model (Olsson et al. 2019). The GNSS dominated weighted mean values have clearly lower uncertainty than the land uplift model, whereas the other techniques' uncertainties are two of threefold. The uncertainty of GNSS time series has been studied a lot and it has been suggested that the true uncertainty might be in the order of 0.2–0.3 mm/year (Lahtinen et al. 2019). This requires still more investigation.

Also the elastic component of the present day melting studied by Simon et al. (2018) could affect both the measurements and modelling. They did not use terrestrial gravity measurements so the effect to this type of study would require more studies.

3.3 Future

The future scenarios from Pellikka et al. (2018) modified using our uplift rates are shown in Table 2. To show the effect on coastline, Fig. 2 shows the future scenarios for the Björkö island in Kvarken as an example, The topography is much steeper on the High Coast side and does not really show much change even with the higher sea level rise scenarios (Huuskonen 2020). The relative sea level fall could not be visualized, because we have no bathymetry data with adequate resolution available for the area.

The future of the Kvarken and High Coast in the coming 100 years remains uncertain as the future scenarios are uncertain. It looks very likely that the land uplift will compensate for the globally rising sea levels, although, depending on the rate of the future sea level rise, land subsidence is possible even in the maximum land uplift area. The sea level change will affect the coast line on the east and west coast of the Bothnian Bay differently, due to different topography of

the landscape. The Hight Coast has very steep topography, which will not be affected much by the rising sea level. In Kvarken the situation is quite the opposite and the landscape is very flat and prone to flooding. Rising sea levels can easily drown landmarks and call for adjustments e.g. in infrastructure. We are taking a closer look at this topic on another publication that is under preparation (Huuskonen and Nordman in prep).

4 Conclusions

We have studied the land uplift in the Bothnian Bay of the Baltic Sea. We have computed the current rates of uplift from geodetic observations, and probabilistic scenarios for future relative sea level rise in the area. The maximum land uplift rates in Fennoscandia are in the Bothnian Bay, where the maximum values are currently on the order of 10 mm/year with respect to the geoid. During the last 100 years, the land has risen from the sea by up to 80 cm in this area. Estimates of future relative sea-level change have considerable uncertainty, with values for the year 2100 ranging from 75 cm of sea-level fall (land emergence) to 30 cm of sea-level rise.

The different techniques we have used give similar results, as could be expected. There are some discrepancies, most likely due to different time periods used for different techniques as well as effects of other phenomena, e.g. changes in local hydrology. Combining the techniques produces some challenges, as all techniques studied here measure different part of the land uplift phenomenon. This can also be seen as an advantage and can be used to distinguish between different parts, namely uplift, mass changes and their relation. Some issues in the differences can be addressed to the theoretical and empirical relations between the measurement types.

The most uncertain part in the current study is the future. The future scenarios cover a wide range, and the uncertainty

Table 2 The probabilistic future scenarios of relative sea level rise for stations in our study area, showing the relative land uplift for the past 100 years from tide gauge time series (first column, see also Table 1) for the tide gauge stations, and change to year 2100 for 5%, 50% and 95% probabilities. The unit is cm

Station	1919–2018 (cm)	2000–99: 5%	2000–99: 50%	2000–99: 95%
Pietarsaari	71 ± 2.6	−25	27	70
Vaasa AA	70 ± 2.3	−27	25	68
Vaasa AB	70 ± 2.3	−35	17	60
Kaskinen	63 ± 2.8	−44	11	55
Skellefteå/Furuögrund	79 ± 2.6	−24	28	71
Bjuröklubb		−27	25	68
Ratan	77 ± 2.5	−26	26	69
Holmsund		−27	25	68
Umeå		−23	29	72
Kramfors		−25	27	70
Sundsvall		−37	18	62

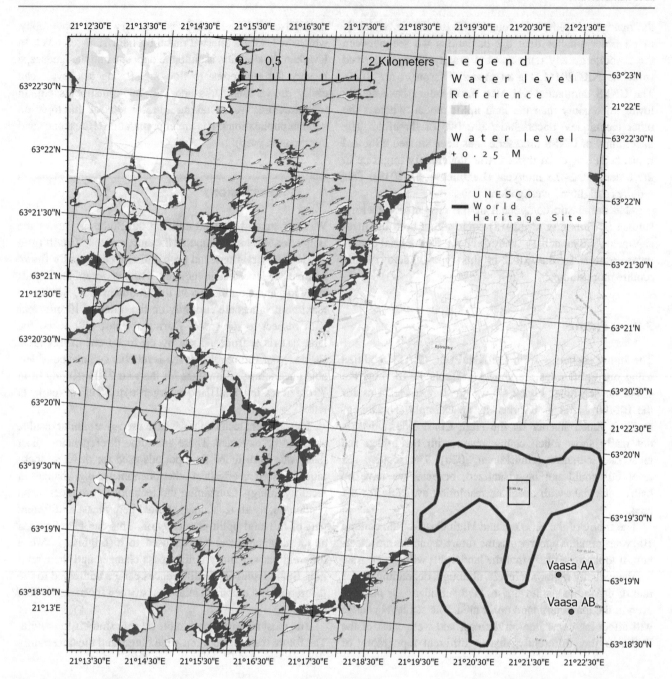

Fig. 2 The future scenario of Table 2 plotted for Kvarken's Björkö. The light blue shows the present coast line, and the dark blue the 95% probability scenario (sea level rise of 25 cm)

of the projections into year 2100 is large, even if it is fitted to local conditions as in Pellikka et al. (2018). Also the effect of the changing coast line is uncertain and should be studied further.

Acknowledgements Aleksi Peltola was funded by European Union Interreg Botnia-Atlantica project Lystra. Maaria Nordman was funded by the Academy of Finland (project number 299626). Erja Huuskonen is kindly acknowledged for Figs. 1 and 2.

References

Dangendorf S, Hay C, Calafat FM, Marcos M, Piecuch CG, Berk K, Jensen J (2019) Persistent acceleration in global sea-level rise since the 1960s. Nat Clim Chang 9(9):705–710

Ekman M (1996) A consistent map of the postglacial uplift of Fennoscandia. Terra Nova 8(2):158–165

Ekman M (1999) Climate changes detected through the world's longest sea level series. Glob Planet Chang 21(4):215–224. https://doi.org/10.1016/S0921-8181(99)00045-4

Ekman M (2009) The changing level of the Baltic Sea during 300 years: a clue to understanding the earth. The Summer Institute for Historical Geophysics, Åland Islands, pp 1–155. ISBN 978-952-92-5241-1

Ekman M, Mäkinen J (1996) Recent postglacial rebound, gravity change and mantle flow in Fennoscandia. Geophys J Int 126(1):229–234. https://doi.org/10.1111/j.1365-246X.1996.tb05281.x

Faller JE, Guo YG, Gschwind J, Niebauer TM, Rinker RL, Xue J (1983) The JILA portable absolute gravity apparatus. BGI Bull Inf 53:87–97

Holgate SJ, Matthews A, Woodworth PL, Rickards LJ, Tamisiea ME, Bradshaw E, Foden PR, Gordon KM, Jevrejeva S, Pugh J (2013) New data systems and products at the permanent service for mean sea level. J Coast Res 29(3):493–504. https://doi.org/10.2112/JCOASTRES-D-12-00175.1

Huuskonen E (2020) Sea-level rise and land uplift in Kvarken and high coast area in 2300. Aalto University, Espoo

Huuskonen E, Nordman M (in prep) Relative sea level rise in the Kvarken archipelago and the High Coast in 2300

Johansson MM, Kahma KK, Boman H (2003) An improved estimate for the long-term mean sea level on the Finnish coast. Geophysica 39(1–2):51–73

Johansson MM, Kahma KK, Boman H, Launiainen J (2004) Scenarios for sea level on the Finnish coast. Boreal Environ Res 9:153–166. ISSN: 1239-6095

Johansson MM, Pellikka H, Kahma KK, Ruosteenoja K (2014) Global Sea level rise scenarios adapted to the Finnish coast. J Mar Syst 129:35–46. https://doi.org/10.1016/j.jmarsys.2012.08.007

Kierulf HP, Steffen H, Simpson MJR, Lidberg M, Wu P, Wang H (2014) A GPS velocity field for Fennoscandia and a consistent comparison to glacial isostatic adjustment models. J Geophys Res Solid Earth 119(8):6613–6629. https://doi.org/10.1002/2013JB010889

Lahtinen S, Häkli P, Jivall L, Kempe C, Kollo K, Kosenko K, Pihlak P, Prizginiene D, Tangen O, Weber M, Parseliunas E, Baniulis R, Galinauskas K (2018) First results of the Nordic and Baltic GNSS analysis Centre. J Geodetic Sci 8:34–42. https://doi.org/10.1515/jogs-2018-0005

Lahtinen S, Jivall L, Häkli P, Kall T, Kollo K, Kosenko K, Galin- auskas K, Prizginiene D, Tangen O, Weber M, Nordman M (2019) Densification of the ITRF2014 position and velocity solution in the Nordic and Baltic countries. GPS Solutions 23(4):95. https://doi.org/10.1007/s10291-019-0886-3

Lambert A, Courtier N, Sasagawa GS, Klopping F, Winester D, James TS, Liard JO (2001) New constraints on Laurentide postglacial rebound from absolute gravity measurements. Geophys Res Lett 28(10):2109–2112

Lambert A, Courtier N, James TS (2006) Long-term monitoring by absolute gravimetry: tides to postglacial rebound. J Geodyn 41:307–317. https://doi.org/10.1016/j.jog.2005.08.032

Lidberg M, Johansson JM, Scherneck HG, Milne GA (2009) Recent results based on continuous GPS observations of the GIA process in Fennoscandia from BIFROST. J Geodyn 50(2010):8–18. https://doi.org/10.1016/j.jog.2009.11.010

Mikolaj M, Meurers B, Mojzeš M (2015) The reduction of hydrology-induced gravity variations at sites with insufficient hydrological instrumentation. Stud Geophys Geod 59:1–xxx. https://doi.org/10.1007/s11200-014-0232-8

Niebauer TM, Sasagawa GS, Faller JE, Hilt R, Klopping F (1995) A new generation of absolute gravimeters. Metrologia 32:159–180

Olsson P-A, Milne G, Scherneck H-G, Ågren J (2015) The relation between gravity rate of change and vertical displacement in previ- ously glaciated areas. J Geodyn 83:76–84. https://doi.org/10.1016/j.jog.2014.09.011

Olsson P-A, Breili K, Ophaug V, Steffen H, Bilker-Koivula M, Nielsen E, Oja T, Timmen L (2019) Postglacial gravity change in Fennoscan- dia - three decades of repeated absolute gravity observations. Geo- phys J Int 217:1141–1156. https://doi.org/10.1093/gji/ggz054

Ophaug V, Breili K, Gerlach C, Gjevestad JGO, Lysaker DI, Omang OCD, Pettersen BR (2016) Absolute gravity observations in Norway (1993–2014) for glacial isostatic adjustment studies: the influence of gravitational loading effects on secular gravity trends. J Geodyn 102:83–94

Pálinkáš V, Lederer M, Kostelecký J, Šimek J, Mojzeš M, Ferianc D, Csapó G (2013) Analysis of the repeated absolute gravity mea- surements in the Czech Republic, Slovakia and Hungary from the period 1991–2010 considering instrumental and hydrological effects. J Geod 87:29–42. https://doi.org/10.1007/s00190-012-0576-1

Pellikka H, Leijala U, Johansson MM, Leinonen K, Kahma KK (2018) Future probabilities of coastal floods in Finland. Cont Shelf Res 157:32–42. https://doi.org/10.1016/j.csr.2018.02.006

Peltola A (2019) Korkearannikon ja Merenkurkun saariston maailman- perintökohteen maanpinnan kohoamisnopeuden havaitseminen eri menetelmillä. (The land uplift rate at the world heritage site high coast and Kvarken archipelago observed with different methods). Aalto University, Espoo. (in Finnish)

Poutanen M, Steffen H (2014) Land uplift at Kvarken Archipelago/High Coast UNESCO world heritage area. Geophysica 50(2):49–64

Simon KM, Riva RE, Kleinherenbrink M, Frederikse T (2018) The glacial isostatic adjustment signal at present day in northern Europe and the British Isles estimated from geodetic observations and geophysical models. Solid Earth 9(3):777

Steffen H, Wu P (2011) Glacial isostatic adjustment in Fennoscandia – a review of data and modeling. J Geodyn 52(3–4):169–204

Timmen L, Gitlein O, Müller J, Strykowski G, Forsberg R (2008) Absolute gravimetry with the Hannover meters JILAg-3 and FG5- 220, and their deployment in a Danish-German cooperation. ZFV - Zeitschrift fur Geodasie, Geoinformation und Landmanagement 133(3):149–163

Vestøl O, Ågren J, Steffen H, Kierulf H, Tarasov L (2019) NKG2016LU: a new land uplift model for Fennoscandia and the Baltic region. J Geod 93(9):1759–1779, 1–21

Estimation of Lesser Antilles Vertical Velocity Fields Using a GNSS-PPP Software Comparison

Pierre Sakic, Benjamin Männel, Markus Bradke, Valérie Ballu,
Jean-Bernard de Chabalier, and Arnaud Lemarchand

Abstract

Vertical land motion in insular areas is a crucial parameter to estimate the relative sea-level variations which impact coastal populations and activities. In subduction zones, it is also a relevant proxy to estimate the locking state of the plate interface. This motion can be measured using Global Navigation Satellite Systems (GNSS), such as the Global Positioning System (GPS). However, the influence of the processing software and the geodetic products (orbits and clock offsets) used for the solution remains barely considered for geophysics studies.

In this study, we process GNSS observations of Guadeloupe and Martinique network (Lesser Antilles). It consists of 40 stations over a period of 18 years for the oldest site. We provide an updated vertical velocity field determined with two different geodetic software, namely EPOS (Gendt et al, GFZ analysis center of IGS–Annual Report. IGS 1996 Annual Report, pp 169–181, 1998) and GINS (Marty et al, GINS: the CNES/GRGS GNSS scientific software. In: 3rd International colloquium scientific and fundamental aspects of the Galileo programme, ESA proceedings WPP326, vol 31, pp 8–10, 2011) using their Precise Point Positioning modes. We used the same input models and orbit and clock offset products to maintain a maximum of consistency, and then compared the obtained results to get an estimation of the time series accuracy and the software influence on the solutions. General consistency between the solutions is noted, but significant velocity differences exist (at the mm/yr level) for some stations.

Keywords

Lesser Antilles · Guadeloupe · Martinique · GNSS · Vertical velocity field · Subsidence · Processing comparison · Time series analysis

P. Sakic (✉) · B. Männel · M. Bradke
GFZ German Research Centre for Geosciences, Helmholtz-Zentrum Potsdam, Potsdam, Germany
e-mail: pierre.sakic@gfz-potsdam.de

V. Ballu
Littoral Environnement et Sociétés, CNRS & University of La Rochelle, La Rochelle, France

J.-B. de Chabalier · A. Lemarchand
Institut de Physique du Globe de Paris, Université de Paris, Paris, France

1 Introduction

Nowadays, the Global Navigation Satellite Systems (GNSS) have become an indispensable tool to monitor the Earth crust motion. Nevertheless, their use in some part of the world remains challenging. It is the case of the Lesser Antilles Subduction Zone, at the convergence of the Nord-American Plate under the Caribbean Plate. This subduction is singular on several aspects: it is one of the slowest in the world (∼2cm/yr), the lack of emerged lands prevents the determination of a complete deformation profile like it

J. T. Freymueller, L. Sanchez (eds.), *Beyond 100: The Next Century in Geodesy*,
International Association of Geodesy Symposia 152, https://doi.org/10.1007/1345_2020_101

can be done in other areas. Moreover, the islands of the volcanic arc are located too far from the trench, prohibiting most often the detection of significant velocity gradients with respect to the stable plate. Because of these reasons, the locking state deduced from GNSS observations, and thus megathrust risk is uncertain. Symithe et al. (2015) estimated with GNSS observations all along the volcanic arc a low coupling rate but did not exclude a megathrust possibility either. Since the estimation of a horizontal deformation rate is a difficult task for this area, vertical motion observations can then become a proxy to help the assessment of a potential strain accumulation (e.g. Mouslopoulou et al. 2016). Moreover, island areas are also threatened by the sea level rise, and extra subsidence can be an aggravating factor (Ballu et al. 2011). For these two reasons, measuring vertical motion in the Lesser Antilles is crucial. Some vertical movement assessments in this area were performed in the past. Paleo-geodesy based on coral reef growth tends to show a subsidence trend in Martinique and Les Saintes Islands (south of Guadeloupe archipelago) (Weil-Accardo et al. 2016; Leclerc and Feuillet 2019). This subsidence is corroborated by GNSS observations for a few stations within the vertical velocity ULR6 solution (Santamaría-Gómez et al. 2017). However, an uplift with decreasing rate for la Désirade Island (West of the Guadeloupe Archipelago) was measured (Léticée et al. 2019). Even though GNSS exploitation in this area is challenging, the islands are generally well instrumented, especially the two islands of Guadeloupe and Martinique.

Therefore, this study has a double objective, technical and scientific: on one hand, we estimate a denser vertical velocity field using a maximum of GNSS data in the area. On the other hand, we compare and quantify the differences between the results (coordinates and velocities inferred from the time series) obtained with two GNSS processing software using homogeneous inputs.

2   Data

We considered the GNSS observations provided by three different station networks, namely the IPGP, the IGN/SONEL, and the ORPHEON networks. Maps of these networks on the two islands are represented in Fig. 1. All three have been deployed for different purposes. The IPGP (*Institut de Physique du Globe de Paris*) network, maintained by the two local volcanological and seismological observatories has been deployed since the early 2000s (the first stations, HOUE and SOUF, were installed in 2000) for geodynamic purposes. Some stations are installed in the vicinity of the volcanic domes to monitor the volcanic activity and the others are deployed to observe potential subduction induced deformations. This network gathers nowadays 28

stations, 20 in Guadeloupe and 8 in Martinique. Due to the remote and/or extreme conditions for some sites, the network is heterogeneous regarding the time series completeness and the continuity in time of the equipment used. The IGN/SONEL (*Institut national de l'information géographique et forestière/Système d'observation du niveau des eaux littorales*) maintains 4 stations in the area (2 in Guadeloupe and 2 in Martinique) mainly for reference frame geodesy, with an application to sea level monitoring for the two stations PPTG and FFT2 (Wöppelmann et al. 2011; Santamaría-Gómez et al. 2017). These stations are installed in the vicinity of tide gauges to monitor their vertical motion. The first station of this network is ABMF and was installed in 2008. Finally, the ORPHEON network which consists in 8 stations (5 in Guadeloupe and 3 in Martinique) was installed in 2013/2014 for RTK surveying purposes, but the notable continuity of the time series without any hardware change make them suitable candidates for geophysics use. These stations are provided within the framework of RENAG (*Réseau national GNSS permanent*, e.g. Walpersdorf et al. 2018; Rabin et al. 2018). We processed the data from May 2000 (deployment of HOUE and SOUF) to end of August 2018 for the IPGP stations and end of November 2018 for the other networks. The timeline of the used observations is represented in Fig. 2.

3   Processing

The observation set described below was processed using two different GNSS processing software, namely EPOS (Gendt et al. 1998; Uhlemann et al. 2015) and GINS (Marty et al. 2011; Loyer et al. 2012), but using a similar strategy, equivalent models and identical product inputs. The underlying idea is to quantify the intrinsic differences due to different software. We used a Precise Point Positioning approach with float ambiguity resolution. A PPP processing is the most suitable one for this area because the reduced number of IGS reference stations prevents efficient differential processing. Moreover, those reference stations can be affected by the same tectonic processes as the geophysics stations. Thus, IGS stations in the area (namely ABMF and LMMF) are used here as "regular" stations. We considered only GPS observations since most of the stations recorded only this constellation during most of the period considered. The orbits and clock offset products have been generated beforehand by the GFZ Analysis Center in preparation of the IGS *Repro3* campaign (Männel et al. 2020). These products are consistent with ITRF2014 (Altamimi et al. 2016). Regarding the models used, we kept a consistency between the two software configurations. The same antenna eccentricities are used for both processings based on station site logs.

Fig. 1 Maps of the stations in Guadeloupe and Martinique Islands used in this study. The colors represent the three different networks: blue for IPGP, orange for IGN/SONEL, green for ORPHEON/RENAG. (**a**) Guadeloupe Archipelago. (**b**) Martinique Island

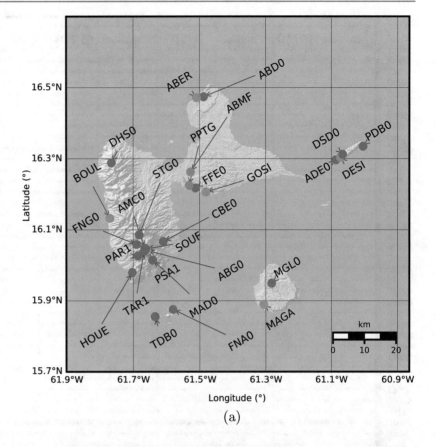

(a)

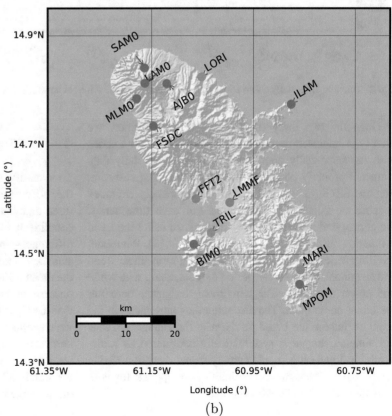

(b)

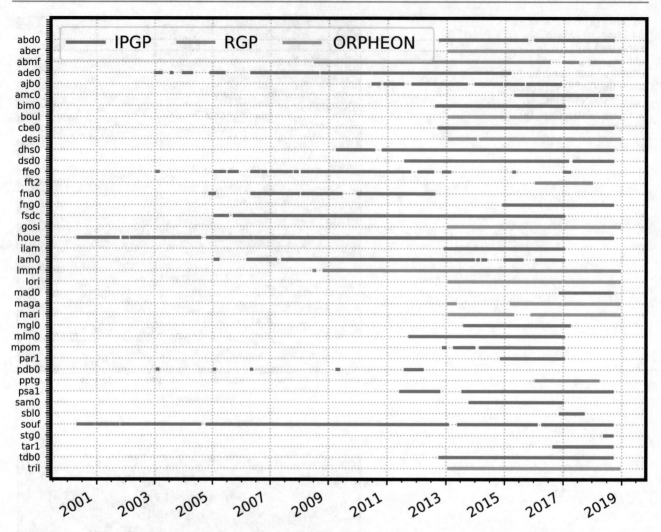

Fig. 2 Timeline of the daily observations processed from the three different networks available

Once the two daily coordinate sets are obtained, we select the intersection of both to get equivalent time series with the same daily coordinates. Indeed, some daily data were not properly computed by one or the other software. Stations STG0, SBL0 and PDB0 are completely excluded because of a lack of observations. For each time series, the corresponding velocities are determined using the trend estimation software HECTOR (Bos et al. 2013). We model the time series as combinations of a long term linear trend and an annual+semi-annual periodic signal, along with white and power-law noise. The term *trend* designates hereafter the linear component. The discontinuities introduced in the trend estimation are based on the material change site logs (an antenna change is systematically considered as a discontinuity) and on a supplementary visual detection (Sakic et al. 2019). The same discontinuities are applied for both equivalent EPOS and GINS solutions.

4 Coordinate and Velocity Differences

To quantify the impact of the processings, we compute the differences between the two coordinate sets for the three topocentric components and the planimetric distance (Euclidean norm of East and North components). These differences are represented as a histogram in Fig. 3 and the statistical indicators are given in Table 1. We remark that the mean difference for the three components is not centered on zero but is shifted by some millimeters. We also remark that the Up difference doesn't respect a Normal distribution, which reflects the fact that the Up component remains the hardest component to estimate with GNSS technique, mainly because this geometrical parameter is highly correlated to the clock offsets and tropospheric delay parameters. For the planimetric distance, the mean difference is 12.33 mm

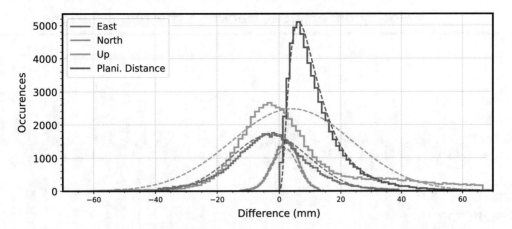

Fig. 3 Differences of the three topocentric coordinates and the planimetric distance for the common daily points of the two estimated solutions

Table 1 Mean, median and standard deviation in millimeters for the three topocentric coordinates and the planimetric distance for the common daily points of the two estimated solutions

Component	Mean	Std. dev.	Median
East	−2.89	12.69	−2.92
North	1.96	4.17	2.08
Up	4.20	19.70	0.15
Plani. distance	12.37	9.91	9.66

but respects a Gamma distribution. Regarding the standard deviations σ, σ_{North} is three times smaller than the σ_{East}, which can be explained by the better resolution of this component due to the general North-South trajectory of the satellites.

We compute also the differences between the two sets of estimated velocities obtained at the end of the processing, represented in Fig. 4. For the absolute differences (Fig. 4a), we observe notable differences at the mm/yr level for the planimetric components, and for some stations a difference over ±1 mm/yr for the vertical component. Regarding the relative differences (Fig. 4b), we note differences on the order of ±10% for the planimetric component, but for the vertical component, these differences can vary by more than a factor of two (stations over 100%). Four stations have a negative relative variation, which means that they have an opposite velocity trend. Since the stations are sorted from the most complete set of data to the sparsest one, we observe no significative relation with the length of the time series.

5 Vertical Velocity Results

The vertical velocity values obtained for the two processings are given in Table 2 and represented in Fig. 5 for the Guadeloupe Archipelago, Fig. 6 for the specific area of the Soufrière Volcano, and Fig. 7 for the Martinique Island.

We note general subsidence for both islands. This tendency is consistent between the two solutions. For the Guadeloupe Archipelago, the subsidence is visible for most of the stations, nevertheless, a more complex behavior around the Soufrière area can be remarked, which might be related to local volcanic deformation but also the frequent hardware change due to the harsh conditions in the area (humidity, corrosion and frequent thunderstorms). On the Marie-Galante Island, South-West of the main island, the MGL0 station has a positive trend. Moreover, the four stations TDB0, ABD0 GOSI and DSD0, have opposite trend depending on the solution. For GOSI and DSD0, the velocities estimated in both cases are very close to zero, which make this opposite trend not significant. For GOSI, the different estimated velocities are still close to each other (with overlapping formal sigmas) but a clear velocity tendency for this station is also non-significant. The case of TDB0 is remarkable, since the time series is long and almost complete but the difference between the two solutions is important (4.6 mm/yr). A detailed view of the raw time series and the estimated tendencies are shown in Fig. 8. We observe that the different scatter for both time series lead to a completely different estimation of the trend (using the strategy we selected). Station DHS0 has a lot of corrupted raw data, which lead to a reduced amount of usable observation and an overestimated vertical velocity of almost −2 cm/yr for one solution. A similar statement can be made for FFE0 station on the main island, where several gaps in the time series along with several hardware changes might explain the positive trend estimated.

The mean velocity rate measured for all stations on the archipelago, with the volcano area excluded, is −1.60 ± 1.54 mm/yr (1σ) using EPOS solution, and −2.17 ± 1.23 mm/yr (1σ) using GINS solution.

For the Martinique Island, the two solutions are also consistent and general subsidence is observed except for SAM0 station. The mean velocity rate measured is

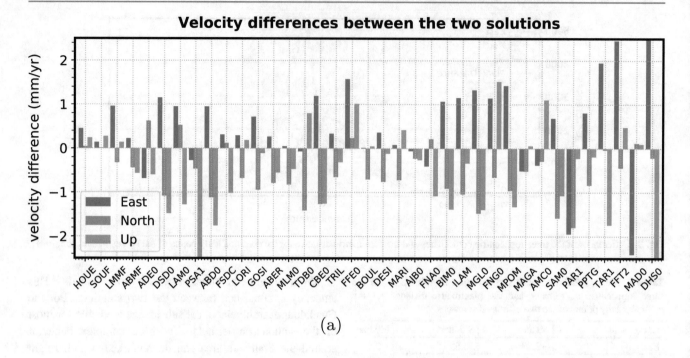

(a)

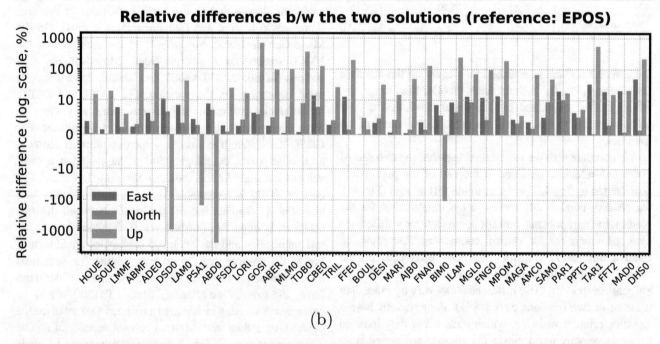

(b)

Fig. 4 Velocity difference for the three topocentric coordinates of the two estimated solutions. The stations are sorted from the most complete to the sparsest one. (**a**) Absolute velocity differences. (**b**) Relative velocity differences

Table 2 Vertical velocities estimated for the Guadeloupe and Martinique network, for both EPOS and GINS solutions

Station	Purpose	Lat.	Long.	Start	End	Total days	Used days	Ratio (%)	Discont. (#)	EPOS solution (mm/yr)		GINS solution (mm/yr)	
										V_{Up}	σV_{Up}	V_{Up}	σV_{Up}
ABD0	T	16.47	298.51	2012-10-11	2018-08-30	2149	1843	85.76	1	−0.81	0.99	0.38	0.70
ABER	S	16.47	298.49	2013-01-30	2018-11-25	2125	1792	84.33	0	−1.27	0.39	−0.67	0.51
ABMF	R	16.26	298.47	2008-07-15	2018-11-25	3785	2926	77.31	2	−0.84	0.40	−0.39	0.42
ADE0	T	16.30	298.91	2003-01-31	2015-02-24	4407	2504	56.82	1	−0.96	0.62	−0.37	0.58
AJB0	V	14.81	298.88	2010-06-29	2016-11-14	2330	1525	65.45	1	−0.89	1.33	−0.35	1.36
AMC0	V	16.05	298.33	2015-05-12	2018-08-30	1206	1086	90.05	0	−0.58	0.84	−1.52	0.97
BIM0	T	14.52	298.93	2012-08-29	2016-12-30	1584	1399	88.32	0	−0.17	0.59	1.12	0.71
BOUL	S	16.13	298.23	2013-01-31	2018-11-25	2124	1680	79.10	0	−3.40	0.93	−3.46	0.95
CBE0	T	16.07	298.39	2012-09-30	2018-08-30	2160	1740	80.56	0	−2.17	1.34	−0.93	1.45
DESI	S	16.30	298.93	2013-02-04	2018-11-25	2120	1657	78.16	0	−0.41	0.42	−0.36	0.51
DHS0	T	16.29	298.23	2009-04-16	2018-08-30	3423	336	9.82	2	−18.20	2.41	−8.29	3.80
DSD0	T	16.31	298.93	2011-08-11	2018-08-30	2576	2253	87.46	2	−0.95	0.83	0.04	0.73
FFE0	T	16.22	298.49	2003-01-26	2017-03-13	5160	1711	33.16	3	0.56	1.28	1.44	1.42
FFT2	R/T	14.60	298.94	2016-01-27	2017-12-12	685	642	93.72	0	−2.84	1.28	−3.32	1.56
FNA0	T	15.88	298.42	2004-11-26	2012-07-25	2798	1486	53.11	2	−2.13	1.11	−0.78	0.99
FNG0	T	16.06	298.31	2014-12-17	2018-08-30	1352	1173	86.76	1	0.09	1.91	−1.29	1.44
FSDC	T	14.73	298.85	2005-02-04	2016-12-30	4347	1831	42.12	3	−5.05	0.92	−3.70	1.12
GOSI	S	16.21	298.52	2013-01-30	2018-11-25	2125	1796	84.52	0	−0.16	0.37	0.03	0.44
HOUE	T	15.98	298.30	2000-05-14	2018-08-30	6682	4543	67.99	3	−0.98	0.26	−1.08	0.30
ILAM	T	14.77	299.12	2012-12-19	2016-12-30	1472	1308	88.86	0	−0.55	0.66	−0.09	0.66
LAM0	V	14.81	298.84	2005-02-02	2016-12-30	4349	2045	47.02	6	−4.60	1.02	−3.27	0.80
LMMF	R	14.59	299.00	2008-11-09	2018-11-25	3668	3247	88.52	2	−2.40	0.46	−2.41	0.41
LORI	S	14.82	298.95	2013-02-06	2018-11-25	2118	1797	84.84	0	−1.02	0.35	−1.17	0.39
MAD0	V	16.01	298.36	2016-11-22	2018-08-30	646	610	94.43	0	0.92	1.78	1.47	2.52
MAGA	S	15.89	298.69	2013-02-02	2018-11-25	2122	1099	51.79	0	−0.93	0.53	−1.14	0.50
MARI	T	14.47	299.14	2013-02-06	2018-11-25	2118	1568	74.03	0	−2.43	0.40	−2.73	0.45
MGL0	T	15.95	298.72	2013-08-20	2017-03-13	1301	1209	92.93	0	0.88	1.41	1.98	0.94
MLM0	V	14.78	298.82	2011-10-06	2016-12-30	1912	1776	92.89	0	−0.85	0.93	−0.44	1.24
MPOM	T	14.44	299.14	2012-11-28	2016-12-30	1493	1170	78.37	0	−1.91	0.76	−0.86	0.70
PAR1	V	16.03	298.31	2014-11-21	2016-12-30	770	748	97.14	4	0.65	4.85	1.01	3.81
PPTG	R/T	16.22	298.47	2016-01-27	2018-03-09	772	718	93.01	0	−2.79	1.12	−2.83	1.06
PSA1	V	16.04	298.33	2011-06-16	2018-08-30	2632	1970	74.85	9	−0.80	1.14	2.07	1.13
SAM0	V	14.84	298.84	2013-10-30	2016-12-21	1148	1055	91.90	0	1.28	1.17	2.31	1.22
SOUF	V	16.04	298.34	2000-05-13	2018-08-30	6683	4427	66.24	3	−1.23	0.31	−1.59	0.34
TAR1	V	16.04	298.33	2016-09-09	2018-08-30	720	659	91.53	10	−2.95	2.31	−1.80	2.97
TDB0	T	15.85	298.36	2012-10-18	2018-08-30	2142	1762	82.26	2	3.10	3.34	−1.51	2.64
TRIL	S	14.54	298.97	2013-02-05	2018-11-25	2119	1732	81.74	0	−1.42	0.36	−1.17	0.37

The *used days* column refers to the common number of days correctly processed in both solutions, and used for the velocity estimation. The main purpose of each station in mentioned in the second column: we distinguish stations located in the vicinity of the volcano domes, and installed for volcanic deformation monitoring (V), stations for tectonic deformation monitoring (T), stations for reference frame definition and orbit determination (R), and stations for RTK surveying (S)

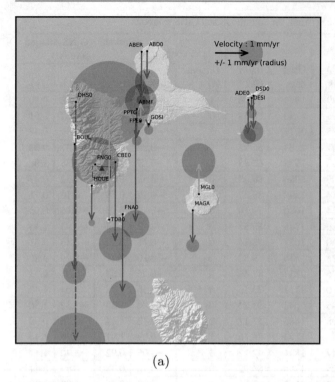

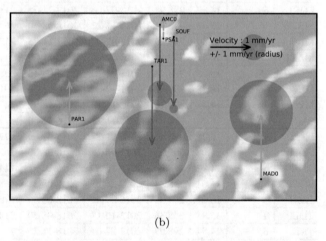

Fig. 5 Vertical velocity field obtained for the two solutions processed for stations located in Guadeloupe. A green arrow indicates an observed uplift and a red arrow an observed subsidence. The red rectangle indicates the area around the Soufrière Volcano, presented in detail in Fig. 6. The volcano summit is represented with a brown triangle. Dashed arrows have been shortened to stay in the frame. (**a**) EPOS solution. (**b**) GINS solution

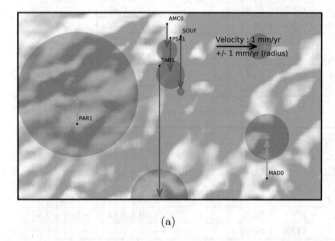

Fig. 6 Vertical velocity field obtained for the two solutions processed for stations located in the vicinity of the Soufrière Volcano (Basse-Terre, southern main Island of Guadeloupe). Dashed arrow has been shortened to stay in the frame. (**a**) EPOS solution. (**b**) GINS solution

-1.80 ± 1.36 mm/yr (1σ) using EPOS solution, and -1.68 ± 1.23 mm/yr (1σ) using GINS solution.

6 Comparison with Existing Solutions

To validate the consistency of our results, we compare the vertical velocities we determined with existing solutions for stations LMMF and ABMF (belonging to IGS network). We consider the velocities provided by SONEL for ULR6a (Santamaría-Gómez et al. 2017), NGL14 (Blewitt et al. 2018), JPL14 (Heflin et al. 2019) and ITRF14 (Altamimi et al. 2016) solutions. Values are given in Table 3. For LMMF, the estimated velocities are consistent with each other ($\sigma_V = 0.44$ mm), while for ABMF we remark more important differences between each solution ($\sigma_V = 0.79$ mm), just like we have also differences between the values of this work's solutions. Nevertheless, the negative trend remains significant.

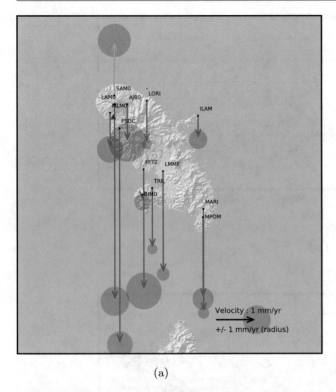

(a)

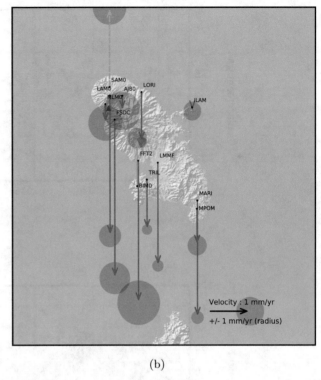

(b)

Fig. 7 Vertical velocity field obtained for the two solutions processed for stations located in Martinique. A green arrow indicates an observed uplift and a red arrow an observed subsidence. The volcano summit is represented with a brown triangle. Dashed arrow has been shortened to stay in the frame. (**a**) EPOS solution. (**b**) GINS solution

7 Discussion

Using two different solutions but based on the same geodetic products and homogeneous models, we obtain significant disparities in terms of coordinates difference repeatability, especially on the East and Up components with a standard deviation at the centimeter level. Regarding the estimated vertical velocities using the same set of points and the same discontinuities, the differences are also significant. This result tends to motivate investigation on velocity combination strategies between different processing centers, as suggested and tested by Ballu et al. (2019) for instance, where a joint least square modeling is developed to combine equivalent time series from different Analysis Centers. A combination based on a maximum likelihood estimation would be also an relevant method.

Nevertheless, for the studied area of the Guadeloupe and Martinique Islands, a negative velocity trend on the Up component is observed for most of the stations, which might suggest generalized subsidence of the area. This tendency is clear for the Martinique Island, but more complex trends for the Guadeloupe Archipelago can be observed, especially in the area around the Soufrière Volcano. This result can also be nuanced, since some stations have a positive trend, which might be due to local effects. A positive trend can also be

due to an important number of discontinuities over the time series period, like the stations PAR1 (furthermore located inside the volcano area) and FFE0 (outside the volcano area). On another hand, a large number of discontinuities for the same station seem to lead also to an overestimated negative trend, like for instance the station LAM0, with a velocity estimated over −3 mm/yr for six discontinuities referenced. This statement reveals the necessity to maintain networks with a minimum of hardware discontinuities, i.e. by reducing the number of antenna changes. MGL0 station, located on the Marie-Galante Island, presents a singular behavior. It is the only station clearly uplifting, with a quasi-complete time-series of 3.5 years, without any visible discontinuity, while the other station on Marie-Galante (MAGA) presents a subsiding trend. Unfortunately, since this station belong to the commercial ORPHEON network, we have only a few metadata that prevent us to explain clearly this behavior.

We corroborate the paleo-geodesy studies carried out in the region. The coral reef records in Martinique (Weil-Accardo et al. 2016) and Les Saintes (Leclerc and Feuillet 2019) indicate also a subsidence but with a smaller order of magnitude of few tenths of a millimeter per year, which can be explained by the difference in the observation time spans (only a few years for GNSS, *ca.* one century for the coral records). According to those studies, long term subsidence can have multiple origins: volcanic activity, crustal faulting,

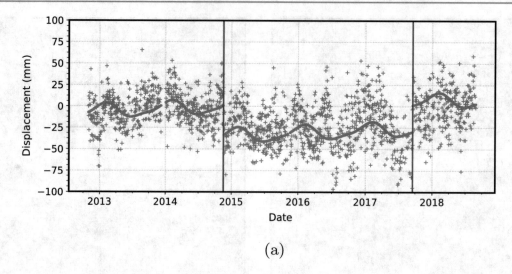

(a)

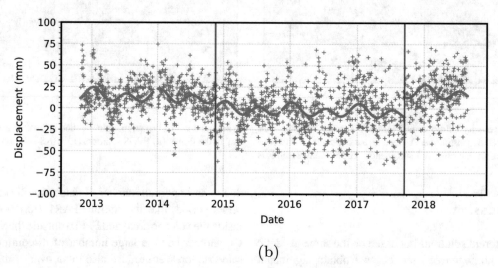

(b)

Fig. 8 Up component time series for the station TDB0, where an opposite velocity trend is visible. Blue dots represent the raw component determined by both software, and green dots the corresponding esti-mated trend. Red vertical bars represent the discontinuities considered. (**a**) EPOS. (**b**) GINS

Table 3 Comparison of vertical velocities for LMMF and ABMF of this study with existing solutions

mm/yr	LMMF	ABMF	Solution end
EPOS	−2.40 ± 0.46	−0.84 ± 0.40	Nov. 2018
GINS	−2.41 ± 0.41	−0.39 ± 0.42	Nov. 2018
ULR6a	−3.55 ± 0.48	N/A	Dec. 2014
NGL14	−2.70 ± 1.33	−2.37 ± 1.15	Apr. 2019
JPL14	−2.49 ± 0.54	−1.74 ± 0.88	Dec. 2019
ITRF14	−2.54 ± 0.21	−0.92 ± 0.37	Dec. 2014

subduction of the Tiburon ridge for the Saintes Islands (Leclerc and Feuillet 2019), and a potential deep interseismic loading for Martinique (Weil-Accardo et al. 2016).

We used only one software for velocity estimation since we mainly focussed on the GNSS processing itself, but some other velocity estimator software are available (e.g. Blewitt

et al. 2016; Santamaría-Gómez 2019). The impact of the velocity estimation software on solutions have been analysed for instance by Mazzotti et al. (2020).

8   Conclusion

This work brings a comparison of the coordinate time series obtained for the same dataset with two different software but using consistent parameters. New homogeneously calculated vertical velocity fields are made available for geophysical modeling, with unprecedented density for the two Martinique and Guadeloupe Islands. A general subsidence trend is observed for both islands.

Acknowledgements We thank the different scientific institutes for the continuously operating GNSS acquisitions, the instrumental maintenance and for providing publicly and freely the observations. Data of the IPGP network, operated on site by the Guadeloupe and Martinique seismic and volcanologic observatories, are available on volobsis.ipgp. fr. Data of the IGN and the SONEL networks are available on rgp. ign.fr and sonel.org respectively. ORPHEON data were provided to the authors for a scientific use in the framework of the agreement between the GEODATA company (orpheon-network.fr) and the RESIF-RENAG network maintained by the CNRS-INSU (renag.resif.fr). We acknowledge the CNES (Centre national d'études spatiales) for its geodetic processing software GINS. We thank A. Walpersdorf & an anonymous reviewer along with the editors J. Freymueller & L. Sanchez, for their constructive comments which improved the content of this paper.

References

Altamimi Z, Rebischung P, Métivier L, Collilieux X (2016) ITRF2014: a new release of the international terrestrial reference frame modeling nonlinear station motions. J Geophys Res Solid Earth 121(8):6109–6131. https://doi.org/10.1002/2016JB013098. http://doi.wiley.com/10.1002/2016JB013098

Ballu V, Bouin MN, Siméoni P, Crawford WC, Calmant S, Boré JM, Kanas T, Pelletier B (2011) Comparing the role of absolute sea-level rise and vertical tectonic motions in coastal flooding, Torres Islands (Vanuatu). Proc Nat Acad Sci 108(32):13019–13022. https://doi.org/10.1073/pnas.1102842108. http://www.pnas.org/lookup/doi/10.1073/pnas.1102842108

Ballu V, Gravelle M, Wöppelmann G, de Viron O, Rebischung P, Becker M, Sakic P (2019) Vertical land motion in the Southwest and Central Pacific from available GNSS solutions and implications for relative sea levels. Geophys J Int 218(3):1537–1551. https://doi.org/10.1093/gji/ggz247. https://academic.oup.com/gji/article/218/3/1537/5499028

Blewitt G, Kreemer C, Hammond WC, Gazeaux J (2016) MIDAS robust trend estimator for accurate GPS station velocities without step detection. J Geophys Res Solid Earth 121(3):2054–2068. https://doi.org/10.1002/2015JB012552. http://doi.wiley.com/10.1002/2015JB012552

Blewitt G, Hammond W, Kreemer C (2018) Harnessing the GPS data explosion for interdisciplinary science. Eos 99. https://doi.org/10.1029/2018EO104623. https://eos.org/project-updates/harnessing-the-gps-data-explosion-for-interdisciplinary-science

Bos MS, Fernandes RMS, Williams SDP, Bastos L (2013) Fast error analysis of continuous GNSS observations with missing data. J Geod 87(4):351–360. https://doi.org/10.1007/s00190-012-0605-0. http://link.springer.com/10.1007/s00190-012-0605-0

Gendt G, Dick G, Söhne W (1998) GFZ analysis center of IGS–Annual Report. IGS 1996 Annual Report, pp 169–181

Heflin M, Moore A, Murphy D, Desai S, Bertiger W, Haines B, Kuang D, Sibthorpe A, Sibois A, Ries P, Hemberger D, Dietrich A (2019) Introduction to JPL's GPS time series GPS time series. https://sideshow.jpl.nasa.gov/post/tables/GNSS_Time_Series.pdf

Leclerc F, Feuillet N (2019) Quaternary coral reef complexes as powerful markers of longterm subsidence related to deep processes at subduction zones: Insights from Les Saintes (Guadeloupe, French West Indies). Geosphere 15(4):983–1007. https://doi.org/10.1130/GES02069.1

Léticée JL, Cornée JJ, Münch P, Fietzke J, Philippon M, Lebrun JF, De Min L, Randrianasolo A (2019) Decreasing uplift rates and

Pleistocene marine terraces settlement in the central lesser Antilles fore-arc (La Désirade Island, 16 deg. N). Quat Int 508(March):43–59. https://doi.org/10.1016/j.quaint.2018.10.030. https://doi.org/10.1016/j.quaint.2018.10.030

Loyer S, Perosanz F, Mercier F, Capdeville H, Marty JC (2012) Zero-difference GPS ambiguity resolution at CNES–CLS IGS Analysis Center. J Geod 86(11):991–1003. https://doi.org/10.1007/s00190-012-0559-2. http://link.springer.com/10.1007/s00190-012-0559-2

Männel B, Brandt A, Bradke M, Sakic P, Brack A, Nischan T (2020) Status of IGS reprocessing activities at GFZ. In: International Association of Geodesy Symposia. Springer, Berlin. https://doi.org/10.1007/1345_2020_98

Marty JC, Loyer S, Perosanz F, Mercier F, Bracher G, Legresy B, Portier L, Capdeville H, Fund F, Lemoine JM, Biancale R (2011) GINS: the CNES/GRGS GNSS scientific software. In: 3rd international colloquium scientific and fundamental aspects of the Galileo programme, ESA proceedings WPP326, vol 31, pp 8–10

Mazzotti S, Déprez A, Henrion E, Masson C, Masson F, Menut JL, Métois M, Nocquet JM, Rolland L, Sakic P, Socquet A, Santamaria-Gomez A, Valty P, Vergnolle M, Vernant P (2020) Comparative analysis of synthetic GNSS time series - bias and precision of velocity estimations. Research report, RESIF. https://hal.archives-ouvertes.fr/hal-02460380

Mouslopoulou V, Oncken O, Hainzl S, Nicol A (2016) Uplift rate transients at subduction margins due to earthquake clustering. Tectonics 35(10):2370–2384. https://doi.org/10.1002/2016TC004248. http://doi.wiley.com/10.1002/2016TC004248

Rabin M, Sue C, Walpersdorf A, Sakic P, Albaric J, Fores B (2018) Present-day deformations of the Jura Arc inferred by GPS surveying and earthquake focal mechanisms. Tectonics 37(10):3782–3804. https://doi.org/10.1029/2018TC005047. http://doi.wiley.com/10.1029/2018TC005047

Sakic P, Mansur G, Kitpracha C, Ballu V (2019) The GeodeZYX toolbox: a versatile Python 3 toolbox for geodetic-oriented purposes. https://doi.org/10.5880/GFZ.1.1.2019.002. http://dataservices.gfz-potsdam.de/panmetaworks/showshort.php?id=escidoc:4754924

Santamaría-Gómez A (2019) SARI: interactive GNSS position time series analysis software. GPS Solutions 23(2):1–6. https://doi.org/10.1007/s10291-019-0846-y. http://dx.doi.org/10.1007/s10291-019-0846-y

Santamaría-Gómez A, Gravelle M, Dangendorf S, Marcos M, Spada G, Wöppelmann G (2017) Uncertainty of the 20th century sea-level rise due to vertical land motion errors. Earth Planet Sci Lett 473:24–32. https://doi.org/10.1016/j.epsl.2017.05.038. http://dx.doi.org/10.1016/j.epsl.2017.05.038

Symithe S, Calais E, de Chabalier JB, Robertson R, Higgins M (2015) Current block motions and strain accumulation on active faults in the Caribbean. J Geophys Res Solid Earth 120(5):3748–3774. https://doi.org/10.1002/2014JB011779. http://doi.wiley.com/10.1002/2014JB011779

Uhlemann M, Gendt G, Ramatschi M, Deng Z (2015) GFZ global multi-GNSS network and data processing results. In: International association of geodesy symposia, vol 12. Springer, New York, pp 673–679. https://doi.org/10.1007/1345_2015_120

Walpersdorf A, Pinget L, Vernant P, Sue C, Deprez A (2018) Does long-term GPS in the Western Alps finally confirm earthquake mechanisms? Tectonics 37(10):3721–3737. https://doi.org/10.1029/2018TC005054. http://doi.wiley.com/10.1029/2018TC005054

Weil-Accardo J, Feuillet N, Jacques E, Deschamps P, Beauducel F, Cabioch G, Tapponnier P, Saurel JM, Galetzka J (2016) Two hundred thirty years of relative sea level changes due to climate and megathrust tectonics recorded in coral microatolls of Martinique (French West Indies). J Geophys Res Solid Earth 121(4):2873–2903. https://doi.org/10.1002/2015JB012406. http://doi.wiley.com/10.1002/2015JB012406

Wöppelmann G, Testut L, Créach R (2011) La montée du niveau des océans par marégraphie et géodésie spatiale: contributions françaises à une problématique mondiale. Annales Hydrographiques 6ème Série 8(777):11–14

Time Variations of the Vertical Component in Some of Japanese GEONET GNSS Sites

S. Shimada, M. Aichi, T. Harada, and T. Tokunaga

Abstract

We analyze the vertical component of GEONET GNSS measurements in Central Japan and clarify in some of the sites the origin of large annual time variations, as well as the secular variations. Many of these vertical movements may be attributable to the use of groundwater for agriculture, for snow melting, industrial, and hospital usages, etc. and the pumping up of the groundwater mining for refining natural gas and iodine at the production area of natural gas dissolved in water. For this reason, highly accurate monitoring of vertical variations by GNSS observations can provide new observation methods for understanding of not only geodynamics but also hydrology through monitoring groundwater fluctuation, and natural gas and oil resource development through monitoring ground movements caused by mining.

Keywords

GEONET · GNSS · Groundwater · Vertical component

1 Introduction

In GNSS observation, ground deformation due to human activity is observed in addition to crustal deformation due to geodynamics such as plate motion, seismic and volcanic activities. In particular, the vertical component is sensitive to the variations due to human activities such as groundwater usage and natural resource mining in the shallow area near the observation site. Regarding these human activities, there are many research subjects in hydrology and resource engineering, and GNSS measurements can contribute to these research fields.

In fact, recently even the vertical component of GNSS permanent station position time series has reached millimeter accuracy, and significant ground movements caused by groundwater usage have been detected at some GEONET observation sites, the CORS sites in Japan (Miyazaki et al. 1998). Surface deformation due to atmospheric, oceanic and hydrological seasonal loading may affect the GNSS solutions on vertical component. For example, continental hydrology loading deformations can be computed using the Global Land Data Assimilation System GLDAS (2020) model (Rodell et al. 2004). The EOST Loading Service (2015) provides the map of the annual vertical deformation where it can be noticed that hydrological loading effect is small over Japan. Recent major sophisticated GNSS analyzing software packages, such as GAMIT/GLOBK software (Herring et al. 2018), have already involved enough accurate models of seasonal atmospheric and ocean tidal loading which has the largest power among those loadings over Japan (Dong et al. 2002). Among those models, for instance in the case of GAMIT/GLOBK software, NAO99b by Matsumoto et al. (2000) includes Sa, Ssa, Mm, and Mf long period tides and is accurate especially near and around Japanese islands because tide gauge observations in the region are assimilated in the model. We use GOTIC2 program (Matsumoto et al. 2001) to calculate amplitude and phase of solid earth and ocean load tides for all Japanese and most of global IGS sites.

S. Shimada (✉) · M. Aichi · T. Tokunaga
Graduate School of Frontier Sciences, the University of Tokyo, Kashiwa-Shi, Chiba Prefecture, Japan
e-mail: shimada@envsys.k.u-tokyo.ac.jp

T. Harada
JX Nippon Oil & Gas Exploration, Tokyo, Japan

Heki (2001), Heki (2004) and Munekane (2010) study on various factors that influence the annual variations of GEONET sites. Munekane (2010) quantitively evaluates subsidence caused by groundwater pumping for agricultural usage in Tsukuba GSI Campus, but carefully checking of those papers, none of them evaluates quantitively subsidence caused by snow loading from the viewpoint of hydrology. Moreover, Geospatial Information Authority of Japan (GSI) in Heki (2001) adopted a specific processing that lead to erroneous vertical annual displacements. Although Tsukuba GSI Campus is showing significant annual variations by the nearby groundwater pumping for agricultural usage (Tobita et al. 2004; Munekane et al. 2004, 2010), the coordinates of Tsukuba IGS/GEONET site were fixed to ITRF96 (Boucher et al. 1998) and the coordinate solution of each GEONET network site then derived by applying the backbone/cluster approach (Miyazaki et al. 1997; Hatanaka et al. 2003; Nakagawa et al. 2009). Thus, Heki (2001) shows time series of vertical coordinate solutions at Naruko and Shizukuishi GEONET sites, but most of the annual variations at those sites are fake motions brought by the erroneous analysis strategy of F1 solution. However the more accurate horizontal coordinates of GEONET sites bring realistic estimations of snow depth distribution in Japanese Island.

In this paper, we report secular and annual variations of the vertical components due to the pumping of groundwater for agricultural and snow melting usages and the extraction of water-soluble natural gas, observed at the GEONET sites in central Japan. As mentioned above, the annual vertical deformations predicted by classical hydrological model over Japan are equal or less than 3 mm (EOST Loading Service 2015), far smaller compared with those amplitude due to the pumping of groundwater for agricultural, snow melting, and water-soluble natural gas we report. The displacement due to atmospheric loading is evaluated in the GAMIT/GLOBK program applying the ECMWF global model (Tregoning and Herring 2006).

2   Data

We analyze more than 100 GEONET sites in Kanto area and more than 50 sites in Niigata region both in Central Japan, using GAMIT/GLOBK program with ITRF2014 reference frame, using IGS final and repro2 precise ephemerides with around 30 IGS fiducial sites in and around East Asia. The reference sites are the following: ARTU, BJFS, CHAN, DAEJ(TEAJ), GUAM, GUUG, HYDE, IRTJ(IRKT), KIT3, KOKB, KWJ1, LHAZ(LHAS), MAG0, MCIL, MKEA, NRIL, NVSK, PETP, PIMO, POL2, SHAO, SUWN, TIXI, TNML, TSKB, ULAB, URUM, USUD, WHIT, WUHN, YAKT, YELL. The reference sites are chosen from the sites whose coordinates and velocities are accurately determined

in ITRF2014 reference frame (Altamimi et al. 2016). Some of them especially Japanese domestic sites are excluded after 2011 Eastern Japan Great Earthquake, because of large postseismic movements. The area where the reference points locate is almost the same as in Fig. 80.2 of the former paper (Shimada 2012).

We estimate simultaneously site coordinates, independent ambiguities, hourly ZTD, and four-hourly atmospheric gradient at every sites Japanese and IGS sites mentioned above for the daily GAMIT analysis. Then we estimate every site coordinates and daily ZTD in the daily GLOBK analysis, applying a 30day-window Kalman filtering to remove short period noise from daily coordinate solutions. The uncertainties of vertical component are usually significantly large compared with horizontal component, but after applying a 30day-window Kalman filtering the one sigma uncertainties of vertical component are less than 3 mm for all Japanese sites.

Because groundwater usages induce mainly vertical ground movements, we evaluate the time series of vertical components in three area in Central Japan. Those areas are the followings (Fig. 1): (a) central Kanto plain for the agricultural usages of groundwater, (b) Kujukuri area in Boso Peninsula in southeast Kanto region for groundwater mining for natural gas and iodine, and (c) the Niigata plain in North part of Central Japan for snow melting usage of groundwater.

3   Result and Discussion

3.1   Agricultural Usage of Groundwater

It can be observed that the annual vertical variations clearly depend on the site location. Figure 2 shows the time series of the vertical component of some of the GEONET sites in the Kanto region as well as the location of those sites. In central Kanto plain, we detect vertical movements that may be due to agricultural usage of ground water at Tsukuba and Sanwa GEONET sites. At Tsukuba GEONET site (VLBI colocation site) in the campus of GSI (Geospatial Information Authority of Japan), annual variations are observed likely caused by the groundwater pumping of the rise paddy field around the site, in harmony with VLBI and groundwater level observations (Munekane et al. 2004). Figure 3 shows the annual variation of Tsukuba GEONET site (92110 site) during 1997 and 2009, as well as the VLBI vertical component analyzed by BKG (Federal Agency for Cartography and Geodesy) in Germany (IVS Combination Center 2017), and annual variation of nearby groundwater level in 1997 (Tobita et al. 2004). For those three observations, the phases of annual motion coincide because groundwater pumping is used to irrigate the paddy fields surrounding

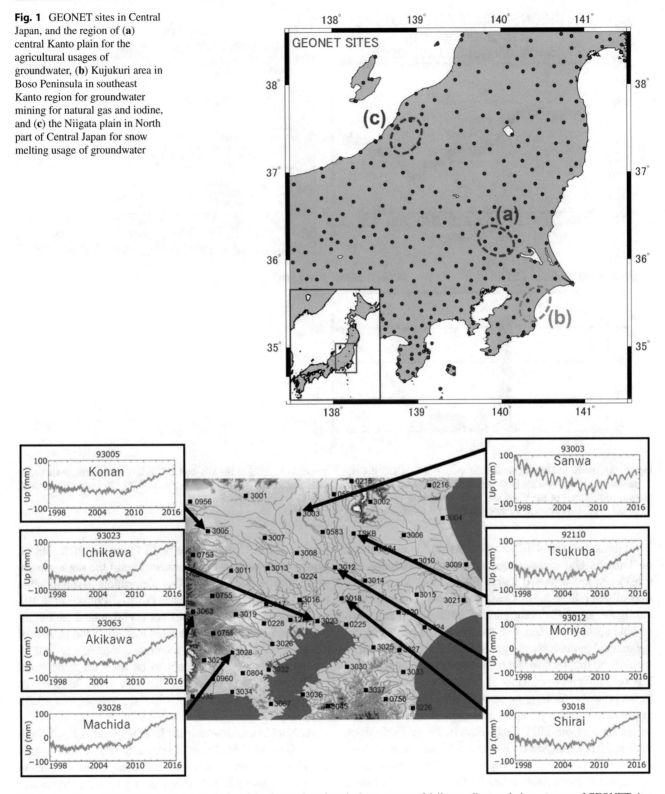

Fig. 1 GEONET sites in Central Japan, and the region of (**a**) central Kanto plain for the agricultural usages of groundwater, (**b**) Kujukuri area in Boso Peninsula in southeast Kanto region for groundwater mining for natural gas and iodine, and (**c**) the Niigata plain in North part of Central Japan for snow melting usage of groundwater

Fig. 2 Location of GEONET sites (central map) and the time series of vertical component of daily coordinate solutions at some of GEONET sites in Kanto area

the GSI campus mainly in spring and summer seasons (Munekane et al. 2010). Comparing the amplitudes of annual variations of GNSS and VLBI measurements, VLBI shows smaller amplitude because the foundation pile of VLBI is driven in the midst of aquifer and deeper than that of GNSS.

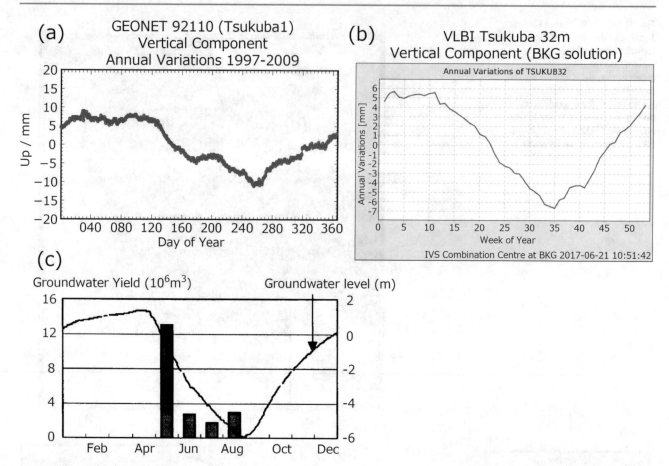

Fig. 3 (**a**) Annual variation of vertical component at the GEONET site in the campus of GSI during 1997 and 2009. (**b**) Annual variation of vertical components of the VLBI site in the campus of GSI obtained by BKG (IVS Combination Center 2017). (**c**) Annual variation of groundwater level of the W4 well at GSI in 1997 and the groundwater yield per month for irrigation in Tsukuba City in 1996 (Tobita et al. 2004)

At the Sanwa GEONET site (93003 site) 35 km northwest of Tsukuba GSI campus significant annual variation and subsidence are observed before 2011 (Fig. 4). Coseismic step of 2011 East Japan Great Earthquake is removed in the figure. After the occurrence of 2011 earthquake, post-seismic uplift is overlaid with the subsidence caused by the agricultural groundwater usage. In Fig. 2 the slope for Sanwa seems to be smaller than that of the other sites after 2011. But it is very difficult to conclude if the smaller slope is caused by the groundwater usage or the deviation of the post-seismic movements. Besides the uplift movements occur prior to the origin time of the 2011 Great Eastern Japan Earthquake. Such uplift is widely seen in Kanto and Niigata areas we have analyzed (see Figs. 2 and 8), not related with the existence of the annual deformations caused by the groundwater usage. The author guesses the movements are the precursory uplift of the 2011 event, but such movement is not significant in the horizontal components. We have not yet surveyed for the sites in the Tohoku region nearest to the epicenter of the earthquake, and also we do not have the concrete image of the mechanism of the precursor, thus the movement

prior to the 2011 earthquake is the future research subject. The leveling survey (Fig. 5) shows almost the same amount of secular subsidence at Nogi Uruushima leveling point 6.2 km northwest of Sanwa site (Tochigi Prefecture 2010), but once-per-year survey does not reveal annual variations. Sanwa and Nogi Uruushima sites located in the middle of the area are well known to undergo subsidence caused by groundwater pumping thought to be used for agricultural groundwater usage (Seki and Koyama 1998; Ministry of the Environment Water and Air Environment Bureau 2020).

At Nogi Uruushima near the leveling point in Fig. 5 there is the Ground Subsidence Observatory that reveals annual variations of groundwater level and compaction by ground subsidence gage. We calculate the long period groundwater level and ground subsidence at gage observations (Tokunaga 2015; Aichi 2008) which reveal annual variations of ground subsidence as well as secular subsidence (Fig. 6). However, a leveling survey realized once per year is totally insufficient to confirm the model as shown in the figure. Contributions of GNSS measurements provide temporal information

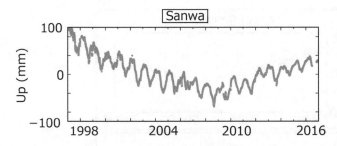

Fig. 4 Time series of vertical component of daily coordinate solution at Sanwa GEONET site during 1997 and 2016

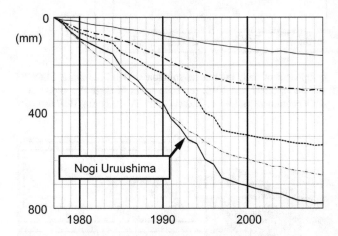

Fig. 5 Time variation of height by leveling survey at Sano Funazukawa, Ashikaga Kenmachi, Oyama Otome, Fujioka Shimomiya, and Nogi Uruushima during 1977 and 2009. Nogi Uruushima indicates the largest subsidence (Tochigi Prefecture 2010)

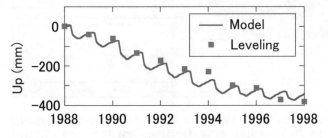

Fig. 6 Time variations of ground subsidence calculated value by the coupled hydrogeological/deformation model applying long period groundwater level and ground subsidence gage observations, and the height by leveling survey both at Nogi Uruushima (Tokunaga 2015)

to validate coupled hydrogeological/deformation simulation of ground subsidence instead of once per year leveling survey.

3.2 Groundwater Mining for Natural Gas and Iodine

A large amount of subsidence is observed at two GEONET sites in harmony with the level survey (Chiba Prefecture

Environmental Life Department 2016) and the InSAR observations (Fig. 7). The subsidence in the area is widely considered to be caused by the groundwater mining of soluble natural gas and iodine, and the land subsidence is suppressed to some extent by the regulation of the groundwater mining by local government (Nojo et al. 2015; Chiba Prefecture Environmental Life Department 2016; Ministry of the Environment Water and Air Environment Bureau 2020). In Fig. 7 two GEONET sites, Oami Shirasato (93027) and Chosei (93033), secular subsidence is observed from the beginning of the observation in 1997 to 2010 before the occurrence of 2011 East Japan Great Earthquake. After the earthquake, post-seismic uplift is overlaid with the subsidence caused by the groundwater mining. Coseismic step of 2011 earthquake is removed in the figure. In the Kujukuri area, southeast Kanto region, many authors investigated the subsidence applying various methodologies; leveling survey, GNSS, InSAR, GIS, and comparison of those technique (Nojo et al. 2015; Deguchi and Rokugawa 2010; Chen et al. 2015).

3.3 Groundwater for Snow Melting

The recent work based on InSAR (Morishita et al. 2020) measurements clarifies that large annual movement in the Niigata area could be caused by the groundwater usage for snow melting. Inland Niigata prefecture is famous for suffering heavy snow in winter season. In the area groundwater is used for snow melting on roads, and on roofs and gardens in private houses, and the groundwater usage for snow melting causes significant subsidence (Kayane 1980; Morishita et al. 2020). At many GEONET sites in Niigata prefecture, annual vertical subsidence is observed where the subsidence is widespread inland the prefecture (right three sites in Fig. 8). The result is confirmed by a levelling survey (Niigata Prefecture Life and Environment Department 2019), although the sites near coastal area where there are less snowfalls do not show significant annual movements (left three sites in Fig. 8). The peaks of the subsidence are in winter season in consistency with the snow-melting usage of groundwater, different from the summer season in the Kanto region where groundwater is pumping and used for agriculture.

For the period after 2011 East Japan Great Earthquake, the post-seismic deformations are widely significant in eastern Japan including the Niigata prefecture. The deformations are described with exponential and/or logarithmic functions with the origin time of earthquake occurrence (Tobita 2016). Niigata prefecture locates near the border of the post-seismic uplift/subsidence of the 2011 earthquake. Thus although the sites in Fig. 8 show slight uplift deformations (Geospatial Information Authority of Japan 2019), the linear trends are

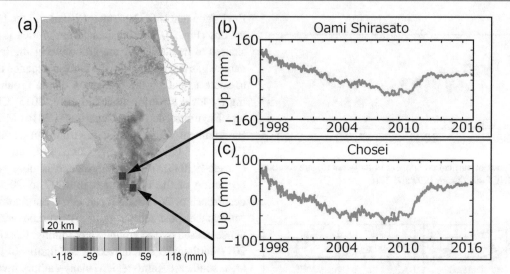

Fig. 7 (**a**) Subsidence rate at Kujukuri area derived from the InSAR observation (GSI 2020) and the location of Oami Shirasato and Chosei GEONET sites. The map was obtained by ALOS mission and time interval of 2008/02/11 and 2011/02/19. (**b**) Time series of vertical component of daily coordinate solution at Oami Shirasato GEONET site during 1997 and 2016. (**c**) Time series of vertical component of daily coordinate solution at Chosei GEONET site during 1997 and 2016

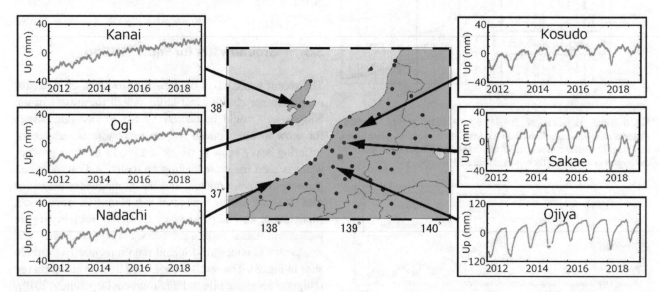

Fig. 8 Location of GEONET sites in Niigata Prefecture (central map) and the time series of vertical component of daily coordinate solutions at GEONET sites, Kanai (960565), Ogi (950235), Nadachi (950243) (from upper to lower in left figures), Kosudo (960571), Sakae (970810), and Ojiya (95240) (from upper to lower in right figures) during 2012 and 2019. The square on the central map shows the location of Nagaoka City

mostly caused by the post-seismic deformation of the 2011 earthquake and not related to anthropogenic effects.

Niigata Prefecture Life and Environment Department (2019) shows that in the year of heavy snowing groundwater level observations become lower than usual year, and also subsidence gauges show rapid subsidence. Recently 2018, for instance, is well known as the heavy snowing year especially during February 5th and 8th (Niigata Prefecture 2020). In Fig. 8, a significant large amount of subsidence is noticed in 2018 at the Sakae site (970810). If it is caused by groundwater pumping around the site, the deformation that follows could be partially attributed to non-reversible deformations as well as the elastic reversible deformations, but at the Ojiya site (950240) the non-reversible deformations seems to be not significant. Both kind of deformations could depend on the geology of sites. For the case of the Ojiya site, because of rich groundwater the deformations are mostly within the sandy aquifer. But for the case of Sakae site, for years that present larger snowfalls than usual years and significant pumping of groundwater, the deformations may also reach the clay layer which may lead to significant non-reversible subsidence.

GNSS measurements must also contribute to sustainable groundwater usage for snow melting by better understanding the nature of annual and secular subsidence caused by the snow melting usage of groundwater.

Civil engineering requires monitoring of river and coastline bank heights to prevent flood and high tide damages. Monitoring of annual and secular subsidence using GNSS measurements must provide accurate and temporally dense information of subsidence compared with traditional leveling survey, and will become a standard ground subsidence measurement alternative to leveling survey.

4　Conclusion

Conventionally, leveling survey has generally been used for geodetic monitoring of land subsidence. However, because leveling survey are realized only once-per-year in the most frequent cases, data of groundwater level and ground subsidence gage are also used for establishing hydrogeological models coupled with crustal deformations in order to monitor ground subsidence caused by groundwater usage. Continuous GNSS monitoring are more interesting in terms of frequency and immediacy compared with leveling survey. They are also more interesting in terms of cost and they directly measure the ground subsidence compared with ground subsidence gage. Finally, they are more interesting in terms of installation/setup compared with other types of geodetic measurements and measurements that require a dedicated well. This paper demonstrates that mm level accuracy daily monitoring of GNSS measurements is useful for studying ground subsidence, and for establishing coupled hydrogeological/deformation simulation of ground subsidence.

Acknowledgements On the annual and semi-annual tidal components, Dr. Tamura of NAO Mizusawa let us know useful information. The Geospatial Information Authority of Japan provides GEONET RINEX files and PCV models. This research is supported JSPS KAKENHI Grant Number 18K03774 and the research program "Development of Underground Water Utilization System in Case of Disaster" of SIP (Cross-ministerial Strategic Innovation Promotion Program) research fund of Japanese Government. We use GMT program by Wessel and Smith (1998) to draw figures.

References

Aichi M (2008) Coupled groundwater flow/deformation modelling for predicting land subsidence. In: Takizawa S (ed) Groundwater management in Asian cities: technology and policy for sustainability. Springer-Verlag, Tokyo, pp 105–124. (Chap. 5)

Altamimi Z, Rebischung P, Métivier L, Collilieux X (2016) ITRF2014: a new release of the International Terrestrial Reference Frame modeling nonlinear station motions. J Geophys Res 121:6109–6131. https://doi.org/10.1002/2016JB013098

Boucher C, Altamimi Z, Sillard P (1998) Results and analysis of the ITRF96. International Earth Rotation Service (IERS) Technical Note #24

Chen HL, Ito Y, Sawamukai M, Su T, Tokunaga T (2015) Has land subsidence changed the flood hazard potential? A case example from the Kujukuri Plain, Chiba Prefecture, Japan. Proc IAHS 372:157–161. https://doi.org/10.5194/piahs-372-157-2015

Chiba Prefecture Environmental Life Department (2016) Current status of land subsidence in Chiba prefecture. (in Japanese) https://www.pref.chiba.lg.jp/suiho/jibanchinka/torikumi/documents/h27gaikyou.pdf. Accessed 02 May 2020

Deguchi T, Rokugawa S (2010) Monitoring of land subsidence around Kanto Plains of Japan by DInSAR and time series analysis. EUSAR 2010. VDE Verlag GmbH. European Conference on Synthetic Aperture Radar (EUSAR), 2010-06-07 - 2010-06-10, Aachen, Deutschland. ISBN 978-3-8007-3272-2

Dong D, Fang P, Bock Y, Cheng MK, Miyazaki S (2002) Anatomy of apparent seasonal variations from GPS-derived site position time series. J Geophys Res 107:2075–2081. https://doi.org/10.1029/2001JB000573

EOST Loading Service (2015). http://loading.u-strasbg.fr/displ_maps.php. Assessed 14 May 2021

Geospatial Information Authority of Japan (2019) Eight years after from 2011 Tohoku Earthquake. https://www.gsi.go.jp/kanshi/h23touhoku_8years.html. Accessed 29 June 2019

Geospatial Information Authority of Japan (2020). https://maps.gsi.go.jp/#10/35.660644/140.210266/&base=english&ls=english%7Calos_subsidence_kujyukuri_20080211-20110219. Accessed 02 May 2020

Global Geophysical Fluid Center (2020). https://ldas.gsfc.nasa.gov/gldas. Assessed 14 May 2021

Hatanaka Y, Iizuka T, Sawada M, Yamagiwa A, Kikuta Y, Johnson JM, Rocken C (2003) Improvement of the analysis strategy of GEONET. Bull Geogr Surv Inst 49:11–34. https://www.gsi.go.jp/common/000001182.pdf. Accessed 06 Sep 2020

Heki K (2001) Seasonal modulation of interseismic strain buildup in Northeastern Japan driven by snow loads. Science 293:89–92. https://doi.org/10.1126/science.1061056

Heki K (2004) Dense GPS array as a new sensor of seasonal changes of surface loads. In: Sparks RSK, Hawkesworth CJ (eds) The state of the planet: frontier and challenges in geophysics, Geophys. Monogr., vol 150. American Geophysical Union, Washington, pp 177–196. https://doi.org/10.1029/150GM15

Herring TA, King RW, Floyd MA, McClusky SC (2018) Documentation for the GAMIT/GLOBK GPS analysis software. Department of Earth, Atmospheric, and Planetary Science, Massachusetts Institute of Technology

IVS Combination Center (2017) Annual variations of TSUKUB32. https://ccivs.bkg.bund.de/index.php?uri=quarterly/annual. Accessed 21 June 2017

Kayane I (1980) Groundwater use for snow melting on the road. GeoJournal 4:173–181. https://doi.org/10.1007/BF00705524

Matsumoto K, Takanezawa T, Ooe M (2000) Ocean tide models developed by assimilating TOPEX/POSEIDON altimeter data into hydrodynamical model: a global model and a regional model around Japan. J Oceanogr 56:567–581. https://doi.org/10.1023/A:1011157212596

Matsumoto K, Sato T, Takanezawa T, Ooe M (2001) GOTIC2: a program for computation of oceanic tidal loading effect. J Geod Soc Jpn 47:243–248. https://doi.org/10.11366/sokuchi1954.47.243

Ministry of the Environment Water and Air Environment Bureau (2020) Overview of subsidence areas nationwide 2018FY. http://www.env.go.jp/water/jiban/gaikyo/gaikyo30.pdf. Accessed 06 Sep 2020

Miyazaki S, Saito T, Sasaki M, Hatanaka Y, Iimura Y (1997) Expansion of GSI's nationwide GPS array. Bull Geogr Surv Inst 43:23–34

Miyazaki S, Hatanaka Y, Sagiya T, Tada T (1998) The nationwide GPS array as an Earth observation system. Bull Geogr Surv Inst 44:11–22

Morishita Y, Lazecky M, Wright TJ, Weiss JR, Elliott JR, Hooper A (2020) LiCSBAS: an open-source InSAR time series analysis package integrated with the LiCSAR automated Sentinel-1 InSAR processor. Remote Sens 12:424–452. https://doi.org/10.3390/rs12030424

Munekane H (2010) On improving precision of GPS-derived height time series at GEONET stations. Bull Geospatial Inf Auth Jpn 58:39–46. https://www.gsi.go.jp/common/000058372.pdf. Accessed 06 Sep 2020

Munekane H, Tobita M, Takashima K (2004) Groundwater-induced vertical movements observed in Tsukuba, Japan. Geophys Res Lett 31:L12608. https://doi.org/10.1029/2004GL020158

Munekane H, Kuroishi Y, Hatanaka Y, Takashima K, Ishimoto M (2010) Groundwater-induced vertical movements in Tsukuba revisited: installation of a new GPS station. Earth Planets Space 62:711–715. https://doi.org/10.5047/eps.2010.08.001

Nakagawa H, Toyofuku T, Kotani K, Miyahara B, Iwashita C, Kawamoto S, Hatanaka Y, Munekane H, Ishimoto M, Yutsudo T, Ishikura N, Sugawara Y (2009) Development and validation of GEONET new analysis strategy (version 4). Bull Geogr Surv Inst 118:39–46. (in Japanese) https://www.gsi.go.jp/common/000054716.pdf. Accessed 06 Sep 2020

Niigata Prefecture (2020). http://micos-sc.jwa.or.jp/niigatayuki/contents_4.html. Accessed 12 May 2021

Niigata Prefecture Life and Environment Department (2019) Subsidence in Nagaoka area. (in Japanese) http://npdas.pref.niigata.lg.jp/kankyotaisaku/5c87637485a5e.pdf. Accessed 02 May 2020

Nojo M, Waki F, Akaishi M, Muramoto Y (2015) The investigation of a new monitoring system using leveling and GPS. Proc IAHS 372:539–542. https://doi.org/10.5194/piahs-372-539-2015

Rodell M, Houser PR, Jambor U, Gottschalck J, Mitchell K, Meng C-J, Arsenault K, Cosgrove B, Radakovich J, Bosilovich M, Entin JK, Walker JP, Lohmann D, Toll D (2004) The Global Land Data Assimilation System. Bull Am Meteor Soc 85(3):381–394. https://doi.org/10.1175/BAMS-85-3-381

Seki Y, Koyama J (1998) New findings on ground subsidence in the central and northern regions of the Kanto Plain (Groundwater level fluctuation-ground fluctuation cycle). Chishitsu News no. 531, 52–64. (in Japanese) https://www.gsj.jp/data/chishitsunews/98_11_06.pdf. Accessed 06 Sep 2020

Shimada S (2012) Comparison of the coordinates solutions between the absolute and the relative phase center variation models in the dense regional GPS network in Japan. Geodesy Planet Earth:651–656. https://doi.org/10.1007/978-3-642-20338-1_80

Tobita M (2016) Combined logarithmic and exponential function model for fitting postseismic GNSS time series after 2011 Tohoku-Oki earthquake. Earth Planets Space 68:41–52. https://doi.org/10.1186/s40623-016-0422-4

Tobita M, Munekane H, Kaidzu M, Matsuzaka S, Kuroishi Y, Masaki Y, Kato M (2004) Seasonal variation of groundwater level and ground level around Tsukuba. J Geod Soc Jpn 50:27–37. https://doi.org/10.11366/sokuchi1954.50.27

Tochigi Prefecture (2010) Report of ground deformation and groundwater level survey in Tochigi prefecture. Utsunomiya Tochigi Japan. (in Japanese) http://www.pref.tochigi.lg.jp/d03/eco/kankyou/hozen/documents/21jiban.pdf. Accessed 02 May 2020

Tochigi Prefecture (2018) Report of ground deformation and groundwater level survey in Tochigi prefecture. Utsunomiya Tochigi Japan. (in Japanese) http://www.pref.tochigi.lg.jp/d03/eco/kankyou/hozen/jiban_houkoku_2017.html. Accessed 06 Sep 2020

Tokunaga T (2015) Groundwater management after the cessation of land subsidence: lessons learnt from the Tokyo Metropolitan Area. J Assoc Groundwater Hydrol 57:37–43. (in Japanese). https://doi.org/10.5917/jagh.57.37

Tregoning P, Herring TA (2006) Impact of a priori zenith hydrostatic delay errors on GPS estimates of station heights and zenith total delays. Geophys Res Lett 33:L23303. https://doi.org/10.1029/2006/GL027706

Wessel P, Smith WHF (1998) New, improved version of the generic mapping tools released. EOS Trans AGU 79:579. https://doi.org/10.1029/98EO00426

An Approximate Method to Simulate Post-Seismic Deformations in a Realistic Earth Model

He Tang, Jie Dong, and Wenke Sun

Abstract

The geodetic observations of static deformations, including gravity perturbations and displacement fields due to huge earthquakes, are understood and explained using recent dislocation theories. Due to multiple possible mechanisms for the post-seismic phase of earthquakes, the dominant mechanism may change at different spatiotemporal ranges for different earthquake types. Accurate forward and inverse modeling of post-seismic deformations is valuable and needed information for geoscience communities. The existing methods for calculating gravitational viscoelastic relaxation can be improved or simplified to make them more suitable for more realistic Earth models and/or to overcome the poor convergence performance and/or overflow risks during numerical calculations. In this study, a simple and effective method for calculating the post-seismic relaxation deformations is proposed. This method is different from previous methods, such as the normal mode summation and rectangle integration methods. The proposed method consists of a rational functional approximation of the integral kernel and a transformation of the numerical inverse Laplace transform (NILT) into an alternating series summation using the residual theorem. Then the intrinsic oscillation and overflow risks are thoroughly suppressed. The accuracy of the calculated Green's functions can be easily controlled by choosing a suitable parameter. In addition, the proposed method also has applicability in different Earth models with linear rheological profiles.

Keywords

Green's function · Inverse Laplace transform · Post-seismic deformation · Rheology structure

Key Points

(a) The viscoelastic Green's functions are expressed as a weighted summation of the complex ones.
(b) The oscillation and overflow risk of the inverse Laplace transform is exhaustively avoided.
(c) The proposed method can also be used for other linear rheological models.

H. Tang · W. Sun (✉)
Key Laboratory of Computational Geodynamics, University of Chinese Academy of Sciences, Beijing, China
e-mail: sunw@ucas.ac.cn

J. Dong
Chinese Academy of Surveying and Mapping, Beijing, China

1 Introduction

An entire earthquake process may contain three major phases: the inter-seismic accumulation, the co-seismic transient deformation, and the long-term post-seismic adjustment. The viscoelastic relaxation deformation, as a major mechanism of the post-seismic phase, can be captured using modern geodetic technologies for megathrust earthquakes (e.g., Wang et al. 2012; Sun et al. 2014; Li et al. 2018; Qiu et al. 2019; Agata et al. 2019). To explain the multiple observations and investigate the physical mechanisms, accurate modeling for such processes is a necessity and attractive to geophysicists and geodesists.

© The Author(s) 2020
J. T. Freymueller, L. Sanchez (eds.), *Beyond 100: The Next Century in Geodesy*,
International Association of Geodesy Symposia 152, https://doi.org/10.1007/1345_2020_96

Some dislocation theories based on a flat viscoelastic Earth model have been well developed in recent decades. For instance, post-seismic deformations due to a point source and a finite fault have been studied by Rundle (1982) using a two-layer viscoelastic-gravitational model. The gravity effect, however, was not considered correctly in this and other previous papers. Wang (2005) presented a consistent method to include the gravity effect in viscoelastic Earth models. His group contributed an open source Fortran code PSGRN/PSCMP for calculating co- and post-seismic deformations on a layered half-space Earth model (Wang et al. 2006).

\looseness1{}Many scientists have also proposed dislocation theories in spherical Earth models. For example, the normal mode summation method was applied for calculating the global deformations in a stratified, self-gravitating, and incompressible viscoelastic Earth model by Sabadini et al. (1984), Pollitz (1992), Piersanti et al. (1995), and others. However, the continuous structure and compressibility of the Earth cannot be considered simultaneously due to difficulties in finding the innumerable mode. The first improvement was proposed by Tanaka et al. (2006, 2007) where a closed path integration method in a complex plane was used to bypass this fault in the normal mode method. Once a suitable integration path was determined, the deformation Love numbers (DLNs) (or Green's functions) could be obtained from the complex ones using numerical integration along a rectangle path. An alternative method was also proposed by Spada and Boschi (2006) that modified the normal mode formulation and used the Post-Widder Laplace inversion formula (Gaver 1966). This method is beneficial because it bypasses the numerical Laplace integral and root-finding procedure at the same time.

These two major methods have been successful in simulating post-seismic deformations; however, their distinct features and applicability scope needs to be discussed. The accuracy and reliability of the closed path integration depend on a rectangle path and a sampling scheme along the path. As pointed out by Tanaka et al. (2006), two trade-off conditions need to be considered: a smooth variation of the integrand and low oscillation of the integral kernel. These considerations result in the utilization of a root-finding scheme to locate the positions of the smallest and largest poles for a given viscoelastic Earth model. This may limit its application in inverse problems, such as estimating the viscosity of the asthenosphere. The modified normal mode method proposed by Spada and Boschi (2006) has suffered due to its slow convergence performance of the series summation formula. In addition, it requires a multiple-precision computer system. Therefore, alternative methods to calculate post-seismic deformations in a realistic Earth model are still required.

Tang and Sun (2019) confirmed that the approximate inverse Laplace transform method presented by Valsa and Brančík (1998) can be applied to simulate post-seismic deformations. However, the mathematical process of this method is complicated and not easy to follow. In this paper, a straightforward approach is proposed as a simplification of the previous study. The integral kernel is directly approximated as a rational function, and the residual theorem is then applied to obtain a similar formula. This new approach is concise and clear, focusing on the nature of the problem and avoiding lengthy non-essential processes. Moreover, the modified method might be considered as common one that can be extended to deal with other similar problems. That is, the kernel of an integral transform (such as the Laplace and Mellin transforms) can be replaced by approximate functions that result in a series of new algorithms.

In the following sections, this simple and effective algorithm for calculating the post-seismic deformations is presented in Sect. 2. A numerical verification of the method on the Love numbers is given in Sect. 3. Finally, the discussions and conclusions are presented.

2　Complex DLNs in the Laplace Domain

The focus of this research is on the time-dependent post-seismic deformations in a spherically symmetric, viscoelastic, and self-gravitating Earth. There are three equations that govern the geodynamic process: the equation of the equilibrium, the stress-strain relationship of the viscoelastic martial, and Poisson's equation (Takeuchi and Saito 1972). Because the stress-strain relationship is time-dependent, the correspondence principle (McConnell 1965) is utilized to transform the viscoelastic problem into an equivalent elastic problem via the Laplace transform. Then all of the equations have the same form as they are in the elastic case, and the only difference is that the Lamé parameters are functions of the Laplace variable, s, for viscoelastic material. After a vector spherical harmonic expansion of the transformed equations, the equations system is uncoupled into two systems: the spheroidal portion involving a 6-D displacement-stress vector, and the toroidal part involving a 2-D one. The boundary conditions of these two systems at the center of the Earth, the source location, the free surface, and the internal discontinuities need to be satisfied. Then, the Runge-Kutta scheme is used to solve the transformed deformation problem to obtain DLNs, similar to what Sun and Okubo (1993) did in treating the co-seismic deformation in the pure elastic Earth model.

To transform the DLNs at a radius, r, in the Laplace domain into the time domain is quite difficult compared with other processes. In addition, the full analytical method, such as used by Tang and Sun (2018a), Tang and Sun (2018b), is restricted in this more realistic Earth model. A numerical method might be the only possible choice. Following the definition of the inverse Laplace transform, the time-dependent DLNs can be written as:

$$\mathbf{Y}(r,t) = \frac{1}{2\pi i} \int_{c-i\infty}^{c+i\infty} \mathbf{Y}_h(r,s)\, e^{st}\, ds, \quad (1)$$

where $\mathbf{Y}_h(r,s) = \mathbf{Y}(r,s)/s$; $\mathbf{Y}(r,s)$ is the complex DLNs; $\mathbf{Y}(r,t)$ is the DLNs in the time domain; i is the imaginary unit of the complex number; and c is a real number larger than all of the real parts of the singularities of the integrand. Here, $1/s$ appears because the focus is on the relaxation deformation of the main shock, i.e., the relaxation of a heaviside slip is considered here.

The complex DLNs change smoothly along the vertical line $\mathrm{Re}(s) = c$; however, the integral kernel, e^{st}, is an oscillating function along this path. More explicitly, the integral kernel can be written as $e^{st} = e^{ct}[\cos(\mathrm{Im}(s)t) + i \sin(\mathrm{Im}(s)t)]$, and it reveals the intrinsic oscillational feature with respect to $\mathrm{Im}(s)t$, which may trouble the inverse Laplace transform.

This problem is avoided using an approximation of the integral kernel (Valsa and Brančik 1998):

$$e^{st} \approx \frac{e^{st}}{1 - e^{-2(a-st)}}, a > \mathrm{Re}(s)t. \quad (2)$$

Then the inverse Laplace transform of $\mathbf{Y}_h(r,s)$ is approximately written as:

$$\mathbf{Y}_a(r,t) = \frac{1}{2\pi i} \int_{c-i\infty}^{c+i\infty} \mathbf{G}(r,s)\, ds, \mathbf{G}(r,s) = \frac{\mathbf{Y}_h(r,s)\, e^{st}}{1 - e^{-2(a-st)}}. \quad (3)$$

To calculate the integral of $\mathbf{G}(r,s)$, another curve integral is added along a semicircle centered at $(c, 0)$ with an infinite radius to obtain a closed path integral (Fig. 1b):

$$\mathbf{Y}_a(r,t) = \frac{1}{2\pi i} \left[\int_{c-i\infty}^{c+i\infty} \mathbf{G}(r,s)\, ds + \int_{arc} \mathbf{G}(r,s)\, ds \right]. \quad (4)$$

It has been shown that the complex DLNs approach a real constant as the Laplace variable $|s|$ approaches infinity [see Eq. (23) of Cambiotti et al. (2009)]. Thus, it is clear that $\mathbf{Y}_h(r,s) \to 0$ as $s \to \infty$. This fact results in $\mathbf{G}(r,s) \to 0$ as $s \to \infty$. Hence, the integration of the second portion of Eq. (4) is zero, and it contributes nothing to the time-dependent Love numbers.

For this closed path integral, the residual theorem can be directly applied. For simplicity, we define the variable $\mathbf{P}(s) = \mathbf{Y}_h(r,s)e^{st}$ and $Q(s) = 1 - e^{-2(a-st)}$, then Eq. (4) is rewritten as:

$$\mathbf{Y}_a(r,t) = -\sum_{k=-\infty}^{+\infty} \mathrm{Res}\,(\mathbf{P}(s)/Q(s), q_k)$$
$$= -\sum_{k=-\infty}^{+\infty} \mathrm{Res}\left(\mathbf{P}(s)/Q'(s)\Big|_{s=q_k} \right), \quad (5)$$

where all of the zero points of $Q(s)$ are $q_k = (a - i\pi k)/t$ and $Q'(s)$ denotes the derivative of it. After some simple algebraic operations, Eq. (5) can be written as:

$$\mathbf{Y}_a(r,t) = \frac{e^a}{2t} \sum_{k=-\infty}^{+\infty} e^{-i\pi k} \mathbf{Y}_h\left(\frac{a}{t} - \frac{i\pi k}{t} \right)$$
$$= \frac{e^a}{2t} \left[\sum_{k=0}^{+\infty} 2(-1)^k \mathbf{Y}_h\left(\frac{a}{t} - \frac{i\pi k}{t} \right) - \mathbf{Y}_h\left(\frac{a}{t} \right) \right], \quad a/t > c. \quad (6)$$

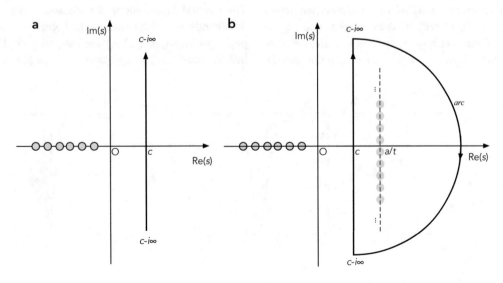

Fig. 1 Diagram of the inverse Laplace transform using two different integration paths. All of the poles (grey points) are located on the real axis of s. The Bromwich integral path (**a**) is along a vertical line $\mathrm{Re}(s) = c$. Our proposed contour integration (**b**) is performed along a closed path: the vertical line and a semicircle with an infinite radius. The grey points along the dashed vertical blue line, $\mathrm{Re}(s) = a/t$, denote all of the poles of the approximated integral kernel

Here, the identical equation $\mathbf{Y}_h{}^* = \mathbf{Y}_h(s^*)$ is used during simplification, and * denotes a conjugate operator. Note that the above formula will result in a real value, although it involves a complex series.

The real DLNs can be found by approximately expressing them as a weighted sum of the samples of the complex counterpart at some special points along a vertical path in the complex plane. To speed up the convergence of the series and maintain high accuracy, the first portion of the series with $0 \leq k < k_c$ should be summed directly, and the Euler's transforms can be applied for the other portions with $k > k_c$. Note that the above formula only holds for post-seismic deformations when $t > 0$. The co-seismic DLNs at $t = 0$ can be obtained from the complex DLNs by setting a large enough $|s|$.

3 Real DLNs in the Time Domain via NILT

3.1 The Oscillational Integral During NILT

The oscillational inverse Laplace transform in Eq. (1) is approximately rewritten as an alternating series of DLNs sampled on a special vertical path without any oscillational terms. Hence, the overflow risk due to the integral kernel is completely avoided, and an accurate calculation of the post-seismic deformations can be achieved using the current method. The new method can be applied to calculate post-seismic relaxation deformations due to four point sources (labeled as 12, 32, 22, and 33) as defined by Sun and Okubo (1993) in the realistic spherical model with a continuous stratification.

To show this point in a straightforward manner, a complex DLN of $k^{32}(r = R, s)$ was used as an example on the surface of the Earth (set $r = R$, the mean radius) at time $t = 5$ years with degree $n = 20$ due to a point dip-slip

source with depth of 20 km calculated using this proposed method (the parameter was set to $a = 6$) and a common numerical integration. The elastic parameters of the PREM (Dziewonski and Anderson 1981) were used. The inner core within a radius $r < 1,221.5$ km and the crust with a thickness of 50 km were elastic. In addition, the viscosity of the mantle ($r > 5,871$ km) was set at 10^{18} Pa·s, and that of the other portion was set at 10^{19} Pa·s. In Fig. 2, the $k^{32}(s)$ approaches the elastic DLN and $k^{32}(s)/s$ approaches zero when the norm of the Im(s) is large. The integrand $e^{st}k^{32}(s)/s$ then oscillated quickly, which was the trouble source of the inverse Laplace transform. In the following subsection, the solution of this problem will be demonstrated using our proposed method.

3.2 Convergence of the Proposed Method

The time-dependent DLN $k^{32}(t)$ was computed using the proposed method and is plotted in Fig. 3. A simple rectangular integration along the path Re(s) = $a/t = 6/t = 1.2$ year^{-1} shows a very poor convergence performance (Fig. 3a). The final series is truncated at various values of the maximum number of k. The summation of the first kth terms are denoted as S_k (Fig. 3b). It was found that S_k exhibited regular oscillations around the convergence value (-1.73E-5, calculated using Euler's accelerated sum in Fig. 3c). In Fig. 3c, the first 21 terms were directly summed and the following 19 terms were accelerated using Euler's transform. It is clear that a satisfactory convergence performance was achieved using approximately 40 terms of the complex DLNs if a suitable convergence acceleration of the series was utilized.

The direct integration of the inverse Laplace transform was too rough to be available for post-seismic simulations. The integral kernel along a vertical line was proportional to sin(Im(s)t), and this means the kernel oscillated with period of $T_0 = 2\pi/t$ along the integral path. If we assume that the maximum sample step length is $T_0/20$ and truncate

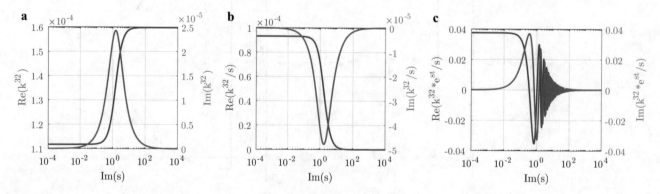

Fig. 2 Numerical visualizations of $k^{32}(s)$ and $k^{32}(s)/s$ along the path Re(s) = a/t = 6/t year^{-1} with a of degree $n = 20$ in the Laplace domain at time $t = 5$ years due to a point dip-slip source with a depth of 20 km. (**a**) The variation of $k^{32}(s)$ over the Laplace variable s: real part (blue line), imaginary part (pink line). (**b**) Similar to (**a**), but for $k^{32}(s)/s$. (**c**) Similar to (**a**), but for the integrand $e^{st}k^{32}(s)/s$. Units: 1/year for each horizontal axis

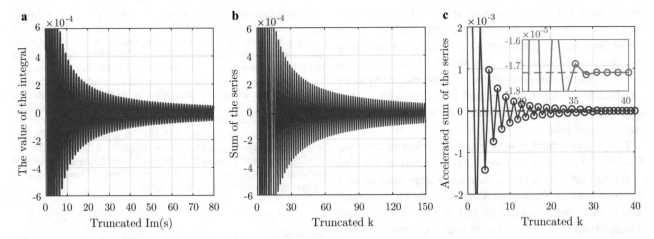

Fig. 3 Comparison of the inverse Laplace transform of $k^{32}(s)/s$ using the direct integration and the series summation (proposed method) along the path $\text{Re}(s) = 1.2$ year^{-1}. (**a**) The integration value of Eq. (1) truncated at various maximum $|\text{Im}(s)|$ (blue line). (**b**) Truncated series S_k of Eq. (6) with different k. (**c**) Similar to (**b**), but for S_k after a direct summation of 21 terms and Euler's transform of the following 19 terms. S_{40} with Euler's transform was set as the convergence value (dashed pink lines)

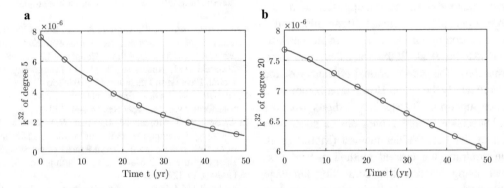

Fig. 4 Normalized time-dependent DLNs $k^{32}(t)$ of degree 5 (**a**) and degree 20 (**b**) for a vertical dip-slip fault. Here, the normalization factor of the DLNs is the same as in Sun and Okubo (1993). The solid lines denote those calculated using the previous method (Tanaka et al. 2006), and the cycles denote the results calculated using the proposed method

the integral at a length of 100 periods, then it requires at least 2,000 samples. In fact, a numerical integration cannot achieve a stable value after approximately 2,000 samplings along the vertical line in Fig. 3a. However, only 40 terms resulted in an acceptable accuracy of the result using this proposed method and Euler's transform to accelerate the convergence speed (Fig. 3c).

3.3 Verifying Using a Comparison with a Previous Method

The time-dependent DLNs were then calculated using the proposed method to verify them. Only k^{32} was used as an example, and the result is shown here (other variables are also verified but are not shown here). Most of the parameters of the Earth model were used, as discussed in the above Sects. 3.1 and 3.2. In addition, the crust was set to 30 km, and the inner core was kept elastic. The degrees of 5 and 20 DLNs of the dip-slip source with depths of 1 km were calculated using Tanaka's method (Tanaka et al. 2006) and our proposed method [Eq. (6)] in the time range from 0 to 50 years. The results are plotted in Fig. 4. The results of the two schemes agree well, and this confirms the correctness of this method.

4 Discussion and Conclusions

In this study, an approximate approach was presented to calculate the post-seismic deformations due to point sources in a realistic viscoelastic spherical Earth model with the continuous stratification, compressibility, and self-gravitation was proposed. The good agreement of the time-dependent dislocation Love numbers calculated using a previous method (Tanaka et al. 2006) and by our proposed method demonstrated that this new approximation scheme had high accuracy. The idea of the method was to approximately treat the integral kernel (Valsa and Brančik 1998) and to implement the residue theorem in a semicircle. The location of this semicircle, i.e., the parameter, can be adjusted according to a specific problem. Then, the numerical oscillation and overflow risk of some previous methods are avoided by using a skillful treatment of the integral kernel. Essentially, the approximate approach is different in comparison with other published methods.

The proposed method is as simple and clear as the Post-Widder method. They have very similar forms of the final expression, i.e., the inversed function is expressed using a weighted sum of the special sampled complex functions. Both of them can be applied for different viscoelastic Earth models because all of the singular points are negative (Plag and Jüttner 1995; Vermeersen and Mitrovica 2000) for a stable Earth model without a heavier layer overlying a lighter one. However, they are different in essence. The Post-Widder method is a differential approach along the positive real axis; while, the current method is an integral approach along the vertical line in complex plane. There are 40 sampling points of the complex function in our method and 16 in the Post-Widder method (Melini et al. 2008) for calculating post-seismic deformations. The computational efficiency of the two methods were nearly the same. However, a commercial Fortran compiler supporting 30 significant digits with the IEEE extended-precision format cannot ensure success in the application of the Post-Widder method (Melini et al. 2008). On the contrary, the current method can be easily implemented using any scientific computing language without an extended precision algebraic library. The less terms in the sampling points of the Post-Widder method primarily benefits from the extended library. In addition, an extended floating-number requirement means large computational memory and performance degradation. If considering computational efficiency and memory consumption together, these two methods have similar performances.

Compared with the rectangle integration method or the normal mode method, the current method focused on how to treat the oscillational integral kernel rather than the poles of a specific problem. The rectangle integration method (Tanaka et al. 2006, 2007) treats the inverse Laplace transform bypassing the innumerable poles. However, there are at least three parameters that should be determined in the numerical computation for a given earth model. In contrast, this proposed method only has one adjustable parameter. In fact, this parameter can be fixed at approximately six because the largest pole of the post-seismic deformation will not be larger than zero in a stable Earth model. In this point of view, the current method is suitable for an inverse problem. In addition, the current method can also be used for other linear rheological models, such as the Kelvin-Voigt material, the Maxwell solid, Burgers body, and their combination.

Acknowledgements We greatly appreciate the constructive comments by the IAG Symposia Series Editor (Jeffrey T. Freymueller), and two anonymous reviewers, which have greatly improved the manuscript. We are grateful to Drs. Gabriele Cambiotti, Fuchang Wang and Tai Liu for the helpful discussions. All data used in this research can be reproduced by the formulae in this paper and all figures are available in a research data repository on Zenodo (https://doi.org/10.5281/zenodo.3520243). We acknowledge financial support by the NNSFC (Natural Science Foundation of China; 41774088, 41974093, 41331066 and 41474059) and the Key Research Program of Frontier Sciences CAS (Chinese Academy of Sciences; QYZDY-SSW-SYS003). J. D. is grateful for funding provided by the NNSFC (National Natural Science Foundation of China; 41604067) and the Basic Research Fund of Chinese Academy of Surveying and Mapping (No. AR1906).

References

Agata R, Barbot SD, Fujita K, Hyodo M, Iinuma T, Nakata R, Ichimura T, Hori T (2019) Rapid mantle flow with power-law creep explains deformation after the 2011 Tohoku mega-quake. Nat Commun 10:1385

Cambiotti G, Barletta VR, Bordoni A, Sabadini R (2009) A comparative analysis of the solutions for a Maxwell Earth: the role of the advection and buoyancy force. Geophys J Int 176:995–1006

Dziewonski AM, Anderson DL (1981) Preliminary reference Earth model. Phys Earth Planet Inter 25:297–356

Gaver DP (1966) Observing stochastic processes, and approximate transform inversion. Oper Res 14:444–459

Li S, Bedford J, Moreno M, Barnhart WD, Rosenau M, Oncken O (2018) Spatiotemporal variation of mantle viscosity and the presence of cratonic mantle inferred from 8 years of postseismic deformation following the 2010 Maule, Chile, Earthquake. Geochem Geophys Geosyst 19:3272–3285

McConnell RK (1965) Isostatic adjustment in a layered Earth. J Geophys Res 70:5171–5188

Melini D, Cannelli V, Piersanti A, Spada G (2008) Post-seismic rebound of a spherical Earth: new insights from the application of the Post-Widder inversion formula. Geophys J Int 174:672–695

Piersanti A, Spada G, Sabadini R, Bonafede M (1995) Global post-seismic deformation. Geophys J Int 120:544–566

Plag HP, Jüttner HU (1995) Rayleigh-Taylor instabilities of a self-gravitating Earth. J Geodyn 20:267–288

Pollitz FF (1992) Postseismic relaxation theory on the spherical earth. Bull Seismol Soc Am 82:422–453

Qiu Q, Feng L, Hermawan I, Hill EM (2019) Coseismic and post-seismic slip of the 2005 Mw 8.6 Nias-Simeulue earthquake: spatial overlap and localized viscoelastic flow. J Geophys Res: Solid Earth 124:7445–7460

Rundle JB (1982) Viscoelastic-gravitational deformation by a rectangular thrust fault in a layered Earth. J Geophys Res: Solid Earth 87:7787–7796

Sabadini R, Yuen DA, Boschi E (1984) The effects of post-seismic motions on the moment of inertia of a stratified viscoelastic earth with an asthenosphere. Geophys J Roy Astron Soc 79:727–745

Spada G, Boschi L (2006) Using the Post-Widder formula to compute the Earth's viscoelastic Love numbers. Geophys J Int 166:309–321

Sun W, Okubo S (1993) Surface potential and gravity changes due to internal dislocations in a spherical earth—I. Theory for a point dislocation. Geophys J Int 114:569–592

Sun T, Wang K, Iinuma T, Hino R, He J, Fujimoto H, Kido M, Osada Y, Miura S, Ohta Y (2014) Prevalence of viscoelastic relaxation after the 2011 Tohoku-oki earthquake. Nature 514:84–87

Takeuchi H, Saito M (1972) Seismic surface waves. In: Methods in computational physics advances in research & applications, vol 11. Academic Press, London, pp 217–295

Tanaka Y, Okuno J, Okubo S (2006) A new method for the computation of global viscoelastic post-seismic deformation in a realistic earth model (I)—vertical displacement and gravity variation. Geophys J Int 164:273–289

Tanaka Y, Okuno J, Okubo S (2007) A new method for the computation of global viscoelastic post-seismic deformation in a realistic earth model (II)-horizontal displacement. Geophys J Int 170:1031–1052

Tang H, Sun W (2018a) Asymptotic co- and post-seismic displacements in a homogeneous Maxwell sphere. Geophys J Int 214:731–750

Tang H, Sun W (2018b) Closed-form expressions of seismic deformation in a homogeneous Maxwell earth model. J Geophys Res-Solid Earth 123:6033–6051

Tang H, Sun W (2019) New method for computing postseismic deformations in a realistic gravitational viscoelastic earth model. J Geophys Res-Solid Earth 124:5060–5080. https://doi.org/10.1029/2019JB017368

Valsa J, Brančik L (1998) Approximate formulae for numerical inversion of Laplace transforms. Int J Numer Modell: Electron Networks Devices Fields 11:153–166

Vermeersen LLA, Mitrovica JX (2000) Gravitational stability of spherical self-gravitating relaxation models. Geophys J Int 142:351–360

Wang R (2005) The dislocation theory: a consistent way for including the gravity effect in (visco)elastic plane-earth models. Geophys J Int 161:191–196

Wang R, Lorenzo-Martín F, Roth F (2006) PSGRN/PSCMP—a new code for calculating co- and post-seismic deformation, geoid and gravity changes based on the viscoelastic-gravitational dislocation theory. Comput Geosci 32:527–541

Wang K, Hu Y, He J (2012) Deformation cycles of subduction earthquakes in a viscoelastic earth. Nature 484:327–332

Geodetic Monitoring of the Variable Surface Deformation in Latin America

Laura Sánchez and Hermann Drewes

Abstract

Based on 24 years of high-level GNSS data analysis, we present a sequence of crustal deformation models showing the varying surface kinematics in Latin America. The deformation models are inferred from GNSS station horizontal velocities using a least-squares collocation approach with empirically determined covariance functions. The main innovation of this study is the assumption of continuous surface deformation. We do not introduce rigid microplates, blocks or slivers which enforce constraints on the deformation model. Our results show that the only stable areas in Latin America are the Guiana, Brazilian and Atlantic shields; the other tectonic entities, like the Caribbean plate and the North Andes, Panama and Altiplano blocks are deforming. The present surface deformation is highly influenced by the effects of seven major earthquakes: Arequipa (Mw8.4, Jun 2001), Maule (Mw8.8, Feb 2010), Nicoya (Mw7.6, Sep 2012), Champerico (Mw7.4, Nov 2012), Pisagua (Mw8.2, Apr 2014), Illapel (Mw8.3, Sep 2015), and Pedernales (Mw7.8, Apr 2016). We see very significant kinematic variations: while the earthquakes in Champerico and Nicoya have modified the aseismic deformation regime in Central America by up to 5 and 12 mm/a, respectively, the earthquakes in the Andes have resulted in changes of up to 35 mm/a. Before the earthquakes, the deformation vectors are roughly in the direction of plate subduction. After the earthquakes, the deformation vectors describe a rotation counter-clockwise south of the epicentres and clockwise north of the epicentres. The deformation model series reveals that this kinematic pattern slowly disappears with post-seismic relaxation. The numerical results of this study are available at https://doi.pangaea.de/10.1594/PANGAEA.912349 and https://doi.pangaea.de/10.1594/PANGAEA.912350.

Keywords

Caribbean · Crustal deformation · Earth surface kinematics · Latin America · SIRGAS · Station velocity model · VEMOS

1 Introduction

Geodetic reference frames comprise coordinates of station positions at a certain epoch and constant velocities describing a secular station motion. In active seismic regions, strong earthquakes cause large displacements of station positions and velocity changes disabling the use of such coordinates over any time periods. The continuous representation of station positions between different epochs requires the computation of reliable station velocity models. Whit these models, we can monitor the kinematics of reference frames, determine transformation parameters between pre-seismic and post-seismic (deformed) coordinates, and interpolate surface motions arising from plate tectonics or crustal deformations in areas where no geodetic stations are established. In the particular case of Latin America, the reference frame is

L. Sánchez (✉) · H. Drewes
Technische Universität München, Deutsches Geodätisches
Forschungsinstitut (DGFI-TUM), München, Germany
e-mail: lm.sanchez@tum.de; h.drewes@tum.de

© The Author(s) 2020
J. T. Freymueller, L. Sanchez (eds.), *Beyond 100: The Next Century in Geodesy*,
International Association of Geodesy Symposia 152, https://doi.org/10.1007/1345_2020_91

called SIRGAS (*Sistema de Referencia Geocéntrico para las Américas*; cf. SIRGAS 1997) and it is a regional densification of the global International Terrestrial Reference Frame (ITRF; Petit and Luzum 2010). The first SIRGAS realisation was established by a GPS (Global Positioning System) observation campaign in May 1995 (SIRGAS 1997). It comprised 58 stations covering all South America. This network was measured again in May 2000 and it was extended to Central and North America including 184 stations (Drewes et al. 2005). Since 2000, the Latin American geodetic reference network is materialised (and frequently extended) by continuously operating GNSS (GPS+GLONASS) stations (Brunini et al. 2012; Sánchez et al. 2013, 2015; Cioce et al. 2018). As the western margin of Latin America is one of the seismically most active regions in the world, the maintenance of the SIRGAS Reference Frame implies the frequent computation of present-day (updated) surface deformation models. Such models were computed in 2003 (Drewes and Heidbach 2005), 2009 (Drewes and Heidbach 2012), 2015 (Sánchez and Drewes 2016), and 2017 (this paper). Here, we present the computation of the deformation model 2017 and its comparison with the previous models to show the very significant variations of the surface kinematics in Latin America during the past 15 years.

2   Surface-Kinematics Modelling Based on GNSS Multi-Year Solutions

Spatial continuous surface deformation may be inferred from pointwise velocities applying geophysical models or geodetic methods based on mathematical interpolation approaches. The approach used in the present study is the least-squares collocation (LSC, e.g., Moritz 1973; Drewes 1978). Previous studies applied also the finite element method used with geophysical models (e.g., Heidbach and Drewes 2003). It has been demonstrated that for the sole representation of the horizontal Earth surface kinematics, the results of both methods are very similar (e.g., Drewes and Heidbach 2005). The vertical deformation is not considered in this work because the station height variations are highly influenced by local effects, and the station distribution (Fig. 1) is too sparse to apply correlations between neighbouring stations. In the modelling of the surface kinematics, we distinguish two components: the velocity field and the deformation model. In the latter one, a secular motion inferred from plate motion estimates (Fig. 1) is removed from the station velocities, and the pointwise residual velocities are interpolated to a regular grid.

The least-squares collocation method is based on the analysis of the correlation of physical quantities between neighbouring points. The vector of the observations (in this case the station velocities) is divided into a systematic part

(trend) and two independent random parts: the signal and the observation error (or noise). The parameters describing the systematic component and the stochastically correlated signals are estimated by minimising the noise. The spatial signal correlation is usually assumed as a function depending on the distance d and, presuming isotropy after trend removal, independent of the direction. The basic LSC formula is given by (Drewes and Heidbach 2005, 2012):

$$\mathbf{v}_{pred} = \mathbf{C}_{new}^{T}(\mathbf{C}_{obs} - \mathbf{C}_{nn})^{-1}\mathbf{v}_{obs} \qquad (1)$$

$\mathbf{v}_{obs}$ contains the station velocities obtained from the GNSS observations at the geodetic stations. $\mathbf{v}_{pred}$ represents the velocities to be predicted at the grid points. $\mathbf{C}_{obs}$ is the correlation matrix between the observed velocities. $\mathbf{C}_{new}$ is the correlation matrix between predicted and observed velocities. $\mathbf{C}_{nn}$ is the noise covariance matrix (it contains the uncertainty of the station velocities obtained within the multi-year solutions). The correlation between the observed velocities v_i, v_k at the (adjacent) geodetic stations i, k is determined under the stationarity condition over a defined domain by

$$C_{obs}(d_{ik}) = E\{v_i \cdot v_k\}, \qquad (2)$$

E is the statistical expectation and d_{ik} is the distance between stations i and k. The C_{obs} values are classified in Δd_j class intervals and the respective cross-covariance $C_{obs}(\Delta d_j)$ and auto-covariance $C_{obs}(d = 0) = C_0$ are determined using:

$$C_{obs}(\Delta d_j) = \frac{1}{n_j}\sum_{i<k}^{j} v_i \times v_k \; ; \quad C_{obs}(d = 0) = C_0 = \frac{1}{n}\sum_{i=1}^{n} v_i^2, \qquad (3)$$

n stands for the number of stations available at the defined domain, while n_j represents the number of stations available at each class interval Δd_j. After estimating the discrete empirical covariance values with Eq. (3), they are approximated by a continuous function $C(d_{ik})$, which here is the exponential function:

$$C(d_{ik}) = a\,e^{-b \cdot d_{ik}} \qquad (4)$$

The function parameters a and b are estimated by a least-squares adjustment. $\mathbf{C}_{obs}$ is symmetrical and its main diagonal ($i = k$) contains the values C_0. Fulfilling the stationarity condition, the elements of $\mathbf{C}_{new}$ are computed using the same Eq. (4) as a function of the distance between the grid node to be interpolated and the geodetic stations. To satisfy the isotropy condition, we estimate a common rotation vector and remove this horizontal motion trend from all station velocities located in the same domain defined by d. Afterwards, we restore the removed trend to the velocities v_{pred} predicted at the grid points (cf. Drewes 1982, 2009).

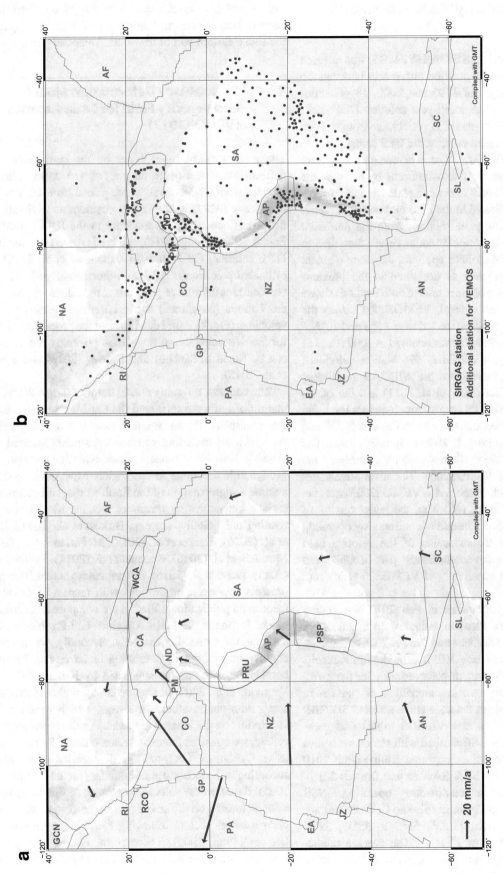

Fig. 1 Generalised tectonic map of Latin America (left) and graphical distribution of the GNSS stations used in this study (right). Blue dots represent stations belonging to the SIRGAS Reference Frame, red stations represent stations provided by UNAVCO. Tectonic plate boundaries after Bird (2003): Grey labels represent the plate name abbreviations, while magenta labels depict the orogeny area abbreviations. Plates: AF Africa, AN Antarctica, AP Altiplano, CA Caribbean, CO Cocos, EA Easter Island, GP Galapagos, JZ Juan Fernandez, NA North America, ND North Andes, NZ Nazca, PA Pacific, PM Panama, RI Rivera, SA South America, SC Scotia. Orogenes: GCN Gorda-California-Nevada, RCO Rivera-Cocos, PRU Peru, PSP Puna-Sierras Pampeanas, WCA West Central Atlantic. On the left map, motion of the large plates after Drewes (2012) and CO, PM, ND, AP and SC after Bird (2003)

3   Existing Velocity Models for SIRGAS (VEMOS)

The first velocity model for SIRGAS (VEMOS) was released in 2003. It is based on the position differences between the two SIRGAS campaigns of 1995 and 2000, 48 velocities derived from the SIRGAS multi-year solution DGF01P01 (Seemüller et al. 2002), and 231 velocities from several geodynamic projects based on episodic GPS campaigns (cf. Drewes and Heidbach 2005). The different data sets were transformed to a common kinematic frame by deriving the rotation vector of the South American plate from the respective station motions located in the rigid part of South America, and reducing these plate motions from the particular data sets. The resulting residual motions were modelled to a continuous deformation field applying the finite element method and LSC approach as described in the previous section. The comparison of the results of both methods shows an agreement in the mm/a level. VEMOS2003 covers the South American area between the latitudes 45°S and 12°N.

The second VEMOS model was released in 2009 (Drewes and Heidbach 2012). It considers 496 station velocities; 95 of them corresponding to the SIRGAS multi-year solution SIR09P01 (Seemüller et al. 2011) and the others derived from repetitive GPS campaigns. It covers the Latin American area between the latitudes 56°S and 20°N and the time-span from January 2, 2000 to June 30, 2009. The continuous surface velocity field was derived applying the same strategies as in VEMOS2003. The main advantages of VEMOS2009 with respect to VEMOS2003 are the increased number of input velocities, the better quality of measurements (due to an increase of continuously operating GNSS stations), and the extension of the velocity field to the Caribbean and the southernmost part of Chile and Argentina. The mean uncertainty of VEMOS2009 is about ±1.5 mm/a.

After the Maule earthquake in Feb 2010, the station velocities in the area between latitudes 30°S and 40°S changed dramatically (Sánchez and Drewes 2016). However, we could not compute a new VEMOS model immediately, because we required 5 years of observations after the earthquake in order to improve the modelling of the strong post-seismic decay signals detected at the affected SIRGAS stations. Consequently, a new VEMOS model was computed in 2015 using the LSC method with station velocities based on GNSS observations captured from March 2010 to March 2015 (VEMOS2015, Sánchez and Drewes 2016). VEMOS2015 is based on continuously operating GNSS stations only; it does not include episodic GPS campaigns. It covers the region from 110°W, 55°S to 35°W, 32°N with a spatial resolution of 1° × 1°. The average prediction uncertainty is ±0.6 mm/a in the north-south direction and ±1.2 mm/a in the east-west direction. The maximum uncertainty (±9 mm/a) occurs in the Maule deformation zone (Chile), while the minimum (±0.1 mm/a) appears in the stable eastern part of the South American plate.

4   Present-Day Deformation Model and Velocity Field for Latin America (VEMOS2017)

The present study concentrates on the computation of a deformation model based on a set of 515 station velocities inferred from GNSS observations gained from January 2014 to January 2017 (Fig. 2). Station positions and velocities are defined at epoch 2015.0 and refer to the IGS14 Reference Frame (Rebischung 2016), which is based on the latest ITRF solution, the ITRF2014 (Altamimi et al. 2016). The estimated precision is ±1.2 mm (horizontal) and ±2.5 mm (vertical) for the station positions at the reference epoch, and ±0.7 mm/a (horizontal) and ±1.1 mm/a (vertical) for the velocities (Fig. 2). More details about the processing strategy for the determination of the station positions and velocities can be found in Sánchez and Drewes (2016) and Sánchez et al. (2015).

The complex on-going crustal deformation in the western margin of Latin America and the Caribbean has been studied intensively. Recent research concentrates on geophysical syntheses including geodetic constraints inferred from GNSS positioning to model tectonic evolution and associated geodynamic processes in this region. Most of these studies assume a segmentation of the Earth's crust and describe the surface kinematics by means of tectonic blocks or slivers rotating individually; see e.g., Brooks et al. (2011), Calais et al. (2016), Franco et al. (2012), McFarland et al. (2017), Mendoza et al. (2015), Nocquet et al. (2014), Symithe et al. (2015), Weiss et al. (2016), and references herein. This paper presents two main innovations with respect to the above-mentioned publications: Firstly, we compute a deformation model for the entire Latin American and Caribbean region and not for isolated areas only. Secondly, we assume a continuous lithosphere deforming under certain kinematic boundary conditions (as suggested by Flesch et al. 2000; Vergnolle et al. 2007; or Copley 2008), without introducing small lithospheric blocks or slivers, which would enforce constraints on the kinematic model. For the collocation procedure, we consider the main tectonic plates South America (SA), Caribbean (CA), and North America (NA) (Fig. 1) according to the tectonic plate boundary model PB2002 (Bird 2003). Based on the velocities obtained in this study for the stations located on the stable part of the plates, we estimate plate rotation vectors following the strategy presented by Drewes (1982, 2009). These plate motions are removed from the pointwise velocities to get the residual velocities, which are interpolated to a continuous deformation model

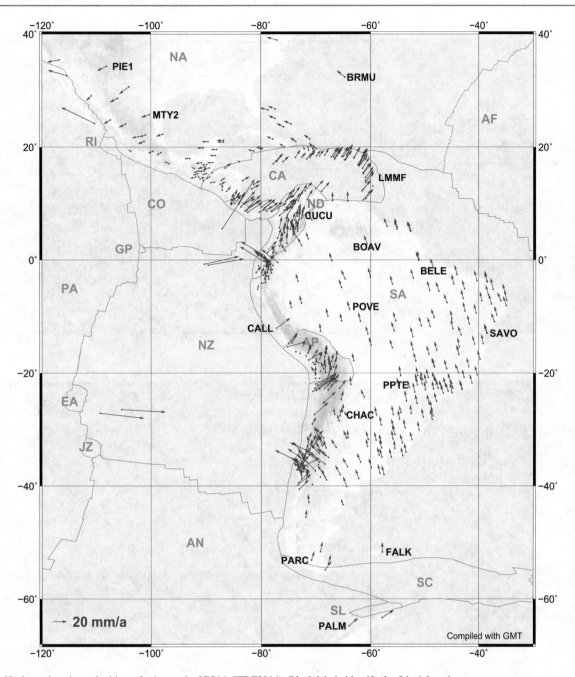

Fig. 2 Horizontal station velocities referring to the IGS14 (ITRF2014). Black labels identify the fiducial stations

using Eqs. (1)–(4). The residual velocities with respect to the Caribbean plate are used for the LSC prediction in Mexico, Central America and the Caribbean (Fig. 3), while the residual velocities with respect to the South American plate are used in South America (Fig. 4). The collocation domain at every grid node is created by selecting the stations located up to a distance of 200 km. If no stations are available at this distance, the LSC is computed using the three nearest stations. In total 2,233 grid points are predicted. Once the LSC prediction is performed, the previously reduced trends (plate rotations) are restored to the interpolated residual velocities at the grid nodes to generate a continuous velocity field referring to the IGS14 (ITRF2014). The average prediction uncertainty is ±1.0 mm/a in the north-south direction and ±1.7 mm/a in the east-west direction. The maximum uncertainty values (up to ±15 mm/a) occur at the zones affected by recent strong earthquakes, not only in the Maule area but also in the northern part of Chile, Ecuador and Costa Rica. The best uncertainty values (about ±0.1 mm/a) are evident in the stable eastern part of the South American plate.

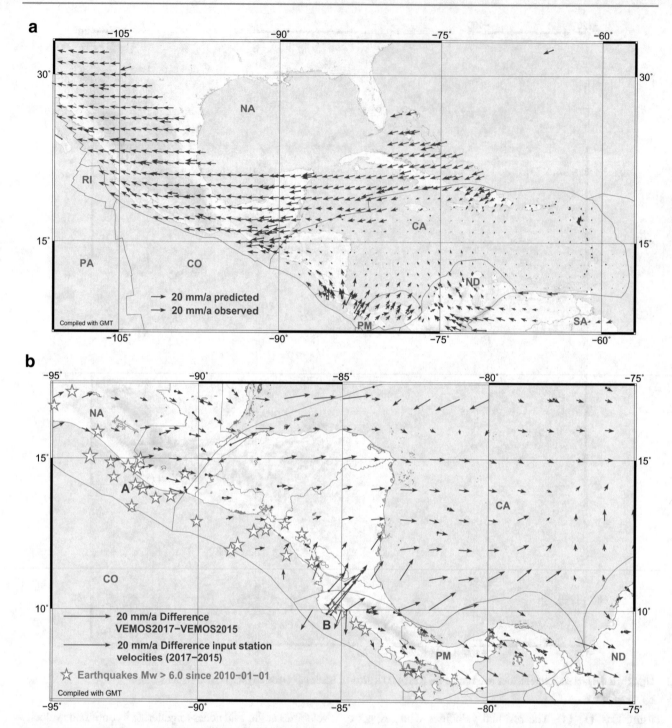

Fig. 3 (**a**) Surface deformation model VEMOS 2017 relative to the Caribbean plate. Blue vectors represent the input station velocities. (**b**) Differences of the input station velocities (blue vectors) and the deformation models (red vectors) VEMOS2015 (Sánchez and Drewes 2016) and VEMOS2017 (this study). Stars represent earthquakes with Mw > 6.0 occurred since Jan 1, 2010. The large discrepancies appear close to the epicentre of the strong earthquakes in Champerico (Mw7.4) on Nov 11, 2012 (marked with A) and Nicoya (Mw7.6) on Sep 5, 2012 (marked with B)

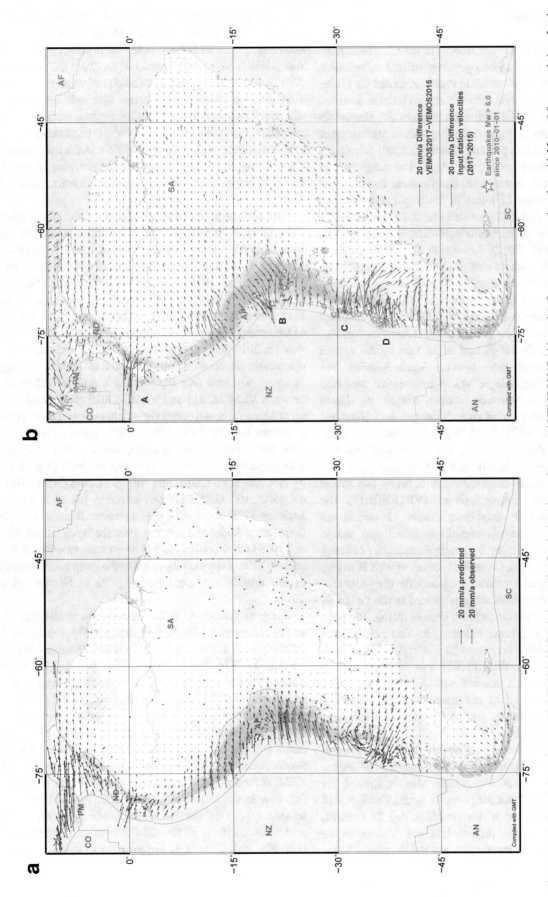

Fig. 4 (a) Surface deformation model VEMOS 2017 relative to the South American plate. Blue vectors represent the input velocities. (b) Differences of the input station velocities (blue vectors) and the deformation models (red vectors) VEMOS2015 (Sánchez and Drewes 2016) and VEMOS2017 (this study). Stars represent earthquakes with Mw > 6.0 occurred since Jan 1, 2010. The large discrepancies appear close to the epicentre of strong earthquakes: (A) Pedernales Mw7.8 on Apr 16, 2016, (B) Pisagua Mw8.2 on Apr 1, 2014, (C) Illapel Mw8.3 on Sep 16, 2015, (D) Maule Mw8.8 on Feb 27, 2010

5   Discussion

The deformation model with respect to the Caribbean plate (Fig. 3a) shows an inhomogeneous surface kinematics. While the deformation vectors in Puerto Rico and the Lesser Antilles show small (less than 0.5 mm/a) relative motions, the direction of the deformation vectors in Hispaniola describes a southward rotation starting with an orientation of S70°W in the northern part and reaching a south orientation in the southernmost part of the island. The magnitude of the vectors also decreases with this rotation: the averaged deformation is about 12 mm/a in the North and less than 1 mm/a in the South. These deformation patterns are in agreement with the GPS results published in earlier studies, e.g.; Benford et al. (2012), Symithe et al. (2015), Calais et al. (2016). In the southern area of Central America (Panama block), we observe horizontal deformations in the range 5–15 mm/a relative to the Caribbean plate. These large magnitudes are dominated in the West by the north-eastward motion of the Cocos plate towards Central America (see Fig. 3a around longitude 84°W) and in the East by the eastern motion of the Nazca plate towards South America (see Fig. 3a around longitude 78°W). A progressive westward rotation of the deformation vectors toward the North American plate is detected over Nicaragua and Honduras (longitudes from 85°W to 90°W), where the very small magnitudes of the deformation vectors suggest that this region moves homogeneously with the Caribbean plate.

Figure 3b presents the differences between this model (VEMOS2017) and the previous one (VEMOS2015). The largest differences in magnitude (about 12 mm/a) are a consequence of post-seismic displacement and station velocity changes caused by the strong earthquake of Nicoya (Mw 7.6, Sep 5, 2012), Costa Rica, (marked with B in Fig. 3b). This earthquake produced co-seismic displacements up to 30 cm at the GNSS stations located in the Peninsula Nicoya (see Fig. 9 in Sánchez and Drewes 2016). The post-seismic relaxation process induces pre- and post-seismic station velocity differences up to 30 mm/a. Another relevant discrepancy between VEMOS2017 and VEMOS2015 is observed in Guatemala (marked with A in Fig. 4). In this case, the difference in the deformation magnitude (about 5 mm/a) is mainly caused by the Champerico earthquake (Mw 7.4, Nov 11, 2012).

The deformation model with respect to the South American plate (Fig. 4a) clearly defines the stable area belonging to the Guiana, Brazilian and Atlantic shields. Indeed, the present VEMOS2017 and the previous model VEMOS2015 are practically identical in this area (Fig. 4b). In contrast, the deformation vectors predicted in the Andean region are characterized by magnitudes up to 30 mm/a. These vectors are roughly parallel to the plate subduction direction and their magnitudes diminish with the distance from the subduction front as already stated by previous publications like Bevis et al. (2001), Brooks et al. (2011), Chlieh et al. (2011), Khazaradze and Klotz (2003) and references herein. However, we observe three zones with *anomalous* vector directions (oriented to the NW): the western part of Ecuador around latitude zero, the north of Chile around latitude 20°S, and the Maule region (around 38°S). As in the case of Central America, these *abnormalities* are also caused by recent strong earthquakes and post-seismic relaxations (Fig. 4b).

The surface deformation predicted for the North Andes (ND) block is characterized by two different kinematic patterns: a north-eastward motion with increasing magnitudes of about 9 mm/a in the southern part of Colombia (latitude 3°N) to 15 mm/a in the northern border area with Venezuela (72°W, 12°N); and opposite oriented deformation vectors in Ecuador (south of latitude 3°S). The latter is a consequence of the strong earthquake occurred in Pedernales (Mw 7.8) on Apr 16, 2016. This earthquake produced co-seismic station displacements up to 80 cm and station velocity changes of about 40 mm/a (see Fig. 4b, mark A). The differences between VEMOS2017 and VEMOS2015 in this area come up to 22 mm/a. South of this region, the poor station coverage in central Peru (latitudes 5°S to 12°S) prevents concluding statements about the deformation pattern in this area; however, our model agrees quite well with the findings published by Nocquet et al. (2014) and Villegas-Lanza (2014). Based on about 100 GNSS stations covering the area between latitudes 12°S and 4.6°N, they conclude that the southern Ecuadorian Andes and northern Peru (between latitudes 5°S and 10°S) move coherently 5–6 mm/a with an orientation of about S70°E. They also suggest that the internal deformation in this area is negligible (see Fig. 2a in Nocquet et al. 2014).

South of latitude 15°S the deformation model (Fig. 4a) and its comparison with the previous one (Fig. 4b) are highly influenced by three major earthquakes: Pisagua (Mw8.2) on Apr 1, 2014, Illapel (Mw8.3) on Sep 16, 2015, and Maule (Mw8.8) on Feb 27, 2010. Before the Pisagua earthquake, the GNSS stations moved about 27 mm/a N45°E; after the earthquake, they are moving 5 mm/a to the North (see ITRF-related station velocities in Fig. 2). This produces an apparent smaller deformation with respect to the South American plate and the differences between both VEMOS models reach magnitudes up to 20 mm/a (mark B in Fig. 4b). In the area Illapel (mark C in Fig. 4b), the post-seismic effects of the 2015 earthquake superimpose the post-seismic effects of the 2010 Maule earthquake (mark D in Fig. 4b). Thus, it is not possible to distinguish their individual contributions to the deformation. As a matter of fact, the complex kinematic pattern south of latitude 25°S described by Sánchez and Drewes (2016, Fig. 18) persists. A large counter clockwise rotation around a point south of the 2010 epicentre (35.9°S, 72.7°W) and a clockwise rotation north of the epicentre are further observed (Fig. 4a). However, magnitude and direction of the deformation vectors considerably differ from those obtained in the previous model VEMOS2015. This is probably a consequence of the post-seismic relaxation process that is bringing the uppermost crust layer to the aseismic NE motion in this zone as suggested by e.g., Bedford et al. (2016), Klein et al. (2016) and Li et al. (2017). The surface kinematics shown in Fig. 4a again makes evident that the deformation regime imposed by the Maule earthquake reaches the Atlantic coast in Argentina. The comparison of the present deformation model with VEMOS2015 in the Maule surroundings presents discrepancies up to 25 mm/a (marks C and D in Fig. 4b). To provide an integrated view of the changing surface-kinematics in the Andean Region, Fig. 5 presents an extract of the models VEMOS2003, VEMOS2009, VEMOS2015 and VEMOS2017.

6 Conclusions and Outlook

This paper presents the surface velocity and deformation models of the entire Latin American and Caribbean region over the time-span 2014–2017 and describes the evolution of the models from previous studies. The effects of the extreme changes in the surface kinematics complicate the long-term stability expected in any reference frame. Therefore, a major recommendation is to materialise the geodetic reference frames by means of a dense network of continuously operating stations and to repeat the velocity computations frequently. This ensures a permanent monitoring of possible reference frame deformations. Nevertheless, a reliable deformation modelling is not yet guaranteed. Some authors suggest the implementation of geodynamic models to predict the pointwise coordinate changes caused by co-seismic and post-seismic effects (see e.g., Snay et al. 2013; Bevis and Brown 2014; Gómez et al. 2015). Since these models rely on hypotheses about the physical properties of the upper Earth crust, different hypotheses produce different results as demonstrated by e.g., Li et al. (2017). We based our analyses on the least-squares collocation as this approach respects the consistency of the geodetic observations and ensures a better agreement with the actual deformation. A problem in the geodetic use of pointwise velocities derived from multi-year solutions is their inconsistency after seismic events, i.e. their short-term validity. In the Andes region, like in any active seismic region of the Earth, there are large discontinuities in the station coordinate time series and considerable variations in the station velocities caused by strong earthquakes. The consequence is that the respective reference frames (e.g., ITRF) cannot be used or have to be frequently updated for geodetic purposes (like SIRGAS). An alternative of using multi-year solutions with station velocities is the release of frequent reference frames (e.g., every week or month). Our recommendation for the SIRGAS national reference frames in seismic active regions is to use the SIRGAS weekly coordinate solutions instead of velocities after seismic events. To consider discontinuities in the coordinates of non-permanently observed points, one has to interpolate them from the coordinate differences in reference stations. In any case, we shall continue the computation of short-period velocity and deformation models for the next future in order to enable the use of coordinates in close alignment with to the IGS reference frames.

7 Supplementary Data

In the preparation of the GNSS data solutions used in this study, we computed a new SIRGAS reference frame solution following the same procedure described in Sánchez and Drewes (2016). This solution, called SIR17P01, covers the time-span from April 17, 2011 to January 28, 2017, contains 345 SIRGAS stations with 502 occupations and is aligned to IGS14, epoch 2015.0. The SIR17P01 station positions and velocities as well as the VEMOS2017 model (velocity and deformation fields) are available at https://doi.pangaea.de/10.1594/PANGAEA.912349 and https://doi.pangaea.de/10.1594/PANGAEA.912350, respectively.

Acknowledgments We are much obliged to SIRGAS for providing us with the weekly normal equations of the SIRGAS reference network (ftp://ftp.sirgas.org/pub/gps/SIRGAS/). They are the primary input of this study. We are also grateful to the International GNSS Service (IGS, Johnston et al. 2017) and UNAVCO for making available some of the invaluable GNSS data sets used in this study (see red dots in Fig. 1). The GNSS data analysis and the computation of the multi-year solution presented in this work were accomplished with the Bernese GNSS Software version 5.2 (Dach et al. 2015). The maps were compiled with the Generic Mapping Tools (GMT) software package version 5.1.1 (Wessel et al. 2013). The topography represented on the maps corresponds to the ETOPO1 dataset (Amante and Eakins 2009).

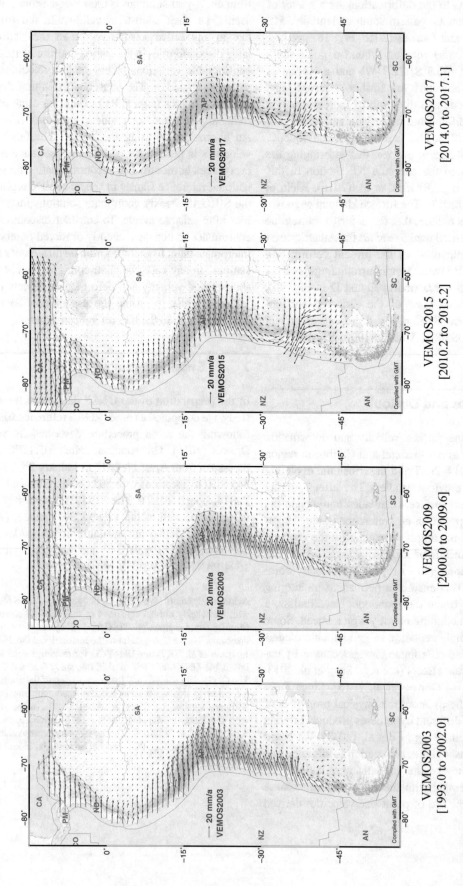

Fig. 5 Deformation model series in Latin America. VEMOS2003 (Drewes and Heidbach 2005), VEMOS2009 (Drewes and Heidbach 2012), VEMOS2015 (Sánchez and Drewes 2016), VEMOS2017 (this study)

References

Altamimi Z, Rebischung P, Métivier L, Collilieux X (2016) ITRF2014: A new release of the International Terrestrial Reference Frame modeling nonlinear station motions. J Geophys Res Solid Earth 121:6109–6131. https://doi.org/10.1002/2016JB013098

Amante C, Eakins BW (2009) ETOPO1 Global Relief Model converted to PanMap layer format. NOAA-National Geophysical Data Center, PANGAEA. https://doi.org/10.1594/PANGAEA.769615

Bedford J, Moreno M, Li S, Oncken O, Baez JC, Bevis M, Heidbach O, Lange D (2016) Separating rapid relocking, afterslip, and viscoelastic relaxation: an application of the postseismic straightening method to the Maule 2010 cGPS. J Geophys Res Solid Earth 121:7618–7638. https://doi.org/10.1002/2016JB013093

Benford B, DeMets C, Calais E (2012) GPS estimates of microplate motions, northern Caribbean: evidence for a Hispaniola microplate and implications for earthquake hazard. Geophys J Int 191(2):481–490. https://doi.org/10.1111/j.1365-246X.2012.05662.x

Bevis M, Brown A (2014) Trajectory models and reference frames for crustal motion geodesy. J Geod 88:283–311. https://doi.org/10.1007/s00190-013-0685-5

Bevis M, Kendrick E, Smalley R Jr, Brooks BA, Allmendinger RW, Isacks BL (2001) On the strength of interpolate coupling and the rate of back arc convergence in the central Andes: an analysis of the interseismic velocity field. Geochem Geophys Geosyst 2:1067. https://doi.org/10.1029/2001GC000198

Bird P (2003) An updated digital model for plate boundaries. Geochem Geophys Geosyst 4(3):1027. https://doi.org/10.1029/2001GC000252

Brooks BA, Bevis M, Whipple K, Arrowsmith JR, Foster J, Zapata T, Kendrick E, Minaya E, Echalar A, Blanco M, Euillades P, Sandoval M, Smalley RJ (2011) Orogenic-wedge deformation and potential for great earthquakes in the central Andean backarc. Nat Geosci 4:380–383. https://doi.org/10.1038/ngeo1143

Brunini C, Sánchez L, Drewes H, Costa SMA, Mackern V, Martinez W, Seemüller W, Da Silva AL (2012) Improved analysis strategy and accessibility of the SIRGAS reference frame. In: Kenyon S, Pacino MC, Marti U (eds) Geodesy for planet earth, vol 136, IAG symposia, pp 3–10. https://doi.org/10.1007/978-3-642-20338-1_1

Calais E, Symithe S, de Lépinay BM, Prépetit C (2016) Plate boundary segmentation in the north-eastern Caribbean from geodetic measurements and Neogene geological observations. C.R. Geosci 348:42–51. https://doi.org/10.1016/j.crte.2015.10.007

Chlieh M, Perfettini H, Tavera H, Avouac J-P, Remy D, Nocquet J-M, Rolandone F, Bondoux F, Gabalda G, Bonvalot S (2011) Interseismic coupling and seismic potential along the Central Andes subduction zone. J Geophys Res 116:B12405. https://doi.org/10.1029/2010JB008166

Cioce V, Martínez W, Mackern MV, Pérez R, De Freitas S (2018) SIRGAS: Reference frame in Latin America. Coordinates XIV(6):6–10

Copley A (2008) Kinematics and dynamics of the southeastern margin of the Tibetan Plateau. Geophys J Int 174(3):1081–1100. https://doi.org/10.1111/j.1365-246X.2008.03853.x

Dach R, Lutz S, Walser P, Fridez P (eds) (2015) Bernese GNSS Software Version 5.2. Astronomical Institute, University of Bern

Drewes H (1978) Experiences with least squares collocation as applied to interpolation of geodetic and geophysical quantities. Proceedings 12th symposium on Mathematical Geophysics. Caracas, Venezuela

Drewes H (1982) A geodetic approach for the recovery of global kinematic plate parameters. Bull Geod 56:70–79. https://doi.org/10.1007/BF02525609

Drewes H (2009) The actual plate kinematic and crustal deformation model APKIM2005 as basis for a non-rotating ITRF. In: Drewes H (ed) Geodetic reference frames, IAG symposia, vol 134, pp 95–99. https://doi.org/10.1007/978-3-642-00860-3_15

Drewes H (2012) Current activities of the International Association of Geodesy (IAG) as the successor organisation of the Mitteleuropäische Gradmessung. Z. f. Verm.wesen (zfv) 137:175–184

Drewes H, Heidbach O (2005) Deformation of the South American crust estimated from finite element and collocation methods. In: Sansò F (ed) A window on the future of geodesy, IAG symposia, vol 128, pp 544–549. https://doi.org/10.1007/3-540-27432-4_92

Drewes H, Heidbach O (2012) The 2009 horizontal velocity field for South America and the Caribbean. In: Kenyon S, Pacino MC, Marti U (eds) Geodesy for planet earth, vol 136, IAG symposia, pp 657–664. https://doi.org/10.1007/978-3-642-20338-1_81

Drewes H, Kaniuth K, Voelksen C, Alves Costa SM, Souto Fortes LP (2005) Results of the SIRGAS campaign 2000 and coordinates variations with respect to the 1995 South American geocentric reference frame. In: A window on the future of geodesy, vol 128, IAG symposia, pp 32–37. https://doi.org/10.1007/3-540-27432-4_6

Flesch LM, Holt WE, Haines AJ, Bingming S-TB (2000) Dynamics of the Pacific-North American Plate Boundary in the Western United States. Science 287(5454):834–836. https://doi.org/10.1126/science.287.5454.834

Franco A, Lasserre C, Lyon-Caen H, Kostoglodov V, Molina E, Guzman-Speziale M, Monterosso D, Robles V, Figueroa C, Amaya W, Barrier E, Chiquin L, Moran S, Flores O, Romero J, Santiago JA, Manea M, Manea VC (2012) Fault kinematics in northern Central America and coupling along the subduction interface of the Cocos Plate, from GPS data in Chiapas (Mexico), Guatemala and El Salvador. Geophys J Int 189(3):1223–1236. https://doi.org/10.1111/j.1365-246X.2012.05390.x

Gómez DD, Piñón DA, Smalley R, Bevis M, Cimbaro SR, Lenzano LE, Barónet J (2015) Reference frame access under the effects of great earthquakes: a least squares collocation approach for non-secular post-seismic evolution. J Geod 90(3):263–273. https://doi.org/10.1007/s00190-015-0871-8

Heidbach O, Drewes H (2003) 3-D finite element model of major tectonic processes in the Eastern Mediterranean. Geol Soc Spec Publs 212:261–274. https://doi.org/10.1144/GSL.SP.2003.212.01.17

Johnston G, Riddell A, Hausler G (2017) The International GNSS Service. In: Teunissen PJG, Montenbruck O (eds) Handbook of global navigation satellite systems, 1st edn. Springer, Cham, pp 967–982. https://doi.org/10.1007/978-3-319-42928-1

Khazaradze G, Klotz J (2003) Short- and long-term effects of GPS measured crustal deformation rates along the south central Andes. J Geophys Res 108(B6):2289. https://doi.org/10.1029/2002JB001879

Klein E, Fleitout L, Vigny C, Garaud JD (2016) Afterslip and viscoelastic relaxation model inferred from the large-scale post-seismic deformation following the 2010 M_w 8.8 Maule earthquake (Chile). Geophys J Int 205(3):1455–1472. https://doi.org/10.1093/gji/ggw086

Li S, Moreno M, Bedford J, Rosenau M, Heidbach O, Melnick D, Oncken O (2017) Postseismic uplift of the Andes following the 2010 Maule earthquake: implications for mantle rheology. Geophys Res Lett 44(4):1768–1776. https://doi.org/10.1002/2016GL071995

McFarland PK, Bennett RA, Alvarado P, DeCelles PG (2017) Rapid geodetic shortening across the Eastern Cordillera of NW Argentina observed by the Puna-Andes GPS Array. J Geophys Res 122:8600–8623. https://doi.org/10.1002/2017JB014739

Mendoza L, Richter A, Fritsche M, Hormaechea JL, Perdomoa R, Dietrich R (2015) Block modeling of crustal deformation in Tierra del Fuego from GNSS velocities. Tectonophysics 651–652:58–65. https://doi.org/10.1016/j.tecto.2015.03.013

Moritz H (1973) Least squares collocation. Dt Geod Komm, Nr. A 75

Nocquet J-M, Villegas-Lanza JV, Chlieh M, Mothes PA, Rolandone F, Jarrin P, Cisneros D, Alvarado A, Audin L, Bondoux F, Martin X, Font Y, Régnier M, Vallée M, Tran T, Beauval C, Maguiña Mendoza JM, Martinez W, Tavera H, Yepes H (2014) Motion of continental slivers and creeping subduction in the northern Andes. Nat Geosci 7:287–291. https://doi.org/10.1038/ngeo2099

Petit G, Luzum B (eds) (2010) IERS Conventions 2010. IERS technical note 36. Verlag des Bundesamtes für Kartographie und Geodäsie, Frankfurt a.M.

Rebischung P (2016) [IGSMAIL-7399] Upcoming switch to IGS14/igs14.atx. https://lists.igs.org/pipermail/igsmail/2016/001233.html. Accessed 9 May 2017

Sánchez L, Drewes H (2016) Crustal deformation and surface kinematics after the 2010 earthquakes in Latin America. J Geodyn 102:1–23. https://doi.org/10.1016/j.jog.2016.06.005

Sánchez L, Seemüller W, Drewes H, Mateo L, González G, Silva A, Pampillón J, Martinez W, Cioce V, Cisneros D, Cimbaro S (2013) Long-term stability of the SIRGAS reference frame and episodic station movements caused by the seismic activity in the SIRGAS region. IAG Symposia 138:153–161. https://doi.org/10.1007/978-3-642-32998-2_24

Sánchez L, Drewes H, Brunini C, Mackern MV, Martínez-Díaz W (2015) SIRGAS Core Network Stability. In: Rizos C, Willis P (eds) IAG 150 years, vol 143, IAG symposia, pp 183–190. https://doi.org/10.1007/1345_2015_143

Seemüller W, Kaniuth K, Drewes H (2002) Velocity estimates of IGS RNAAC SIRGAS stations. In: Drewes H, Dodson A, Fortes LP, Sánchez L, Sandoval P (eds) Vertical reference systems, vol 124, IAG symposia, pp 7–10. https://doi.org/10.1007/978-3-662-04683-8_2

Seemüller W, Sánchez L, Seitz M (2011) The new multi-year position and velocity solution SIR09P01 of the IGS Regional Network Associate Analysis Centre (IGS RNAAC SIR). In: Pacino C et al (eds) Geodesy for planet earth, vol 136, IAG symposia, pp 675–680. https://doi.org/10.1007/978-3-642-20338-1_110

SIRGAS (1997) SIRGAS Final Report; Working Groups I and II IBGE, Rio de Janeiro; 96 p. http://www.sirgas.org/fileadmin/docs/SIRGAS95RepEng.pdf. Accessed 9 May 2017

Snay RA, Freymueller JT, Pearson C (2013) Crustal motion models developed for version 3.2 of the horizontal time-dependent positioning utility. J Appl Geod. https://doi.org/10.1515/jag-2013-0005

Symithe S, Calais E, de Chabalier JB, Robertson R, Higgins M (2015) Current block motions and strain accumulation on active faults in the Caribbean. J Geophys Res 120:3748–3774. https://doi.org/10.1002/2014JB011779

Vergnolle M, Calais E, Dong L (2007) Dynamics of continental deformation in Asia. J Geophys Res 112:B11403. https://doi.org/10.1029/2006JB004807

Villegas-Lanza JC (2014) Earthquake cycle and continental deformation along the Peruvian subduction zone. Dissertation, Universite de Nice Sophia Antipolis – UFR Sciences, Ecole Doctorale des Sciences Fondamentales et Appliquées

Weiss JR, Brooks BA, Foster JH, Bevis M, Echalar A, Caccamise D, Heck J, Kendrick E, Ahlgren K, Raleigh D, Smalley R, Vergani G (2016) Isolating active orogenic wedge deformation in the southern Subandes of Bolivia. J Geophys Res Solid Earth 121:6192–6218. https://doi.org/10.1002/2016JB013145

Wessel P, Smith WHF, Scharroo R, Luis JF, Wobbe F (2013) Generic Mapping Tools: Improved version released. EOS Trans AGU 94:409–410

Progress in GTEWS Ground Displacement Measurements and Tsunami Warning

Marcelo C. Santos

Abstract

Since its early days, GNSS has been employed for the monitoring of sudden ground movements, such as earthquakes. Its use as a tool to enhance tsunami detection was boosted after analysis of data following the December 2004 Great Ocean Indian Tsunami. The contribution of GNSS towards tsunami warning systems is possible due to several factors, such as advances in the measurement of crustal displacement, developments in GNSS methodology, the growing availability of real-time data streams and advances in processing power and communication means. The paper focuses on the progress of Global Navigation Satellite System Tsunami Early Warning Systems (GTEWS) identifying current implementations and future directions and challenges. The discussion leads to the conclusion that the GNSS technology already satisfies requirements of tsunami early warning systems and that the major hurdles are with other aspects, such as optimal network configuration, real-time flow of data, communication infrastructure, and national and international collaboration. The paper ends highlighting the important role that the Global Geodetic Observing System (GGOS) can play to help overcoming those hurdles.

Keywords

Early warning systems · Geohazards monitoring · GGOS · GNNS · Ground-based GNSS · Tsunami

1 Introduction

GTEWS, GNSS Tsunami Early Warning Systems, is a GNSS-based tsunami early warning system, which offers new observables highly complementary to existing ones. We can consider that it came as a response to a massive tragedy, the Great Indian Ocean Tsunami of December 26, 2004 (because of the date, commonly known as *Boxing Day* tsunami), which claimed more than 230,000 lives across 14 countries (ABC, Boxing Day tsunami 2014). The tragedy was partially blamed on the lack of a tsunami warning system in the Indian Ocean. Early warning would not have avoided material loss but it would have helped to decrease human loss. It was also recognized that global seismology failed to accurately diagnose the magnitude of this event in time. It became clear there was a need to strengthen tsunami early warning systems and need to an accurate and rapid estimation of tsunami potential. GNSS emerged as a complementary observation platform.

Until then, tsunami early warning relied on seismically determined earthquake magnitude, source and extent of the earthquake. Even with modern seismometers that can provide earthquake information online a few minutes after the earthquake, the inversion of seismographs for complex models is time consuming, delaying the estimate (Song 2007) even though there exists faster ways to compute metrics of an earthquake source. The main aspect though is that seismic methods cannot accurately sense long period energy, the

M. C. Santos (✉)
Department of Geodesy and Geomatics Engineering, University of New Brunswick, Fredericton, NB, Canada
e-mail: msantos@unb.ca

© The Author(s) 2020
J. T. Freymueller, L. Sanchez (eds.), *Beyond 100: The Next Century in Geodesy*,
International Association of Geodesy Symposia 152, https://doi.org/10.1007/1345_2020_115

larger the magnitude of the earthquake is, something that GNSS can without any concern of saturation (Melgar et al. 2015).

The potential of GNSS was proved with the several post-mission analysis following the 2004 event, when it became evident that 15-min warning was possible if GNSS data were available in real time (Blewitt et al. 2006; Sobolev et al. 2006; Song 2007). For example, Blewitt et al. (2006) demonstrated the use of GPS real-time displacement data would have allowed estimating the moment of the earthquake hours before seismograph estimate was available, therefore, predicting the likelihood of the tsunami and potentially saving thousands of lives. Later, the Tohoku-oki event of 2011, in Japan, was also used to demonstrate that a 25 min accurate early warning was possible (Ohta et al. 2012; Song et al. 2012; Xu and Song 2013; Melgar et al. 2013; Hoechner et al. 2013; Melgar and Bock 2013).

This paper identifies current implementations and future directions of ground-based GNSS tsunami early warning systems, under the light of the IUGG 2015 Resolution 4 (IUGG, Resolution 4 2015) and the UN Sendai Framework for Disaster Risk Reduction (UNISDR 2015). Section 2 overviews the contributions GNSS can provide and the requirements that need to be satisfied for that, Sect. 3 goes over some of the existing and proposed networks, while Sect. 4 indicates GNSS-aided models. Section 5 looks at the recommendations and Sect. 6 ends the paper with a discussion on the status of GNSS for tsunami early detection.

2   The Contribution of GNSS

As mentioned before, there are limitations with seismically determined earthquake magnitude for large earthquakes particularly for tsunami early detection due to possible under-estimation of earthquake's magnitude, source and extent. The positive contribution brought by GNSS displacement measurements is to provide both magnitude and direction of ground motion, a critical information for estimating seafloor displacement and its tsunamigenic potential (Song 2007; Sobolev et al. 2006). Quoting from Song (2007):

> Coastal GPS stations are able to detect continental slope displacements of faulting due to big earthquakes, and that the detected seafloor displacements are able to determine tsunami source energy and scales instantaneously.

In 2007, Sobolev et al. (2007) introduced the idea of the *GNSS Shield*, a near-field GNSS array capable of predictions within 10 min after the occurrence of an earthquake. Such array would follow an optimum distribution derived from a geodynamic numerical model of the local geology, significantly improving the measurement accuracy of crustal displacement and the potential tsunami impact. The GNSS

shield could be formed for specific zones or around the Indo-Pacific Ring of Fire for tsunami early warning.

These early studies proved that GNSS is indeed capable of providing fundamental information for early detection of tsunami, alleviating the chances of false alarm. More recent studies have corroborated and advanced those initial findings.

To fulfill its full potential, GNSS enhancement to tsunami early warning requires:

– real time access to an optimally distributed network of GNSS receivers,
– reliable broadband communications,
– a good number of capable analysis centers,
– and products that can be rapidly assimilated into the existing tsunami early warning systems

The bottleneck is not the GNSS capabilities (including models and processes) but how to make sure there is GNSS-derived information available as soon as possible.

3   GTEWS Networks

What follows is a brief review of GNSS networks used in tsunami early warning systems.

Indonesian Tsunami Early Warning System for the Indian Ocean (InaTEWS) InaTEWS is a nice example of international cooperation between nations, which followed in the aftermath of the 2004 Boxing Day Tsunami (Fig. 1). Initially called German-Indonesian Tsunami Early Warning System for the Indian Ocean (GITNEWS), it started in 2005 and is run by Indonesian Meteorological, Climatological and Geophysical Services (BMKG) since 2014. Babeyko (2017), Harig et al. (2019).

Real-Time GEONET Analysis for Rapid Deformation Monitoring (REGARD) GEONET is a network under the Geospatial Information Authority of Japan (GSI) and Japan's Meteorological Agency (JMA) early warning mandate. GEONET may very well be the world's densest GNSS ground network, with 1200 GPS receivers, providing valuable observations of crustal deformation (Kawamoto et al. 2017). I have been informed that GEONET data may not be generally available (Fig. 2).

Network of the Americas (NOTA) NOTA is a federated network that incorporates three existing major networks: the EarthScope Plate Boundary Observatory (PBO)—a set of 1,100 stations spanning Alaska, the continental US and Puerto Rico; TLALOCNet—40 stations in Mexico; and COCONet—85 stations spanning the Caribbean. NOTA forms a hemispherical-size network with the goal to support a wide range of scientific applications and stakeholders. The intention is to offer high-rate (1 Hz or higher), low-latency (1 s or less) GNSS data, with 25% of the stations

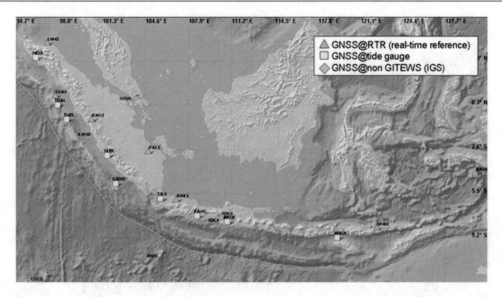

Fig. 1 Indonesian Tsunami Early Warning System for the Indian Ocean

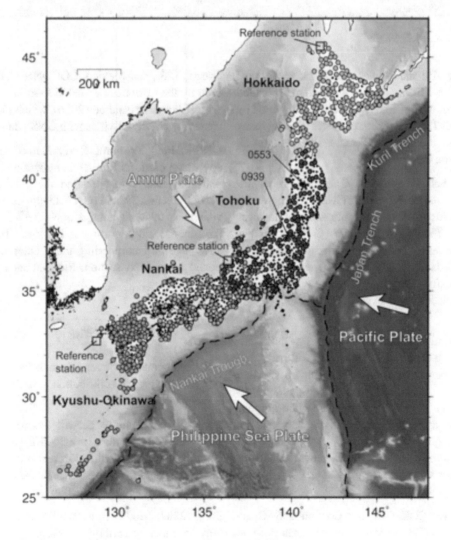

Fig. 2 Real-time GEONET Analysis for Rapid Deformation Monitoring (REGARD)

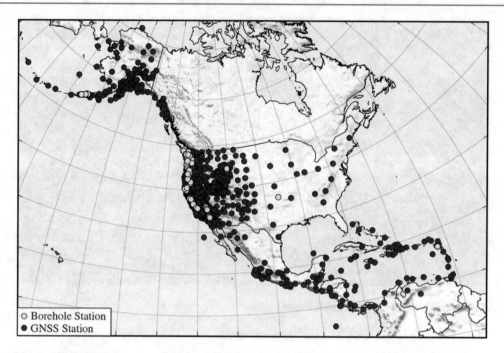

Fig. 3 Network of the Americas (NOTA)

being upgraded to full multi-constellation. A number of GNSS stations have MEMS accelerometers, being called seismogeodetic stations (Fig. 3). GNSS data is being used by NOAA US National Tsunami Warning System and by many other institutions such as CWU, JPL and SIO (Geng et al. 2013a,b; Hodgkinson et al. 2018).

Chilean National Seismic Network The network is composed of nearly 150 stations, operating GNSS, seismometers and strong motion instruments. It is run under the auspices of Chile's *Centro Sismologico Nacional* (CSN). It applies precise point positioning and ambiguity resolution in the treatment of GNSS data (Geng et al. 2013b), and, according to Riquelme et al. (2016), applies W-phase and peak ground displacement models for the analysis of seismogeodetic data (Fig. 4).

Asia-Pacific Reference Frame Network It is a GNSS network of participating countries, with a Central Bureau, co-ordinated by Geoscience Australia, to maintain an accurate geodetic framework serving all types of geospatial applications (Fig. 5). Even though not designed for the purpose of a GTEWS, the availability of real-time data and its geographical coverage, near several important trenches and subduction zones, some of the data collected by this network are appealing to be used for tsunami detection (Asia-Pacific Reference Frame 2019).

Caribbean GTEWS COCONet, the Continuously Operating Caribbean GPS Observational Network, has become part of the federated Network of the Americas (NOTA), which also includes networks spanning Alaska, the con-

tiguous U.S., and Mexico. COCONet GNSS data could be used by the Caribbean Tsunami Warning Program (CTWP) (von Hillebrandt-Andrade 2016), which plans to enhance its tsunami warning activities with GNSS data (Fig. 6).

GNSS-Aided Tsunami Early Detection System Taking advantage of all its infrastructure and availability of GNSS real-time data, NASA started the GNSS-Aided Tsunami Early Detection System (GATED) project (Song et al. 2018), which also takes advantages of NOAA's DART (Deep-ocean Assessment and Reporting of Tsunamis) buoys (Fig. 7). The advantage of incorporating these latter data will be mentioned in the next section. More about DART in Mungov et al. (2013).

4   GNSS-Aided Models

The development of several models have allowed GNSS to offer a fundamental contribution to early tsunami detection. A summary follows.

Blewitt (2006) Model Blewitt (2006) model showed that the correct magnitude of the 2004 Boxing Day earthquake could have been determined in real time by complementing the seismic data with measurements from relatively few GPS tracking sites (Blewitt et al. 2006).

Song (2007) Model Song (2007) model estimates the energy an undersea earthquake transfers to the ocean to generate a tsunami by using data from coastal GPS stations near the epicenter. With these data, ocean floor displacements

Fig. 4 Chilean National Seismic Network

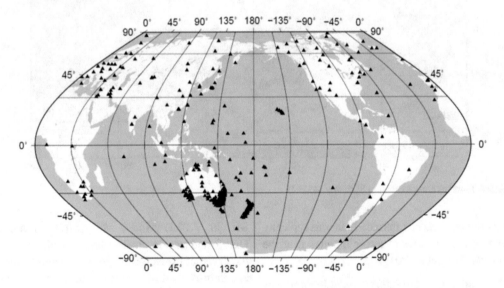

Fig. 5 Asia-Pacific Reference Frame Networks

Fig. 6 COCONet—Continuously Operating Caribbean GPS Observational Network

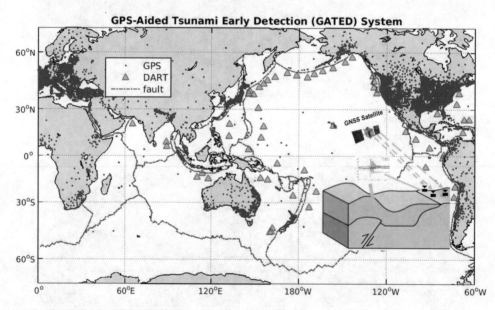

Fig. 7 GNSS-Aided Tsunami Early Detection System (GATED)

caused by the earthquake can be inferred. Tsunamis typically originate at undersea boundaries of tectonic plates near the edges of continents (Song et al. 2017).

Sobolev (2007) Model Sobolev (2007) model incorporates numerical models of regional geology and local infrastructure to derive the deployment of an optimal GNSS network (Sobolev et al. 2007).

Melgar (2013) Model Melgar (2013) model combines GNSS and accelerometer data to estimate seismogeodetic displacement waveforms, providing mm-level 3-D accuracy and improved estimation of coseismic deformation compared to GNSS-only methods (Melgar et al. 2013).

Titov (2016) Model Titov (2016) model combines GNSS data with DART data determining real-time tsunami source energy improving forecast accuracy and early cancellations, decreasing the chance of false alarm (Titov et al. 2016).

Going Mobile During his oral presentation at the IUGG General Assembly, in 2019, Dr. T. Song discussed the development of an App that would provide tsunami warning via the cell phone, derived from the GATED infra-structure (Song 2019). Imagine swimmers accessing that while enjoying the beach! That reminded me of an existing online application called VADASE, the Variometric Approach for Displacements Analysis Standalone Engine, algorithm (Benedetti et al. 2014). VADASE was successfully applied to estimate in a real-time scenario the ground velocities and displacements induced by several earthquakes, and is available on line (Fortunato et al. 2019). It does not offer tsunami warning though.

5   A Look into Recommendations

To further our discussion, let us look at the two documents of interest. In a nutshell, the IUGG Resolution 4 (focus on technology) (IUGG, Resolution 4 2015):

> supports the enhancement of Tsunami Early Warning Systems with GNSS real time technology.

The UN Sendai Framework is a very large document, but here we will summarize their concern with human impact and networking. In a nutshell, it suggests actions to UNISDR (2015):

> decrease the impacts, being human, economic and on infrastructure and disruption of basic services, enhance international collaboration, and increase the availability of and access to multi-hazard early warning systems.

6   Discussion

As it stands, GNSS technology seems to be capable to provide tremendous contribution to early warning systems. Tests showed that tsunami detection (without the risk of false alarms) is possible just a few minutes after the earthquake, provided the array of GNSS receivers are located within a certain range from the epicenter. As said by Babeyko (2017):

> each particular geographical region needs its own strategy for optimal GNSS-based early warning.

Today, there are capable analysis centers and a growing experience with the handling of GNSS data in real time. There is also a growing availability of high-quality real-time products. And it is reasonable to say that further developments are expected as far as geodetic science is concerned.

Things look bright from the perspective of GNSS.

The problem is how to deliver the technology, in terms of:

- optimal network configuration (possible network configuration),
- real-time flow of data,
- communication infrastructure, and
- national and international collaboration (including funding).

All those challenges do not depend on the GNSS technology. The fundamental challenge might be more of coordination rather than with GNSS technology itself.

We conclude the paper emphasizing the important role that GGOS can play helping with coordination and with bringing the multiples organizations and nations together, not just for the optimal use of GNSS technology, but also to help overcome the bottlenecks that prevent the GNSS technology to be used in all its plenitude.

Acknowledgements The author would like to acknowledge financial support from the Natural Sciences and Engineering Research Council of Canada. He also thanks the very useful critique from both anonymous reviewers.

References

ABC, Boxing Day tsunami (2014) How the disaster unfolded 10 years ago. Australian Broadcasting Corporation, The Internet, 23 Dec 2014

Asia-Pacific Reference Frame (APREF) (2019) Retrieved July 5, 2019 from http://www.ga.gov.au/scientific-topics/positioning-navigation/geodesy/asia-pacific-reference-frame

Babeyko A (2017) GITEWS - lessons learned, GNSS tsunami early warning system workshop, July 25–27, Sendai, Japan, 2017

Benedetti E, Branzanti M, Biagi L, Colosimo G, Mazzoni A, Crespi M (2014) Global navigation satellite systems seismology for the 2012mw6.1 emilia earthquake: Exploiting the VADASE algorithm. Seismol Res Lett 85(3):649–656. https://doi.org/10.1785/0220130094

Blewitt G, Kreemer C, Hammond WC, Plag H-P, Stein S, Okal E (2006) Rapid determination of earthquake magnitude using GPS for tsunami warning systems. Geophys Res Lett 33:L11309. https://doi.org/10.1029/2006GL026145

Fortunato M, Sonnessa A, Ravanelli R, Mazzoni A, Crespi M (2019) Validation of GNSS variometric web engine: a new tool for GNSS community. IUGG general assembly, 8–17 July, 2019

Geng J, Melgar D, Bock Y, Pantoli E, Restrepo J (2013a) Recovering coseismic point ground tilts from collocated high-rate GPS and accelerometers. Geophys Res Lett 40:5095–5100. https://doi.org/10.1002/grl.51001

Geng J, Bock Y, Melgar D, Crowell BW, Haase JS (2013b) A seismo-geodetic approach applied to GPS and accelerometer observations of the 2012 Brawley seismic swarm: Implications for earthquake early warning. Geochem Geophys Geosyst 14(7):2124–2142. https://doi.org/10.1002/ggge.20144

Harig S, Immerz A, Griffin WJ, Weber B, Babeyko A, Rakowsky N, Hartanto D, Nurokhim A, Handayani T, Weber R (2019) The tsunami scenario database of the Indonesia Tsunami Early Warning System (InaTEWS): Evolution of the coverage and the involved modeling approaches. Pure Appl Geophys 177:1379–1401. https://doi.org/10.1007/s00024-019-02305-1

Hodgkinson K, Mencin MD, Sievers C, Dittman T, Feaux K, Austin KE, Walls CP, Mattioli GS (2018) A real-time GNSS network of the Americas. In: American geophysical union fall meeting, abstract IN42B-04, 2018

Hoechner A, Ge M, Babeyko AY (2013) Sobolev SV Instant tsunami early warning based on real time GPS – Tohoku 2011 case study. Nat Hazards Earth Syst Sci 13:1285–1292. https://doi.org/10.5194/nhess-13-1285-2013

IUGG, Resolution 4 (2015) International union of geodesy and geophysics general assembly, 22 June - 2 July, 2015, Prague, Czech Republic

Kawamoto S, Ohta Y, Hiyama Y, Todoriki M, Nishimura T, Furuya T, Sato Y, Yahagi T, Miyagawa K, (2017) REGARD: a new GNSS-based real-time finite fault modeling system for GEONET. J Geophys Res 122:1324–1349. https://doi.org/10.1002/2016JB013485

Melgar D, Bock Y (2013) Near-feld tsunami models with rapid earthquake source inversions from land and ocean-based observations: The potential for forecast and warning. J Geophys Res 118:5939–5955. https://doi.org/10.1002/2013JB010506

Melgar D, Crowell BW, Bock Y, Haase JS (2013) Rapid modeling of the 2011 Mw 9.0 Tohoku-oki earthquake with seismogeodesy. Geophys Res Lett 40:2963–2968. https://doi.org/10.1002/grl.50590

Melgar D, Crowell BW, Geng J, Allen RM, Bock Y, Riquelme S, Hill EM, Protti M, Ganas A (2015) Earthquake magnitude calculation without saturation from the scaling of peak ground displacement. Geophys Res Lett 42:5197–5205. https://doi.org/10.1002/2015GL064278

Mungov G, Eblé M, Bouchard R (2013) DART® tsunameter retrospective and real-time data: A reflection on 10 years of processing in support of tsunami research and operations. Pure Appl Geophys 170(9–10):1369–1384. https://doi.org/10.1007/s00024-012-0477-5

Ohta Y, Kobayashi T, Tsushima H, Miura S, Hino R, Takasu T, Fujimoto H, inuma T, Tachibana K, Demachi T, Sato T, Ohzono M, Umino N (2012) Quasi real-time fault model estimation for near-field tsunami forecasting based on RTK-GPS analysis: Application to the 2011 Tohoku-Oki earthquake (Mw 9.0). J Geophys Res. https://doi.org/10.1029/2011JB008750

Riquelme S, Bravo F, Melgar D, Benavente R, Geng J, Barrientos S, Campos J (2016) W-phase source inversion using high-rate regional GPS data for large earthquakes. Geophys Res Lett 43:3178–3185. https://doi.org/10.1002/2016GL068302

Sobolev SV, Babeyko AY, Wang R, Galas R, Rothacher M, Stein D, Schröter J, Lauterjung J, Subarya C (2006) Towards real-time tsunami amplitude prediction, EOS, AGU, 87/37, 374. https://doi.org/10.1029/2006EO370003

Sobolev S, Babeyko A, Wang AR, Hoechner A, Galas R, Rothacher M, Sein DV, Schroter J, Lauterjung J, Subarya C (2007) Tsunami early warning using GPS-shield arrays. J Geophys Res 112:B08415. https://doi.org/10.1029/2006JB004640

Song YT (2007) Detecting tsunami genesis and scales directly from coastal GPS stations. Geophys Res Lett 34:L19602. https://doi.org/10.1029/2007GL031681

Song TY (2019) NASA GNSS tsunami early detection system and its performance in real time. IUGG general assembly, 8–17 July, 2019

Song TY, Fukimori I, Shum CK, Yi Y (2012) Merging tsunamis of the 2011 Tohoku-Okiearthquake detected over the open ocean. Geophys Res Lett 39:L05606. https://doi.org/10.1029/2011GL050767

Song YT, Mothat A, Yim S (2017) New insights on tsunami genesis and energy source. J Geophys Res Oceans 122:4238–4256. http://dx.doi:10.1002/2016JC012556

Song TY, Chen K, Liu Z (2018) Developing GNSS-aided tsunami early detection system. 10th ACES international workshop, Awaji Island, September 25–28, 2018

Titov V, Song T, Tang L, Bernard EN, Bar-Sever Y, Wei Y (2016) Consistent estimates of tsunami energy show promise for improved early warning. Pure Appl Geophys 173(12):3863–3880. https://doi.org/10.1007/s00024-016-1312-1

UNISDR (2015) Sendai framework for disaster risk reduction 2015–2030, UN office for disaster risk reduction, Geneva, Switzerland, 2015

von Hillebrandt-Andrade CG (2016) Caribbean tsunami warning system, National tsunami hazard mitigation program annual meeting, February 3–5, Boulder, Colorado, US, 2016

Xu Z, Song YT (2013) Combining the all-source Green's functions and the GPS-derived source for fast tsunami prediction – illustrated by the March 2011 Japan tsunami. J Atmos Ocean Technol 30:1542–1554. http://dx.doi.org/10.1175/JTECH-D-12-00201.1

Geodesy for Atmospheric and Hydrospheric Climate Research (IAG, IAMAS, IACS, IAPSO)

Characterization of the Upper Atmosphere from Neutral and Electron Density Observations

Andres Calabia and Shuanggen Jin

Abstract

Upper-atmospheric processes under different space weather conditions are still not well understood, and the existing models are far away from the desired operational requirements due to the lack of in-situ measurements input. The ionospheric perturbation of electromagnetic signals affects the accuracy and reliability of Global Navigation Satellite Systems (GNSS), satellite communication infrastructures, and Earth observation techniques. Furthermore, the variable aerodynamic drag, due to variable thermospheric mass density, disturbs orbital tracking, collision analysis, and re-entry calculations of Low Earth Orbit (LEO) objects, including manned and unmanned artificial satellites. In this paper, we use the Principal Component Analysis (PCA) technique to study and compare the main driver-response relationships and spatial patterns of total electron content (TEC) estimates from 2003 to 2018, and total mass density (TMD) estimates at 475 km altitude from 2003 to 2015. Comparison of the first TEC and TMD PCA mode shows a very similar response to solar flux, but annual cycle shown by TEC is approximately one order of magnitude larger. A clear hemispheric asymmetry is shown in the global distribution of TMD, with higher values in the southern hemisphere than in the northern hemisphere. The hemispheric asymmetry is not visible in TEC. The persistent processes including a favorable solar wind input and particle precipitation over the southern magnetic dip may produce a higher thermospheric heating, which results in the hemispheric asymmetry in TMD.

Keywords

Principal Component Analysis · Thermospheric mass density · Total electron content · Upper atmosphere

1 Introduction

The connection between solar drivers, Earth's magnetosphere, and Ionosphere-Thermosphere (IT) phenomena in the upper atmosphere is very complex and dependent on many processes, including energy-absorption, ionization, and dissociation of molecules due to variable X-ray and Extreme Ultra Violet (EUV) solar radiance (Calabia and Jin

A. Calabia · S. Jin (✉)
School of Remote Sensing and Geomatics Engineering, Nanjing University of Information Science and Technology, Nanjing, China
e-mail: andres@calabia.com; sgjin@nuist.edu.cn

2019). In addition, the variable solar wind plasma, combined with a favorable alignment of the Interplanetary Magnetic Field (IMF), can produce aurora particle precipitation at high latitudes, which results in chemical reactions and Joule heating through collisions between electrically-charged and neutral particles.

Consequences of upper-atmospheric conditions on human activity highlight the necessity to better understand and predict IT processes, potentially preventing detrimental effects to orbiting, aerial, and ground-based technologies. Charged particles (mostly free electrons in the ionosphere) are able to influence the propagation of electromagnetic radio waves in Global Navigation Satellite Systems (GNSS), satellite communication infrastructures, and Earth observation tech-

J. T. Freymueller, L. Sanchez (eds.), *Beyond 100: The Next Century in Geodesy*,
International Association of Geodesy Symposia 152, https://doi.org/10.1007/1345_2020_123

niques. For instance, the ionospheric effects on GNSS result in a time-delay of the modulated components from which pseudo-range measurements are made, and a time-advance in carrier-phase measurements (Jin and Su 2020; Jin et al. 2019, 2020). These effects are directly related to the final accuracy and reliability of GNSS onboard satellites, aviation, and ground vehicles, providing unacceptable errors for single-frequency receivers (Hofmann-Wellenhof et al. 2008).

Existing ionospheric total electron content (TEC) models are designed to give a general description of the variations in the ionosphere. These include, for example, the Bent ionospheric model (Bent and Llewellyn 1973), the Param-eterized Ionospheric Model (PIM) (Daniell et al. 1995), the NeQuick model (Radicella 2009), and the International Reference Ionosphere (IRI) model (Bilitza et al. 2011). However, the highly variable ionosphere and the complex processes involved in the IT coupling result in difficulties for accurate modeling, and the existing models only provide monthly averages of behavior for especially magnetically quiet conditions. For instance, IRI and NeQuick are only climatological models, and cannot accurately represent space weather events such as geomagnetic storms.

The variable aerodynamic-drag on LEO satellites is asso-ciated with the upper-atmospheric expansion/contraction in response to variable solar and geomagnetic activity. The drag makes tracking difficult, decelerates LEO orbits, reduces their altitude, and shortens the lifespan of satellite missions. In addition, the exponential increase of space debris (includ-ing the recent destructive events of Fengyun-1C, Iridium, and Mission Shakti) has recently highlighted the importance of orbital tracking and prediction of potential collisions with orbiting satellites. Unfortunately, the existing aerodynamic-drag models used in Precise Orbit Determination (POD) result in positioning errors, far away from the operational requirements (Anderson et al. 2009; Calabia et al. 2020), and several recent studies have exposed limitations to accurately predict the actual neutral density variability (e.g., Müller et al. 2009; Sutton et al. 2009; Doornbos et al. 2010; Emmert and Picone 2010; Liu et al. 2011; Lei et al. 2012; Chen et al. 2014; Cnossen and Förster 2016; Calabia and Jin 2016; Panzetta et al. 2018).

The first empirical models based on orbital decay were introduced by Harris and Priester (1962) and Jacchia (1964). Upper-atmospheric total mass density (TMD) models have since been improving with the use of new techniques, algo-rithms, and proxies, as for example, Bowman et al. (2008) with the JB2008, Bruinsma (2015) with the Drag Temper-ature Model (DTM), and Picone et al. (2002) with the US Naval Research Laboratory Mass Spectrometer and Incoher-ent Scatter-radar Extended 2000 model (NRLMSISE-00).

The exact connection between the ionospheric TEC and thermospheric TMD models is still unclear, but physics-based models such as the Thermosphere-Ionosphere-Electrodynamics General Circulation Model (TIEGCM) (Richmond et al. 1992), which generally use empirical parameterizations and boundary conditions to solve 3-dimensional fluid equations, can provide an approximated solution with various prognostic variables including, but not limited to, TEC, TMD, Joule heating, etc. Compared to empirical models, the physics-based models are far more complex, but can help to better understand the physical mechanisms responsible for the observed IT variability, and have a greater potential for predictions and projections of future occurrences.

Unfortunately, the existing upper-atmospheric models are incapable of predicting the IT variability as required, in spite of the efforts to model variations, anomalies, and climatology over the last half-century. This is largely due to the limited quality and quantity of observations used to better characterize the driver-response relationship of the IT variability, and the lack of comprehensive approaches for calibrating the models. In response to this situation, the international community sought to increase scientific research on upper-atmosphere modeling, develop safeguard schemes, and produce space weather products, centers, and services. As a result, on 22 April 2017, in Vienna, the Focus Area on Geodetic Space Weather Research (FA GSWR) was created within the structure of the Global Geodetic Observing System (GGOS) of the International Association of Geodesy (IAG). Since then, the GGOS FA GSWR has been defining its internal structure and main objectives with the purpose to collate existing initiatives and guide future activity in this area. Since many geodetic tasks depend on upper-atmospheric properties, the development of iono-sphere and neutral density models as GGOS products for direct applications has become the main objective. The committee of the FA GSWR selected the electron and the neutral density as potential candidates for the list of essential geodetic variables (EGV).

In this paper, we aim at the first objective with an innova-tive study based on the Principal Component Analysis (PCA) of large data-sets of globally sparse observations. In this scheme, the main PCA modes derived from different phys-ical variables will be investigated, modeled, and compared in a common space-time frame and the correlations, similari-ties, and differences will provide evidence regarding possible coupling and driving mechanisms. Previous studies using the PCA technique have mapped the data to an orthogonal basis composed of a number of spherical harmonics (SH) functions expressed in the Local Solar Time (LST) and magnetic dip latitude coordinate system (Matsuo and Forbes 2010; Wan et al. 2012). In our method, instead of employing a basis composed of a number of SH functions, we benefit from the full resolution provided by the initial variables and use of the geographical latitude and longitude coordinate system. This coordinate system will reveal possible geographical contributions, including the South Atlantic Anomaly (SAA), or the South magnetic dip, as well as simplify the modeling of the initial variables for practical applications. Other alternative techniques widely employed in spatiotemporal analyses include the use of neural networks (e.g., Gowtam et al. 2019) or wavelet analyses (e.g., Dabbakuti and Ratnam 2016).

Here we present the first results from comparing the main PCA components of TEC and TMD observations from a complete full solar cycle. Section 2 briefly introduces the data and methods employed in our research; PCA results are illustrated and compared in the third section; and finally, conclusions are given in the last section.

2 Data and Methods

2.1 Ionospheric Observations

The 16 year TEC time-series from IGS global ionosphere maps (GIMs) (2003–2018) have been downloaded from the Crustal Dynamics Data Information System (CDDIS) of National Aeronautics and Space Administration (NASA) (https://cddis.nasa.gov/index.html). The data is the result of an integrated set of GNSS dual-frequency tracking data from more than 350 permanent stations. The temporal resolution of each TEC GIM is provided at 2 h LST, and the spatial resolution is 2.5° by 5° in latitude by longitude, respectively.

2.2 Thermospheric Observations

We employ 13 years' (2003–2015) of TMD time-series inferred from accelerometer and POD measurements made by the GRACE (Gravity Recovery and Climate Experiment) mission (Tapley et al. 2004). TMD estimates were obtained from the Information System and Data Center (ISDC) Geo-ForschungsZentrum (GFZ), computed by Calabia and Jin (2016, 2017), and provided at 3-min interval sampling in Calabia and Jin (2019). In order to deal with the different orbit locations and altitudes, the data have been normalized to 475 km altitude and interpolated between orbits. More details on the data processing can be found in Calabia and Jin (2016).

2.3 Space Weather and Geomagnetic Indices

Space weather and geomagnetic indices have been downloaded from the Low-Resolution OMNI (LRO) data set of NASA (http://omniweb.gsfc.nasa.gov/form/dx1. html) and from the International Service of Geomagnetic Indices (ISGI) website (http://isgi.unistra.fr/data_download. php). NASA/ESA (European Space Agency) Solar and Heliospheric Observatory (SOHO) satellite Solar Extreme-ultraviolet Monitor (SEM) measurements are provided by Bowman et al. (2008).

2.4 PCA Modeling

The aim of the PCA technique is to determine a new set of basis vectors that capture the most dominant structures in the data, based on eigenvalue decomposition of the covariance matrix. Detailed analyses and the selection of retained modes can be found in Preisendorfer (1988) and Wilks (1995), and a readily computable algorithm in Bjornsson and Venegas (1997). In our modeling scheme, given a time-series of estimates, the spatial patterns of its variability, temporal variation, and the measure of its importance are presented as a low-dimensional space spanned by a set of modes, which can be parameterized in terms of most representative proxies (e.g., solar flux index, annual cycles, etc.). The measure of importance of each component is provided by the analysis itself, and given in % to the total variance. This is calculated from the eigenvalues (contribution of each mode to the total variability). Our general method (Calabia and Jin 2016) is composed in three basic steps, which include (1) obtaining grids for given time moments; (2) arranging each grid in by columns and finding the eigenvectors (space-dependent components), eigenvalues (contribution of each mode to the total variability), and the projections of the initial matrix on each eigenvector (time-dependent components); and (3) analytical fitting of resulting modes in terms of most representative proxies, including a statistical assessment (i.e., standard deviation, Pearson correlation).

3 Results and Analysis

The results from both TEC and TMD analyses show that the dominant forcing is the solar flux cycle. Figures 1 and 2 show the space and time-dependent components of the first PCA mode, respectively. Secondary and above PCA modes are not included in this manuscript. Concerning the analysis of TMD, the first PCA mode accounts for 92% of the total variability, and its parameterization reaches 96% of the correlation. As for the TEC variability, the first leading mode accounts for 75% of the total variance at each LST case (12 PCA analyses, one each 2 h sampling), and its parameterization reaches 98% of correlation. For both cases, the high values of explained variance indicate marked patterns of variability. The high correlations in the fitting of the PCA modes indicate high accuracy to represent the actual TMD and TEC variability.

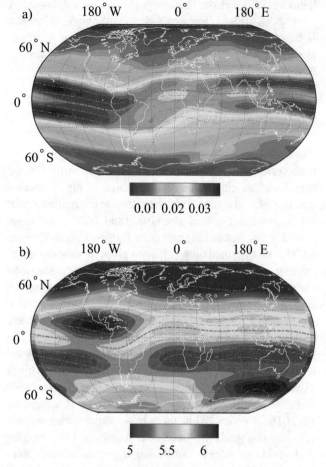

Fig. 1 First spatial PCA component of (**a**) TEC and (**b**) TMD at 475 km altitude. All longitudes provide values at noon (12 h LST). Dimensionless quantities. Dip isoclinic lines are plotted in dash-dotted lines

Each spatial PCA component is interpreted as a set of values in space that vary depending on a coefficient that varies on time (temporal PCA component). High PCA coefficient indicates high variability. Figure 1 shows the spatial components at noontime of the first PCA mode from both the 16 years' time-series of TEC and the 13 years' time-series of TMD (475 km altitude). The corresponding time-expansion coefficients (Fig. 2) are mainly related to the solar flux cycle (analysis presented in the next paragraph). In Fig. 1, we can observe that both values of TEC and TMD show a drop off along the magnetic equator, depicting the characteristic equatorial bulge with a two-crest shape, at about ±20° dip latitude for TEC (namely equatorial ionization anomaly or EIA), and about ±30° dip latitude for TMD (namely equatorial mass anomaly or EMA) (Liu et al. 2009). Minimum values along the magnetic equator are similarly located for both cases at 0°E and at about 90°E, and maximum values in the crests are located at the West, over the eastern Pacific Sea. However, distinct features show no clear spatial correlation between the TEC and TMD and

suggest they originate from different physical mechanisms. For instance, a clear asymmetry is shown only in the global distribution of TMD, with higher values in the southern hemisphere than that shown in the northern hemisphere. The most characteristic discrepancy is the marked enhancement near the southern magnetic pole (dip = 90°). None of these structures are reflected by the ionospheric TEC distribution. In addition, we can observe wider and less marked equatorial crests of TMD when comparing to the TEC crests. The maximum peak of the northern TMD crest is displaced about 40° East from the maximum peak of the TEC crests. The northern TMD crest in the eastern longitudes is very weak when comparing to the southern crest. In general, the TMD crests are very asymmetric, while the TEC crests only show a minor asymmetry over the SAA. Although both TEC and TMD show a common trough along the magnetic equator, these distinct features show no clear spatial correlation.

Both TEC and TMD temporal PCA components show a clear common driver-response relationship to solar flux, plus a small annual modulation. The corresponding time-dependent components to Fig. 1 are presented in Fig. 2, in terms of the $F10.7$ solar radio flux and annual cycle. The annual variations shown in Fig. 2 have been normalized to common solar flux F10.7 = 110 sfu. Note as well that we only investigate the first PCA mode, and the North-South annual variation caused by the angle between the ecliptic and the equatorial plane (including also Earth-Sun distance) is usually reflected in the second PCA components (Calabia and Jin 2016, Fig. 6b). In both TMD and TEC cases, clear quadratic dependencies to the $F10.7$ solar radio flux are seen in Fig. 2a and c. Concerning the annual modulation of the first PCA component, both cases have shown to increase with solar activity (not shown in this manuscript), and during equinox seasons, with higher values in December than in June (Fig. 2b). However, no clear asymmetries are depicted between March and September for any of the both TEC or TMD analysis. The annual modulation in the first PCA components shows a larger contribution for TEC, by approximately one order of magnitude, while both have a very similar shape. The maximum peaks in equinox both show to be approximately 10 days delayed from March equinox (day-of-year 90) and approximately 1 month from September equinox (day-of-year 300). The minimum peaks in solstice both show to be delayed 20 days from June solstice (day-of-year 191) and approximately 1 month from December solstice (day-of-year 15).

4 Conclusions and Discussion

In this paper, we present our first results on the main PCA mode of TEC and TMD variability from 16 years of IGS TEC GIMs and from 13 years of TMD estimates at 475 km

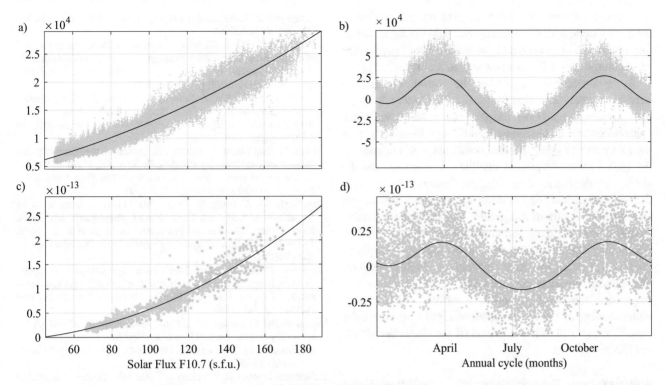

Fig. 2 Solar flux (left panels) and annual (right panels) contributions to the first time-expansion PCA component of TEC (top panels) and TMD at 475 km altitude (bottom panels). Dimensionless quantities. Annual cycles normalized to F10.7 = 110 sfu. Fitting is shown in black solid lines

altitude inferred from GRACE accelerometers and POD. The sparse nature of the initial observations has been successfully synthesized by the PCA to derive the main spatiotemporal patterns that can be compared in a common frame. The dominant patterns of both TEC and TMD variations represented by the first PCA mode show a main dependence on the solar flux with a minor annual modulation. The effect of the varying Earth-Sun distance along the ecliptic plane for a seasonal asymmetry is clearly shown in Fig. 2b and d with higher values in December than in June. Although both TEC and TMD annual variations in the first PCA component show a very similar shape, the annual contribution in TEC is approximately one magnitude larger than that in the TMD variations. These results are in agreement with the common knowledge that neutral species of the upper atmosphere are heated and ionized by radiation from the Sun, resulting in the atmospheric expansion and the creation of electrically charged species, including free electrons and electrically charged atoms and molecules. However, as seen by the spatial analysis, although both EIA and EMA troughs show a clear alignment to the dip magnetic equator, there is no clear spatial correlation between the main spatial structures defined by TEC and TMD (Fig. 1). The EIA crests are clearly located at about ±20° dip latitude, but the EMA lacks such a clear two-crest structure. Moreover, a clear asymmetry is shown in the global distribution of TMD, with higher values in the southern hemisphere than that in the

northern hemisphere. The most characteristic discrepancy is a clear TMD enhancement near the southern magnetic pole (dip = 90°). None of these other structures are reflected by the TEC distribution. These results might suggest the solar irradiance as the main driver of TEC variations, and that TMD variations are strongly driven by solar wind and magnetospheric forcing.

The hemispheric asymmetry in TMD may be caused by a long-term forcing, related to processes under favorable solar wind input and particle precipitation over the southern magnetic dip. These processes would produce a higher thermospheric heating and subsequent TMD increase in the southern hemisphere (Huang et al. 2014). Previous results have shown that the southern hemisphere TMD is more susceptible to magnetospheric driving conditions (Ercha et al. 2012; Calabia and Jin 2019), and a larger conductance in the southern hemisphere also has been reported in many studies (e.g., Sheng et al. 2017; Wilder et al. 2013, Förster and Haaland 2015, Lu et al. 1994). Other studies on thermospheric temperature and winds (Maruyama et al. 2003; Lei et al. 2010; Drob et al. 2015) could suggest that the EMA trough along the magnetic equator is caused by temperature reduction, due to diverging meridional winds (adiabatic cooling due to upward vertical winds) associated with parallel ion drag, and that equatorward winds during elevated solar and magnetospheric activity could modulate ion drag with the consequent increase of temperature and TMD at the EMA.

This could indicate that the long-term forcing related to magnetospheric activity and magnetic field configuration in the southern hemisphere could produce stronger disturbances to thermospheric TMD than that seen to ionospheric TEC.

By systematically improving the overall consistency between models and observations, future research might lead to new geophysical insights and better quantitative characterization of observations. Although the response of the IT system to space weather is still not well understood, the global nature of present satellite observations, and the ability of their advanced scientific instruments, provide the unprecedented opportunity to measure the upper-atmospheric conditions at a greater resolution and accuracy, enabling us to calibrate and constrain the existing models. Monitoring and predicting the Earth's upper-atmospheric processes driven by solar activity is highly relevant to science, industry, and defense. These communities emphasize the need to increase efforts for a better understanding of the IT responses to highly variable solar conditions, as well as how detrimental space weather affects our life and society.

Acknowledgments This work was supported by the National Natural Science Foundation of China-German Science Foundation (NSFC-DFG) Project (Grant No. 41761134092), Jiangsu Distinguished Professor Project and the Startup Foundation for Introducing Talent of NUIST. Great appreciation is extended to ISDC GFZ and IGS for providing the GRACE and TEC data, and to NASA/ESA SOHO for the SEM measurements. Special thanks are given to M Schmidt and K Börger for their invitation to participate in the IUGG 2019 at Montreal, and to IUGG for economical travel support. We are very thankful to the anonymous reviewers and Catherine M Jones for their revisions and suggestions to improve a previous version of the manuscript. There is no conflict of interest regarding the publication of this paper.

References

Anderson RL, George HB, Forbes JM (2009) Sensitivity of orbit predictions to density variability. J Spacecr Rocket 46(6):1214–1230. https://doi.org/10.2514/1.42138

Bent RB, Llewellyn SK (1973) Documentation and description of the Bent ionospheric model. Space and Missile Organisation, Los Angeles

Bilitza D, McKinnell LA, Reinisch B, Fuller-Rowell T (2011) The International Reference Ionosphere (IRI) today and in the future. J Geod 85:909–920. https://doi.org/10.5194/ars-12-231-2014

Bjornsson H, Venegas SA (1997) A manual for EOF and SVD analyses of climatic data, McGill Univ., CCGCR Report No. 97-1, Montréal, Québec, 52 pp

Bowman BR, Tobiska WK, Marcos FA, Huang CY, Lin CS, Burke WJ (2008) A new empirical thermospheric density model JB2008 using new solar and geomagnetic indices. In: AIAA/AAS Astrodynamics Specialist Conference and Exhibit, 18–21 August 2008, Honolulu, Hawaii, n. AIAA 2008–6438

Bruinsma SL (2015) The DTM-2013 thermosphere model. J Space Weather Space Climate 5:A1. https://doi.org/10.1051/swsc/2015001

Calabia A, Jin SG (2016) New modes and mechanisms of thermospheric mass density variations from GRACE accelerometers. J

Geophys Res Space Physics 121(11):11191–11212. https://doi.org/10.1002/2016ja022594

Calabia A, Jin SG (2017) Thermospheric density estimation and responses to the March 2013 geomagnetic storm from GRACE GPS-determined precise orbits. J Atmos Sol Terr Phys 154:167–179. https://doi.org/10.1016/j.jastp.2016.12.011

Calabia A, Jin SG (2019) Solar-cycle, seasonal, and asymmetric dependencies of thermospheric mass density disturbances due to magnetospheric forcing. Ann Geophys 37:989–1003. https://doi.org/10.5194/angeo-37-989-2019

Calabia A, Tang G, Jin SG (2020) Assessment of new thermospheric mass density model using NRLMSISE-00 model, GRACE, Swarm-C, and APOD observations. J Atmos Solar Terr Phys 199:105207. https://doi.org/10.1016/j.jastp.2020.105207

Chen GM, Xu J, Wang W, Burns AG (2014) A comparison of the effects of CIR- and CME-induced geomagnetic activity on thermospheric densities and spacecraft orbits: statistic al studies. J Geophys Res Space Physics 119:7928–7939. https://doi.org/10.1029/2012ja017782

Cnossen I, Förster M (2016) North-south asymmetries in the polar thermosphere-ionosphere system: solar cycle and seasonal influences. J Geophys Res Space Physics 121:612–627. https://doi.org/10.1002/2015ja021750

Dabbakuti JRKK, Ratnam DV (2016) Characterization of ionospheric variability in TEC using EOF and wavelets over low-latitude GNSS stations. Adv Space Res 57(12):2427–2443. https://doi.org/10.1016/j.asr.2016.03.029

Daniell RE, Brown LD, Anderson DN, Fox MW, Doherty PH, Decker DT, Sojka JJ, Schunk RW (1995) Parameterized ionospheric model: a global ionospheric parameterization based on first principles models. Radio Sci 30:1499–1510. https://doi.org/10.1029/95rs01826

Doornbos E, van den IJssel J, Lühr H, Förster M, Koppen-wallner G (2010) Neutral density and cross-wind determination from arbitrarily oriented multiaxis accelerometers on satellites. J Spacecr Rocket 47(4):580–589. https://doi.org/10.2514/1.48114

Drob DP, Emmert JT, Meriwether JW, Makela JJ, Doornbos E, Conde M, Hernandez G, Noto J, Zawdie KA, McDonald SE et al (2015) An update to the Horizontal Wind Model (HWM): the quiet time thermosphere. Earth Space Sci 2:301–319. https://doi.org/10.1002/2014ea000089

Emmert JT, Picone JM (2010) Climatology of globally averaged thermospheric mass density. J Geophys Res 115:A09326. https://doi.org/10.1029/2010ja015298

Ercha A, Ridley AJ, Zhang D, Xiao Z (2012) Analyzing the hemispheric asymmetry in the thermospheric density response to geomagnetic storms. J Geophys Res 117:A08317. https://doi.org/10.1029/2011ja017259

Förster M, Haaland S (2015) Interhemispheric differences in ionospheric convection: cluster EDI observations revisited. J Geophys Res Space Physics 120:5805–5823. https://doi.org/10.1002/2014JA020774

Gowtam VS, Ram ST, Reinisch B, Prajapati A (2019) A new artificial neural network-based global three-dimensional ionospheric model (ANNIM-3D) using long-term ionospheric observations: preliminary results. J Geophys Res Space Physics 124:4639–4657. https://doi.org/10.1029/2019JA026540

Harris I, Priester W (1962) Time-dependent structure of the upper atmosphere. J Atmos Sci 19(4):286–301. https://doi.org/10.1175/1520-0469(1962)019{\mathsurround=\opskip$<$}0286:TDSOTU{\mathsurround=\opskip$>$}2.0.CO;2

Hofmann-Wellenhof B, Lichtenegger H, Wasle E (2008) GNSS-global navigation satellite systems. Springer, Berlin

Huang CY, Su Y-J, Sutton EK, Weimer DR, Davidson RL (2014) Energy coupling during the August 2011 magnetic storm. J Geophys Res Space Physics 119:1219–1232. https://doi.org/10.1002/2013JA019297

Jacchia LG (1964) Static diffusion models of the upper atmosphere with empirical temperature profiles. Smithsonian Astrophysical Observatory Special Report, 170, https://doi.org/10.5479/si.00810231.8-9.213

Jin SG, Su K (2020) PPP models and performances from single-to-quad-frequency BDS observations. Satell Navig 1(1):16. https://doi.org/10.1186/s43020-020-00014-y

Jin SG, Gao C, Li JH (2019) Atmospheric sounding from FY-3C GPS radio occultation observations: first results and validation. Adv Meteorol 2019:1–13. https://doi.org/10.1155/2019/4780143, Article ID 4780143

Jin SG, Gao C, Li J (2020) Estimation and analysis of global gravity wave using GNSS radio occultation data from FY-3C meteorological satellite. J Nanjing Univ Infor Sci Tech (Nat Sci Edn) 12(1):57–67. https://doi.org/10.13878/j.cnki.jnuist.2020.01.008

Lei J, Thayer JP, Forbes JM (2010) Longitudinal and geomagnetic activity modulation of the equatorial thermosphere anomaly. J Geophys Res 115:A08311. https://doi.org/10.1029/2009JA015177

Lei J, Thayer JP, Wang W, Luan X, Dou X, Roble R (2012) Simulations of the equatorial thermosphere anomaly: physical mechanisms for crest formation. J Geophys Res 117:A06318. https://doi.org/10.1029/2012JA017613

Liu H, Yamamoto M, Lühr H (2009) Wave-4 pattern of the equatorial mass density anomaly: a thermospheric signature of tropical deep convection. Geophys Res Lett 36:L18104. https://doi.org/10.1029/2009GL039865

Liu R, Ma S-Y, Lühr H (2011) Predicting storm-time thermospheric mass density variations at CHAMP and GRACE altitudes. Ann Geophys 29:443–453. https://doi.org/10.5194/angeo-29-443-2011

Lu G et al (1994) Interhemispheric asymmetry of the high-latitude ionospheric convection pattern. J Geophys Res 99(A4):6491–6510. https://doi.org/10.1029/93JA03441

Maruyama N, Watanabe S, Fuller-Rowell TJ (2003) Dynamic and energetic coupling in the equatorial ionosphere and thermosphere. J Geophys Res 108(A11):1396. https://doi.org/10.1029/2002JA009599

Matsuo T, Forbes JM (2010) Principal modes of thermospheric density variability: empirical orthogonal function analysis of CHAMP 2001–2008 data. J Geophys Res 115. https://doi.org/10.1029/2009JA015109

Müller S, Lühr H, Rentz S (2009) Solar and geomagnetic forcing of the low latitude thermospheric mass density as observed by CHAMP. Ann Geophys 27:2087–2099. https://doi.org/10.5194/angeo-27-2087-2009

Panzetta F, Bloßfeld M, Erdogan E, Rudenko S, Schmidt M, Müller H (2018) Towards thermospheric density estimation from SLR observations of LEO satellites: a case study with ANDE-Pollux satellite. J Geod 93(3):353–368. https://doi.org/10.1007/s00190-018-1165-8

Picone JM, Hedin AE, Drob DP, Aikin AC (2002) NRLMSISE-00 empirical model of the atmosphere: statistical comparisons and scientific issues. J Geophys Res 107(A12):1468. https://doi.org/10.1029/2002JA009430

Preisendorfer R (1988) Principal component analysis in meteorology and oceanography. Elsevier, Amsterdam

Radicella S (2009) The NeQuick model genesis, uses and evolution. Ann Geophys 52(3-4):417–422. https://doi.org/10.4401/ag-4597

Richmond AD, Ridley EC, Roble RG (1992) A thermosphere/ionosphere general circulation model with coupled electrodynamics. Geophys Res Lett 19(6):601–604. https://doi.org/10.1029/92GL00401

Sheng C, Deng Y, Lu Y, Yue X (2017) Dependence of Pedersen conductance in the E and F regions and their ratio on the solar and geomagnetic activities. Space Weather 15:484–494. https://doi.org/10.1002/2016SW001486

Sutton EK, Forbes JM, Knipp DJ (2009) Rapid response of the thermosphere to variations in Joule heating. J Geophys Res 114:A04319. https://doi.org/10.1029/2008JA013667

Tapley BD, Bettadpur S, Watkins M, Reigber C (2004) The gravity recovery and climate experiment: Mission overview and early results. Geophys Res Lett 31:L09607. https://doi.org/10.1029/2004GL019920

Wan WX, Ding F, Ren ZP, Zhang ML, Liu LB, Ning BQ (2012) Modeling the global ionospheric total electron content with empirical orthogonal function analysis. Sci China Technol Sci 55:1161–1168. https://doi.org/10.1007/s11431-012-4823-8

Wilder FD, Eriksson S, Wiltberger M (2013) Investigation of the interhemispheric asymmetry in reverse convection near solstice during northward interplanetary magnetic field conditions using MHD simulations. J Geophys Res Space Phys 118:4289–4297. https://doi.org/10.1002/jgra.50421

Wilks DS (1995) Statistical methods in the atmospheric sciences. Academic, San Diego

Tropospheric Products from High-Level GNSS Processing in Latin America

María V. Mackern, María L. Mateo, María F. Camisay, and Paola V. Morichetti

Abstract

The present geodetic reference frame in Latin America and the Caribbean is given by a network of about 400 continuously operating GNSS stations. These stations are routinely processed by ten Analysis Centres following the guidelines and standards set up by the International Earth Rotation and Reference Systems Service (IERS) and International GNSS Service (IGS). The Analysis Centres estimate daily and weekly station positions and station zenith tropospheric path delays (ZTD) with an hourly sampling rate. This contribution presents some attempts aiming at combining the individual ZTD estimations to generate consistent troposphere solutions over the entire region and to provide reliable time series of troposphere parameters, to be used as a reference. The study covers ZTD and IWV series for a time-span of 5 years (2014–2018). In addition to the combination of the individual solutions, some advances based on the precise point positioning technique using BNC software (BKG NTRIP Client) and Bernese GNSS Software V.5.2 are presented. Results are validated using the IGS ZTD products and radiosonde IWV data. The agreement was evaluated in terms of mean bias and rms of the ZTD differences w.r.t IGS products (mean bias −1.5 mm and mean rms 6.8 mm) and w.r.t ZTD from radiosonde data (mean bias −2 mm and mean rms 7.5 mm). IWV differences w.r.t radiosonde IWV data (mean bias 0.41 kg/m^2 and mean rms 3.5 kg/m^2).

Keywords

IWV · Radiosonde · SIRGAS · ZTD

M. V. Mackern
Consejo Nacional de Investigaciones Científicas y Tecnológicas, Mendoza, Argentina

Facultad de Ingeniería. Universidad Nacional de Cuyo, Mendoza, Argentina

Facultad de Ingeniería. Universidad Juan Agustín Maza, Mendoza, Argentina
e-mail: vmackern@mendoza-conicet.gob.ar

M. L. Mateo (✉) · M. F. Camisay
Facultad de Ingeniería. Universidad Nacional de Cuyo, Mendoza, Argentina

Facultad de Ingeniería. Universidad Juan Agustín Maza, Mendoza, Argentina
e-mail: laura.mateo@ingenieria.uncuyo.edu.ar

P. V. Morichetti
Facultad de Ingeniería. Universidad Juan Agustín Maza, Mendoza, Argentina

1 Introduction

Integrated Water Vapour (IWV) plays a fundamental role in several weather processes that deeply influence human activities. Retrieving IWV content in the atmosphere can be performed in different ways using independent techniques: from the traditional ones like radiosondes and ground-based microwave radiometers, up to the recent ones based on satellite techniques. In particular, the GNSS-based tropospheric Zenith Total Delay (ZTD) estimates allow inferring IWV values with high accuracy equivalent to that expected from

© The Author(s) 2020
J. T. Freymueller, L. Sanchez (eds.), *Beyond 100: The Next Century in Geodesy*,
International Association of Geodesy Symposia 152, https://doi.org/10.1007/1345_2020_121

direct observational techniques, such as radiosondes and microwave radiometers (Bonafoni et al. 2013; Van Baelen et al. 2005; Calori et al. 2016). Several studies have been devoted to the use of GNSS stations for the estimation of IWV over South America. Bianchi et al. (2016) estimated mean IWV based on GNSS data (IWV_{GNSS}) and its trends during 2007–2013 over more than a hundred GNSS tracking sites from SIRGAS-CON. Calori et al. (2016) analysed a period of 45 days where deep convective processes with hail precipitation took place over Mendoza province, in the Central-Western Argentina (CWA). For this assessment, the authors used IWV_{GNSS} maps to draw insight into the accumulation and influence of humidity over the region. Even fewer studies have performed a validation of the IWV_{GNSS}; for this, Fernández et al. (2010) used radiosonde data from four locations over Central-North Argentina in order to validate IWV estimates from Global Positioning System (GPS) stations during a 1-year period (2006–2007). The authors found an agreement between IWV_{GNSS} and IWV estimated through radiosonde data (IWV_{RS}), with differences as large as 3 kg/m^2. Llamedo et al. (2017) used GPS-derived IWV to analyse moisture anomalies over South America during El Niño-Southern Oscillation phases, finding positive anomalies over northern Argentina during El Niño events.

Camisay et al. (2020) estimated IWV_{GNSS} time series for a 4-year period (2015–2018), to assess the accuracy through a comparison in two GNSS Argentinean stations with radiosonde observations and explore the role of IWV in the development of regional precipitation events over the CWA. The obtained agreement between IWV_{GPS} and IWV_{RS} was close to 2 kg/m^2 in terms of mean absolute error. In Latin-American region, in situ meteorological observations are scarce; therefore, GNSS atmospheric monitoring has significant relevance for the understanding of regional meteorological processes. This kind of information is extremely valuable, and it can be used to achieve a better knowledge of IWV variable in the study region.

The GNSS allows monitoring the IWV from a network that surpasses traditional techniques due to its significant temporal and spacial density. This is of interest to study the regional trends of the climatic variable for which it is necessary to have a long time series by site and region. On the other hand, the ZTD can be estimated in real-time and near real-time mode, in order to be assimilated in regional forecast models.

SIRGAS (Sistema de Referencia Geocéntrico para las Américas) is the geocentric reference frame in Latin America and the Caribbean. It is at present given by a network of about 420 continuously operating GNSS stations (Cioce et

al. 2018) (Fig. 1). These stations are routinely processed by the SIRGAS Analysis Centres (AC), following the guidelines and standards set up by the International Earth Rotation and Reference Systems Service (IERS) and International GNSS Service (IGS). Since 2014, the routine GNSS data processing includes the estimation of hourly ZTD values based on GPS and GLONASS observations (Camisay et al. 2020; Sánchez et al. 2015; Brunini et al. 2012).

Pacione et al. (2017) shows the great potential that a continental GNSS network offers in atmospheric studies. EUREF Permanent Network (EPN) (Bruyninx et al. 2019) had been used as a valuable database for the development of a climate data record of GNSS tropospheric products over Europe. It had been used as a reference in the regional numerical weather prediction reanalyses and climate model simulations and had been used for monitoring IWV trends and variability. Guerova et al. (2016) showed and discussed the advantages of the application of GNSS tropospheric products in operational weather prediction and in the climate monitoring.

In this contribution, we report on the estimation and validation of the ZTD and IWV values in Latin America GNSS stations, using as input data the ZTD values obtained in: (1) the operational processing of the SIRGAS regional reference frame and (2) applying the Precise Point Positioning (PPP) approach, with two softwares, BKG NTRIP Client (BNC) and Bernese v5.2. (BSW52). To assess the reliability of our results (ZTD and IWV values), they are compared with the operational IGS products (ZTD_{IGS}), IWV values extracted from radiosonde profiles (IWV_{RS}) and ZTD estimations inferred from integrate the correspondent radiosonde profile data (ZTD_{RS}).

In Sect. 2, the methodology used in operational SIRGAS processing to estimate ZTD product is reviewed. ZTD_{SIR} internal consistency is presented. ZTD products estimated by PPP in SIRGAS stations are reviewed. Section 3 summarises the ZTD_{SIR} and IWV_{SIR} products validation with respect to ZTD_{IGS} products and IWV radiosonde data. Conclusions, outlook and future work are given in Sect. 4.

2 Methodology

2.1 Estimation of ZTD Values Based on the Operational SIRGAS Processing (ZTD_{SIR})

The ZTD estimations based on the operational SIRGAS GNSS processing (ZTD_{SIR}) are routinely calculated for all the SIRGAS-CON stations (Fig. 1) by the SIRGAS Analysis

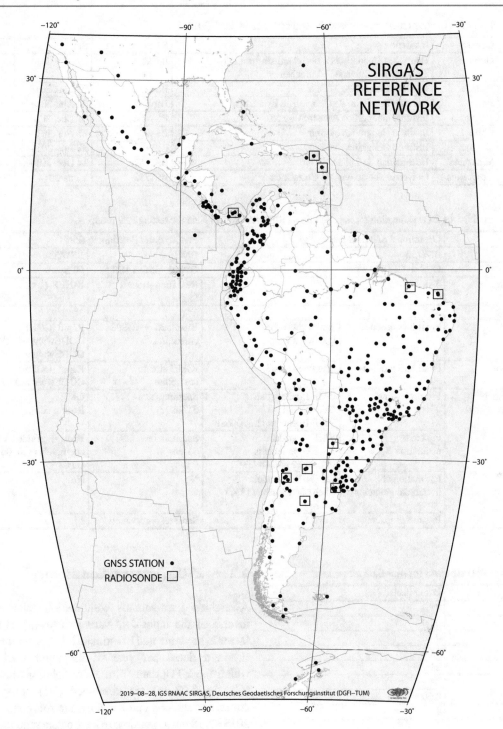

Fig. 1 SIRGAS GNSS stations and radiosonde sites considered in this study

Centres (AC) for a 5-year period (2014–2018). The eight official SIRGAS-AC (Table 1) used Bernese GNSS Software v5.2 (BSW52, Dach et al. 2015).

The SIRGAS operational ZTD products (ZTD_{SIR}) are calculated with the final IGS products (orbits and earth rotation parameters, ERP). Table 2 summarizes the methodology implemented for the operational SIRGAS products and the testing PPP products.

Each SIRGAS-AC processes a different sub-network of SIRGAS GNSS stations. The distribution of the stations considers that each station parameter (ZTD_i) is available in three different solutions, so it is possible to evaluate the

Table 1 SIRGAS Analysis Centres (AC) that estimated ZTD for the period 2014–2018

SIRGAS AC	Country	Institution	Software used	Start	End
DGF	Germany	Deutsches Geodätisches Forschungsinstitut der Technischen Universität München	BSW52	27 Apr. 2014	–
ECU	Ecuador	Instituto Geográfico Militar	BSW52	21 Dec. 2014	–
IBG	Brasil	Instituto Brasileiro de Geografia e Estatistica	BSW52	27 Apr. 2014	–
IGA	Colombia	Instituto Geográfico Agustín Codazzi	BSW52	21 Dec. 2014	–
CHL	Chile	Instituto Geográfico Militar	BSW52	27 Apr. 2014	–
URY	Uruguay	Instituto Geográfico Militar	BSW52	27 Apr. 2014	–
LUZ	Venezuela	Universidad de Zulia	BSW52	14 Dec. 2014	9 Feb. 2019
UNA	Costa Rica	Universidad Nacional de Costa Rica	BSW52	1 Jan. 2014	29 Dec.2018

Table 2 Models used for the ZTD estimation for the operational SIRGAS products and the testing PPP products

	Operational SIRGAS processing		Precise Point Positioning (PPP)	
Software	BSW52		BNC	BSW52
Observations	GPS + GLONASS		GPS + GLONASS	GPS
Sampling interval	30 s		Real time streams (1 s)	RINEX (1 s)
Elevation cut off	3°		3°	3°
Orbits and ERP	Final IGS products	igswwwwD.sp3 igswwww7.erp	Broadcast + IGS03 correction	Rapid (CODE) CODEwwwwD.EPH CODEwwwwD.ERP
Clock correction	Final IGS products	igswwwwD.sp3	IGS03 stream correction	Rapid (CODE) CODEwwwwD.CLK
A-priori troposphere modeling and mapping funtion	Pre-processing	GMF Boehm et al. (2006b) and VMF Boehm et al. (2006a)	Saastamoinen (1973) dT/cos(z)	GMF Boehm et al. (2006b)
	Parameter estimation	VMF + gridded VMF1 coefficients (00, 06 12 and 18 UTC)	Saastamoinen (1973) dT/cos(z)	VMF + gridded VMF1 coefficients (00, 06 12 and 18 UTC)
	Estimation of horizontal gradients	CHENHER model Chen and Herring (1997) (24 h)	No	No
	Parameter spacing	1 or 2 h	Same as observation	1 h

Table 3 Rejected ZTD estimates ($\sigma_{ZTD} > 0.02$ m)

AC	Data rejected (%)
CHL	7
DGF	0.09
ECU	0.10
IBG	0.06
IGA	23
LUZ	22
UNA	0.06
URY	2

internal consistency and generate the final combined ZTD products (ZTD$_{SIR}$).

The ZTD$_i$ variance is used as a filter ($\sigma_{ZTD} > 0.02$ m), prior to the combination. The 5% of the ZTD$_i$ values are rejected in the analysed period. Table 3 shows the number of rejected estimates (in %) for each AC.

2.2 ZTD$_{SIR}$ Internal Consistency

A weighted least-squares combination scheme using the inverse of the input data variances (σ_{ZTD}) as a weighting factor is implemented to estimate ZTD$_{SIR}$ products. Figure 2 shows a detail per year of the number of stations in which the ZTD$_i$ data (3 or more solutions available, with $\sigma_{ZTD} < 0.02$ m) are combined (Nc) compared to the number of stations that had only one solution. For the years 2015–2018 it was possible to have a data redundancy in more than 75% of the stations.

The internal consistency of the ZTD$_{SIR}$ values is evaluated considering the residuals of each contributing ZTD solutions with respect to the combined ZTD value (ZTD$_i$-ZTD$_{SIR}$). After a weighted least squares combination process, rms of each ZTD$_{SIR}$ parameters are determined. A mean rms is calculated per station and per year (Table 4). The mean rms

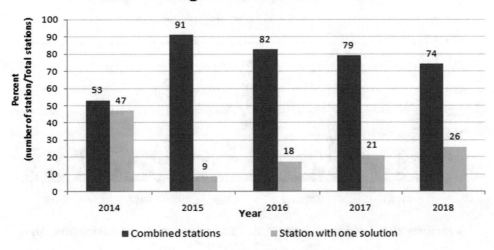

Fig. 2 Number of stations in which ZTD_i data were combined vs stations with one solution

Table 4 Summary of combination process statistics

Year	Nc/Total	%	Mean rms [mm]
2014	180/339	53	0.15
2015	345/378	91	0.09
2016	320/388	82	0.27
2017	308/390	79	0.43
2018	303/409	74	0.54

is less than 1 mm in more than the 84% of the estimated values in the period 2014–2018 (Fig. 3).

2.3 ZTD$_{SIR}$ Validation with IGS Tropospheric Products

For validation, the ZTD final products (ZTD$_{SIR}$) are compared with the operational IGS (Byram et al. 2011; Byun and Bar-Sever 2009) products (ZTD$_{IGS}$) at 15 GNSS stations. Figure 4 shows both ZTD time series in two selected stations, AREQ (16.46 °S; 71.49 °W; 2488.92 m.) and OHI2 (63.32 °S; 57.90 °W; 32.47 m.) in the study period (Jan 2014–Dec 2018).

2.4 ZTD$_{SIR}$ Validation with Radiosonde Data

ZTD$_{SIR}$ also, are compared with ZTD values calculated from data of 10 radiosonde stations (ZTD$_{RS}$). Table 5 details characteristics of the RS used.

The ZTD$_{RS}$ are calculated from the precipitable water for entire sounding (IWV$_{RS}$), data extracted from radiosonde

profiles available at Wyoming Weather Web-University of Wyoming (http://weather.uwyo.edu/upperair/sounding.html). First, ZWD$_{RS}$ values were calculated by Askne and Nordius (1987) with the physical constants for atmospheric refractivity from Rüeger (2002) (Eq. 1). The mean temperature of the atmosphere (Tm) used in (1) is calculated integrating the radiosonde profiles data (temperature and dew-point) in each level profiles up to GNSS station height (Eq. 2). The zenith hydrostatic delay values at the RS sites (ZHD$_{RS}$) are obtained according to Davis et al. (1985) (Eq. 3), where pressure is calculated to the GNSS height (P$_{GNSS}$) from pressure radiosonde data. An adaptation to the standard pressure model of Berg (1948) to correct for the height differences is applied (Eq. 4). Finally, ZTD$_{RS}$ values are calculated by adding ZHD$_{RS}$ to ZWD$_{RS}$

$$ZWD = \frac{\left(22,9744 + \frac{375463}{Tm}\right)0,4614991785}{10^5}IWV \quad (1)$$

$$Tm = \frac{\int_H^\infty e/T\,dz}{\int_H^\infty e/T^2\,dz} \quad (2)$$

$$ZHD = 0,002276738.\frac{P_{GNSS}}{1 - 0,00266.\cos{(2\varphi)} - 0,28.10^{-6}.h_{GNSS}} \quad (3)$$

$$P_{GNSS} = P_{RS}(1 - 0.0000226\,(h_{GNSS} - h_{RS}))^{5.225} \quad (4)$$

2.5 ZTD Estimation Applying PPP

In order to have a product in near real time to be used in numerical weather prediction model, we tested the Precise

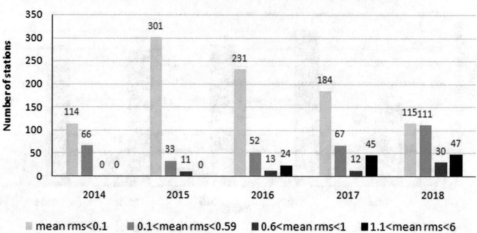

Fig. 3 Distribution of the ZTD_{SIR} mean rms per year (2014–2018)

Point Positioning processing technique (ZTD_{PPP}). Two case of study are analysed

- Case 1: Feb 21 to Mar 27 (36 days), 2016; ten GNSS stations (located in the central-western region of South America).
- Case 2: Jan 1 to Dec. 31, 2019 (365 days); thirty GNSS stations (located in Argentina).

The ZTD_{PPP} values estimated are compared with the corresponding ZTD_{SIR} values.

This estimation approached with two softwares, BNC (Weber et al. 2016) and BSW52, in the first case study. PPP with BSW52 showed better results (not shown). In the second period (year 2019) we decided to estimate ZTD_{PPP} only by BSW52. In both cases of study, with BSW52 PPP, rapid IGS products (orbits, ERP and satellite clock corrections) were used so the ZTD_{PPP} were estimated with 24 h delay. Table 2 summarizes the input data, models and main configuration used for each software.

2.6 Determination of IWV Values from GNSS-Based ZTD Estimates

The GNSS-based ZTD values are used to calculate the IWV applying the ratio of Askne and Nordius (1987) to the wet component of the delay (ZWD), (Eq. 1). In this work, the ZTD_{SIR} and the one from applying PPP (from BSW52) were used. ZWD values were obtained by removing the ZHD, which was calculated according to Davis et al. (1985) (Eq. 3). Sea level pressure values (P_{ref}) were extracted from the ERA-Interim products and were reduced to the height of the GNSS stations (P_{GNSS}) following Berg (1948) (Eq. 5).

$$P_{GNSS} = P_{ref}.\left(1 - 0,0000226.\left(h_{GNSS} - h_{ref}\right)\right)^{5,225} \quad (5)$$

In this case, the weighted mean temperature of the atmosphere (Tm) was calculated in accordance with Mendes (1999) using the surface temperature (Ts) also provided by ERA-Interim. The values for the refractivity constants were taken from Rüeger (2002). Following this strategy, IWV_{SIR} series from a 5 years (2014–2018) period were estimated in each SIRGAS station. We generated four daily IWV maps by Hunter (2007) (at 00:00, 06:00, 12:00 and 18:00 UTC) for the entire SIRGAS region, see some examples in Fig. 5 (24-6-2018).

The IWV_{SIR} values were tested in the 10 radiosonde stations selected (Table 5). The Figs. 6 and 7 show the comparison of IWV_{SIR} (inferred from ZTD_{SIR}) values with values obtained from radiosonde profiles (IWV_{RS}) at two SIRGAS stations: MZAC (located in an arid region) and IGM1 (located in a humid region), respectively.

3 Results

3.1 ZTD_{SIR} Validation

Our results presented a quite good agreement with the IGS products (see Fig. 4). Discrepancies between ZTD_{SIR} and ZTD_{IGS} values are compared at 15 IGS (SIRGAS) stations (Fig. 8). The results present a mean root mean square (rms) value of 6.8 mm (0.29% of the mean ZTD) with a negative mean bias of 1.5 mm (0.07% of the mean ZTD).

The comparison of ZTD_{SIR} w.r.t. ZTD_{RS} is also very promising: discrepancies computed at 10 radiosonde stations (see Fig. 1 and Table 5) have a mean rms of 7.5 mm (0.32% of the mean ZTD) and a negative mean bias of 2 mm (0.09%

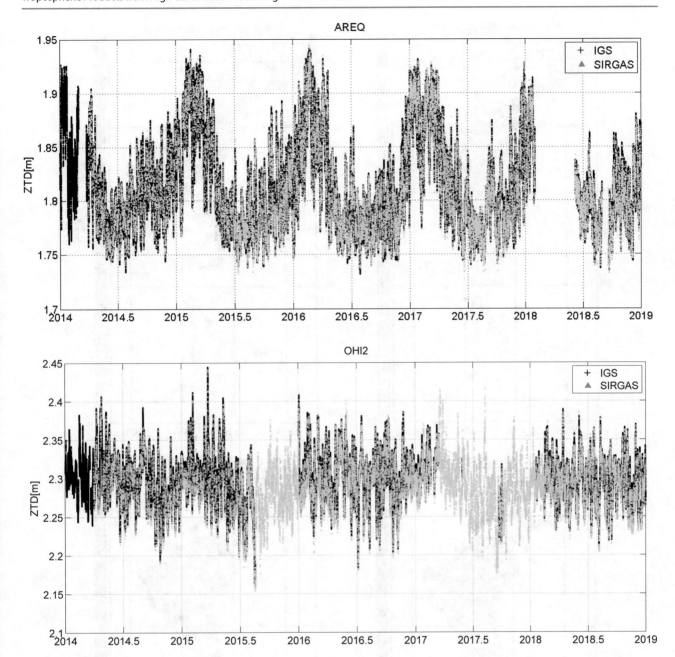

Fig. 4 Time series of $\mathrm{ZTD_{SIR}}$ (grey) and $\mathrm{ZTD_{IGS}}$ (black) values at two selected SIRGAS stations, AREQ (Arequipa, Peru) and OHI2 (O'Higgins, Antartica), period: Jan 2014–Dec 2018

Table 5 Location of ten RS stations used (bold used in $\mathrm{ZTD_{PPP}}$ validation), distance to GNSS sites, and heights (h_{GNSS} and h_{RS})

RS station	GNSS site	Lat. (°)	Long. (°)	h_{RS} (m)	h_{GNSS} (m)	Distance (km)
78866 (TNCM)	SMRT	18.03	−63.09	9	−32.48	3
78897 (TFFR)	ABMF	16.21	−61.41	8	−25.57	12
78807 (MPCZ)	IGN1	8.98	−79.46	19	47.56	13
82280	SALU	−2.53	−44.28	51	18.99	11
82397	CEFT	−3.59	−38.45	19	4.90	15
87155 (SARE)	**CHAC**	**−27.36**	**−59.04**	**52**	**77.95**	**10**
87418 (SAME)	**MZAC**	**−32.83**	**−68.78**	**704**	**859.86**	**13**
87623 (SAZR)	**SRLP**	**−36.57**	**−64.27**	**191**	**223.83**	**7**
87344 (SACO)	**CORD**	**−31.32**	**−64.22**	**474**	**746.83**	**34**
87576 (SAEZ)	**IGM1**	**−34.65**	**−58.42**	**20**	**50.69**	**28**

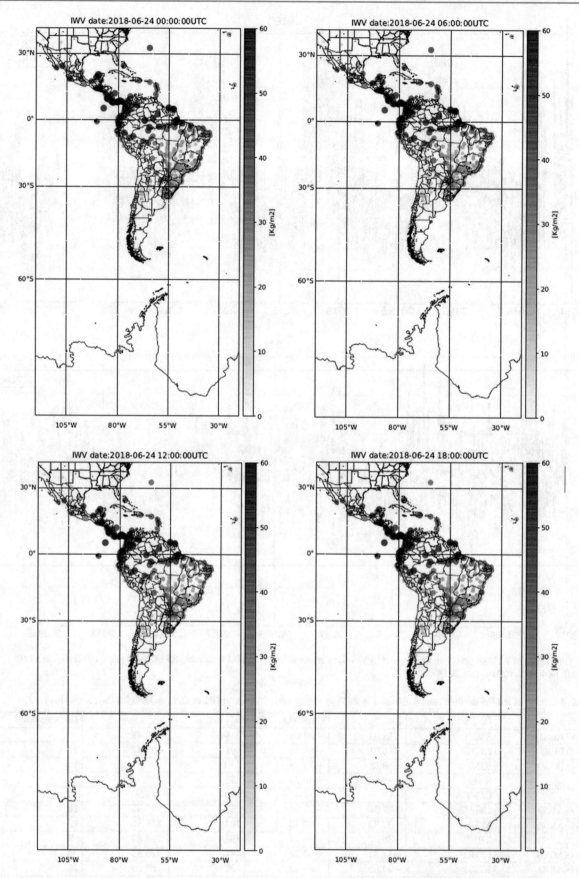

Fig. 5 Maps of IWV inferred from the ZTD estimates produced within the operational SIRGAS processing (24-6-2018; 00,06,12 and 18 hs UTC)

Fig. 6 IWV$_{SIR}$ (MZAC GNSS station) and IWV$_{RS}$ (RS: SAME)

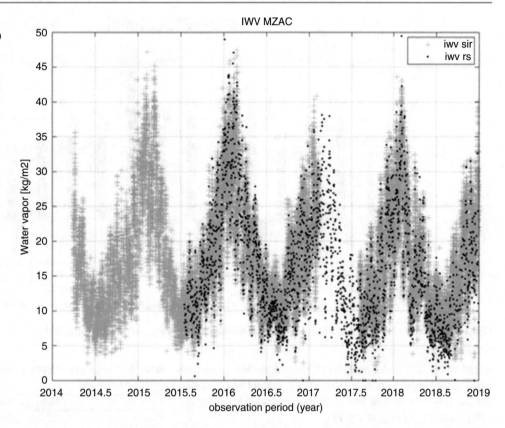

Fig. 7 IWV$_{SIR}$ (IGM1 GNSS station) and IWV$_{RS}$ (RS: SAEZ)

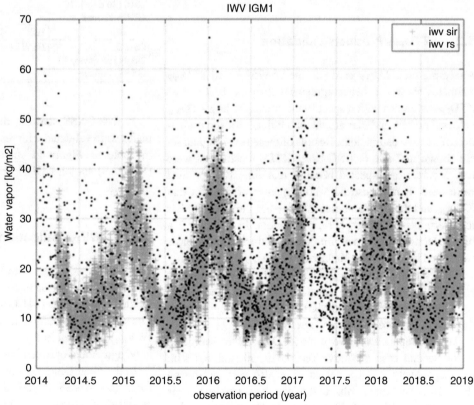

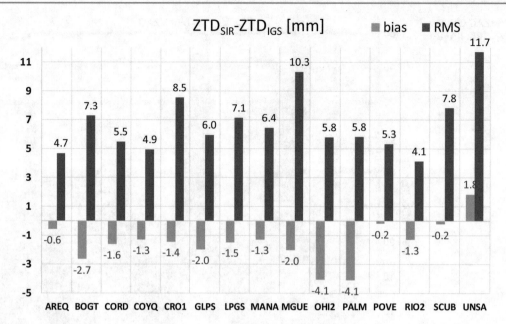

Fig. 8 Comparison of ZTD_{SIR} and ZTD_{IGS} values at 15 selected SIRGAS stations (Jan 2014–Dec 2018)

of the mean ZTD). An analysis of the radiosonde types has been started at each analysed site, which it could be the cause for the negative bias in line with the results of Wang et al. (2007) and Pacione et al. (2017).

3.2 ZTD_PPP Products Validation

Analysing the ZTD_{PPP} products, the BSW52-based ZTD_{PPP} estimates showed a better agreement than the BNC-based ZTD_{PPP} estimates with respect to the corresponding ZTD_{SIR} values. The rms and bias are the two indexes for the evaluation of the two estimations. Results of these two-test data set are shown in Table 6. BNC-based ZTD_{PPP} estimates were less accurate as expected because real time IGS product were used. It may also be a consequence of the fact that ZTD_{SIR} and the BSW52-based ZTD_{PPP} use the same models to determine the tropospheric parameters. In the case 2 a bias-reduction scheme was implemented on a monthly basis as applied in Douša and Vaclavovic (2014).

The comparison of the BSW52-based ZTD_{PPP} estimates and ZTD_{SIR} values at two selected SIRGAS station, EBYP (in a subtropical region) and MGUE (in an arid region), with the data in the case 1, are shown in Fig. 9.

The discrepancies between the ZTD_{PPP} values estimated in the second case of study (Year 2019, 30 stations) with the respectively ZTD_{SIR} values were also very promising (Fig. 10). The mean rms and mean bias per station is shown in the Fig. 10. The 84% of the stations had a mean rms < 28 mm and the rest 16% had a mean rms < 31 mm.

Table 6 Comparison of ZTD_{PPP} values with the operational SIRGAS processing (ZTD_{SIR})

Case	Software	Bias [mm]	rms [mm]
Case 1 2016 (36 days), 10 GNSS stations	BSW52	49 (1.8% of the ZTD)	55
	BNC	118 (4.8% of the ZTD)	125
Case 2 2019 (365 days), 30 GNSS stations	BSW52	2 (0.07% of the ZTD)	22

In five GNSS stations, the BSW52-based ZTD_{PPP} estimates were validated with respect to ZTD_{RS} (detailed in bold in Table 5). Figure 11 shows this comparison in the IGS (SIRGAS) station CORD in the centre of Argentina, as an example.

3.3 IWV_SIR Validation

The IWV_{SIR} validation with IWV_{RS} also showed agreement. The results for a period of 5 years, in 10 RS – GNSS locations yielded a mean bias 0.41 kg/m^2 and a mean rms 3.5 kg/m^2. The correlation coefficient of the two series (IWV_{SIR} and IWV_{RS}) presented in Fig. 12 is 0.94, which indicates a very good agreement between both estimations.

In the other hand, the comparison of IWV_{PPP} (calculated from the BWS52-based ZTD_{PPP} values) with IWV_{RS}, produces discrepancies with a mean rms of 1 kg/m^2, a standard deviation of 0.73 kg/m^2 and a bias of 2.37 kg/m^2 (not shown).

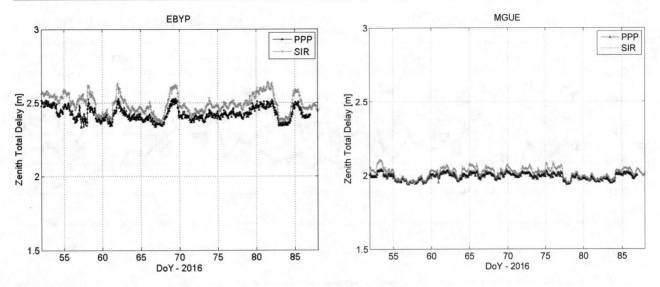

Fig. 9 Comparison of ZTD$_{SIR}$ and BSW52-based ZTD$_{PPP}$ values at two selected SIRGAS stations

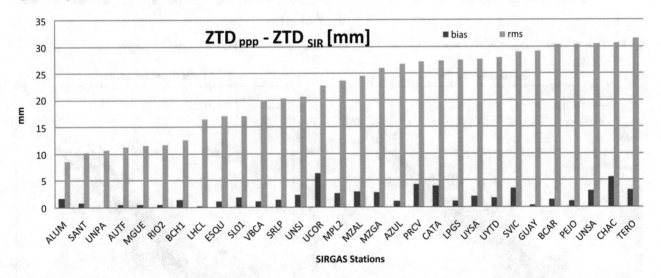

Fig. 10 Comparison of BWS52-based ZTD$_{PPP}$ with ZTD$_{SIR}$ at 30 GNSS station (365 days)

4 Conclusions

Latin America has SIRGAS network, an infrastructure of GNSS stations that generates ZTD (per hour), offering regional and continental coverage that can be used in atmospheric studies.

The internal consistency of the ZTD$_{SIR}$ values, calculated by SIRGAS ACs, have been evaluated for a period of 5 years (2014–2018). An average rms less than 1 mm, in more than the 84% of the values, indicate the rigorous weighted least squares combination process implemented to get the SIRGAS reference products.

The ZTD$_{SIR}$ series for a 5-year period have been validated with two different time series. They agree with the corresponding values of the ZTD series obtained by the IGS

(mean rms 6.8 mm; mean bias −1.5 mm) as well as those from the radiosonde technique (mean rms 7.5 mm; mean bias −2 mm).

The ZTD obtained by PPP with BSW52, using the RAPID CODE products (ephemeris and clock corrections) are validated with respect to the post-processing products ZTD$_{SIR}$. The mean rms of the differences is 22 mm (84% of the stations had a mean rms < 28 mm) for an annual case of study (2019, 30 stations). It remains to continue improving the methodology to increase accuracy and decrease the positive bias that on average resulted in 2 mm (0.07% of the ZTD mean value in the stations evaluated). Anyway, these accuracy of ZTD$_{PPP}$ complies with the threshold requirements for the operational NWP nowcasting – the relative accuracy of 5 kg/m^2 in integrated water vapor (IWV) and 30 mm in ZTD when approximating the conversion factor defined by Bevis

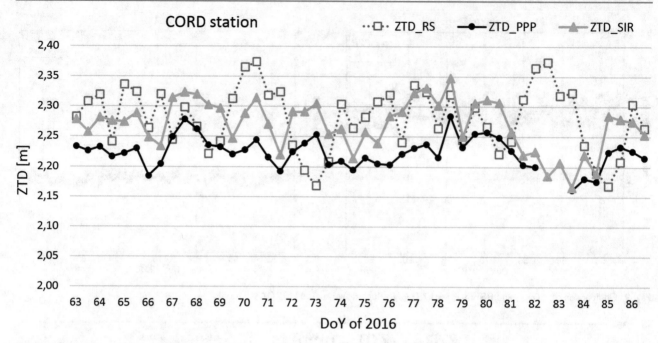

Fig. 11 Time series of ZTD$_{SIR}$, BWS52-based ZTD$_{PPP}$ and ZTD$_{RS}$ values at a GNSS station located in Cordoba (Argentina)

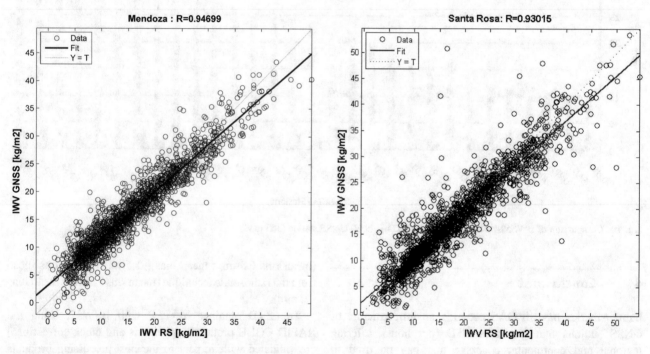

Fig. 12 Scatter plot comparing IWV values inferred from GNSS-based ZTD estimates (IWV$_{SIR}$) and radiosonde profile data (IWV$_{RS}$) at two selected SIRGAS stations (Jan 2014 to Dec 2018)

et al. (1994) and Douša and Vaclavovic (2014). However, we must work to obtain a product in near real time (with 90 min of latency), applying ultra-rapid orbits and clocks, or even better using real-time corrections (Guerova et al. 2016).

The publication of this new product from SIRGAS opens the opportunity for new research topics that can be carried out both continentally and regionally in Latin America. As

an example, it has been shown that SIRGAS ZTD products can be used to calculate the IWV over SIRGAS stations, thus providing IWV with a spatial and temporal density not existing in Latin America by conventional techniques. This variable has also been validated with radiosonde data (mean correlation coefficient 0.89, in 10 compared sites). SIRGAS ZTD products can be used as a reference for different scientific applications (e.g. validation of regional numerical weather prediction reanalyses) and they could be used for monitoring trends and variability in atmospheric water vapour in Latin America region, similar than EUREF Permanent network (Pacione et al. 2017).

Acknowledgments The authors are grateful for the silent task of those responsible for the GNSS stations, the data centers and the SIRGAS analysis centers (CHL, DGF, ECU, IBG, IGA, LUZ, URY and UNA), without which this research could not have been carried out.

The ERA-Interim data used were provided by ECMWF. Radiosonde data were provided by Wyoming Weather Web, University of Wyoming.

References

Askne J, Nordius H (1987) Estimation of tropospheric delay for microwaves from surface weather data. Radio Sci 22:379–386. https://doi.org/10.1029/RS022i003p00379

Berg H (1948) Allgemeine Meteorologie. Dümmler's Verlag, Bonn, p 337

Bevis M, Businger S, Chiswell S, Herring TA, Anthes RA, Rocken C, Ware RH (1994) GPS meteorology: mapping zenith wet delays onto precipitable water. J Appl Meteorol 33:379–386

Bianchi CE, Mendoza LPO, Fernández LI, Natali MP, Meza AM, Moirano JF (2016) Multi-year GNSS monitoring of atmospheric IWV over central and South America for climate studies. Ann Geophys 34:623–639

Boehm J, Niell AE, Tregoning P, Schuh H (2006a) Global mapping function (GMF): a new empirical mapping function based on numerical weather model data. Geophys Res Lett 33:25. https://doi.org/10.1029/2005GL025546

Boehm J, Werl B, Schuh H (2006b) Troposphere mapping functions for GPS and very long baseline interferometry from European Centre for Medium-Range Weather Forecasts operational analysis data. J Geophys Res 111:B02406. https://doi.org/10.1029/2005JB003629

Bonafoni S, Mazzoni A, Cimini D, Montoponi M, Pierdicca N, Basili P, Ciotti P, Carlesimo G (2013) Assessment of water vapor retrievals from a GPS receiver network. GPS Solut 17(4):475–484

Brunini C, Sánchez L, Drewes H, Costa S, Mackern V, Martínez W, Seemuller W, da Silva A (2012) Improved analysis strategy and accessibility of the SIRGAS reference frame. In: Kenyon S, Pacino M, Marti U (eds) Geodesy for planet earth. International association of geodesy symposia, vol 136. Springer, Berlin, pp 3–10

Bruyninx C, Legrand J, Fabian A et al (2019) GNSS metadata and data validation in the EUREF permanent network. GPS Solut 23:106. https://doi.org/10.1007/s10291-019-0880-9

Byram S, Hackman C, Tracey J (2011) Computation of a high-precision GPS-based troposphere product by the USNO. In: Proceedings of the 24th international technical meeting of the satellite division of the institute of navigation (ION GNSS 2011). 2001

Byun SH, Bar-Sever YE (2009) A new type of troposphere zenith path delay product of the international GNSS service. J Geod 83(3–4):1–7

Calori A, Santos JR, Blanco M, Pessano H, Llamedo P, Alexander P, de la Torre A (2016) Ground-based GNSS network and integrated water vapor mapping during the development of severe storms at the Cuyo region (Argentina). Atmos Res 176–177:267–275

Camisay MF, Rivera JA, Mateo ML, Morichetti PV, Mackern MV (2020) Estimation of integrated water vapor derived from global navigation satellite system observations over Central-Western Argentina

(2015–2018). Validation and usefulness for the understanding of regional precipitation events. J Atmos Sol Terr Phys 197:105143. https://doi.org/10.1016/j.jastp.2019.105143. ISSN 1364-6826

Chen G, Herring TA (1997) Effects of atmospheric azimuthal asymmetry on the analysis of space geodetic data. J Geophys Res 102:20,489–20,502

Cioce V, Martínez W, Mackern MV, Pérez R, De Freitas S (2018) SIRGAS: reference frame in Latin America. Coordinates XIV(6):6–10. ISSN 0973-2136

Dach R, Lutz S, Walser P, Fridez P (2015) Bernese GNSS software version 5.2. Astronomical Institute, University of Bern, Bern. https://doi.org/10.7892/boris.72297. ISBN: 978-3-906813-05-9; Open Publishing

Davis JL, Herring TA, Shapiro I, Rogers AE, Elgened G (1985) Geodesy by interferometry: effects of atmospheric modeling errors on estimates of base line length. Radio Sci 20:1593–1607

Douša J, Vaclavovic P (2014) Real-time zenith tropospheric delays in support of numerical weather prediction applications. Adv Space Res 53:1347–1358. https://doi.org/10.1016/j.asr.2014.02.021

Fernández LI, Salio P, Natali MP, Meza AM (2010) Estimation of precipitable water vapour from GPS measurements in Argentina: validation and qualitative analysis of results. Adv Space Res 46:879–894

Guerova G, Jones J, Douša J, Dick G, de Haan S, Pottiaux E, Bock O, Pacione R, Elgered G, Vedel H, Bender M (2016) Review of the state of the art and future prospects of the ground-based GNSS meteorology in Europe. Atmos Meas Tech 9:5385–5406. https://doi.org/10.5194/amt-9-5385-2016

Hunter JD (2007) Matplotlib: a 2D graphics environment. Comp Sci Eng 9(3):90–95

Llamedo P, Hierro R, de la Torre A, Alexander P (2017) ENSO-related moisture and temperature anomalies over South America derived from GPS radio occultation profiles. Int J Climatol 37:268–275

Mendes VB (1999) Modeling the neutral-atmosphere propagation delay in radiometric space techniques. Ph.D. dissertation, Department of Geodesy and Geomatics Engineering Technical Report No 199, Univ. of New Brunswick, Canada

Pacione R, Araszkiewicz A, Brockmann E, Dousa J (2017) EPN-Repro2: a reference GNSS tropospheric data set over Europe. Atmos Meas Tech 10:1689–1705. https://doi.org/10.5194/amt-10-1689-2017

Rüeger JM (2002) Refractive index formula for radio waves. In: Proc. XXII FIG Int. Congr., April 19–26, 2002, Web. http://www.fig.net/resources/proceedings/fig_proceedings/fig_2002/Js28/JS28_rueger.pdf

Saastamoinen J (1973) Contributions to the theory of atmospheric refraction. Bull Geod 107:13–34. https://doi.org/10.1007/BF02521844

Sánchez L, Drewes H, Brunini C, Mackern MV, Martínez-Díaz W (2015) SIRGAS core network stability. In: Rizos C, Willis P (eds) IAG 150 years. International Association of Geodesy Symposia, vol 143. Springer, Cham, pp 183–191

Van Baelen J, Aubagnac JP, Dabas A (2005) Comparison of near-real time estimates of integrated water vapor derived with GPS, Radiosondes, and microwave radiometer. J Atmos Ocean Technol 22:201–210

Wang J, Zhang L, Dai A, Van Hove T, Van Baelen J (2007) A near-global, 2-hourly data set of atmospheric precipitable water dataset from ground-based GPS measurements. J Geophys Res 112:D11107. https://doi.org/10.1029/2006JD007529

Weber G, Mervart L, Stürze A, Rülke A, Stöcker D (2016) BKG Ntrip Client, Version 2.12. Mitteilungen des Bundesamtes für Kartographie und Geodäsie, vol 49. Frankfurt am Main, 2016

Can Vertical GPS Displacements Serve As Proxies for Climate Variability in North America?

Shimon Wdowinski [iD] and Tonie M. van Dam [iD]

Abstract

Vertical crustal displacements induced by atmospheric, hydrological, cryospheric, and oceanic load changes are detectable with sub-cm accuracy by precise continuous GPS measurements. Areas subjected to rapid load changes due to ice sheet melt, drought, massive groundwater extraction, or lake level drop, are characterized by a dominant non-linear vertical signal. Here, we investigate possible relations between vertical crustal movements and climate change by analyzing the relations between observed GPS vertical movements, predicted movements, and climatic indices, where we have long GPS time series (>20 years). Applying our analysis to GPS records from western and eastern North America indicates different load change characteristics. In the western US, the seasonal and climatic signals are dominated by hydrological load changes and, consequently, the GPS signal correlates well with the Palmer Severe Drought Index (PSDI) calculated for the same region. However, vertical crustal movements in eastern North America, as detected by long GPS time series, reveal poor correlation with PSDI and other climatic indices. Our results suggest that long continuous GPS observations of vertical crustal displacements primarily driven by climate related changes in water storage can serve as independent measures of regional-scale climate change in some cases, mainly in western north America.

Keywords

Climate indices · Crustal deformation · GPS · GRACE · Palmer Severe Drought Index

1 Introduction

Displacement of the Earth's crust, mostly in the vertical direction, occurs in response to load changes induced by the atmosphere, hydrosphere, and cryosphere; these components of the Earth system are affected by climate change. The largest load changes occur by sediment deposition and the melting of thick ice sheets and results in tens or even hundreds of meters of vertical crustal movements over periods of thousands of years, as recorded by uplifted shorelines in Fenoscandia and in other near polar regions (Mörner 1979). Smaller load changes induce smaller movements, which are measured nowadays using space geodetic techniques, mainly GPS and InSAR, with sub-cm accuracy level (Wdowinski and Eriksson 2009). GPS observations in Greenland and the northern Atlantic regions revealed non-linear rates of crustal uplift, reflecting the accelerating rate of ice mass loss in the region in response to the changing climate (Jiang et al. 2010; Bevis et al. 2012). Similarly, GPS observations in the western US, mainly in California, detected transient crustal movements, reflecting crustal response to changes in the hydrological load due to changing lake levels, groundwater depletion, and the California drought (Amos et al. 2014; Brosa et al. 2014; Wahr et al. 2013; Hammond et al. 2016). The above two examples demonstrate that observations of the GPS vertical component can be used as a proxy for

S. Wdowinski (✉)
Department of Earth and Environment, Florida International University, Miami, FL, USA
e-mail: shimon.wdowinski@fiu.edu

T. M. van Dam
Université du Luxembourg, Esch-sur-Alzette, Luxembourg

© The Author(s) 2020
J. T. Freymueller, L. Sanchez (eds.), *Beyond 100: The Next Century in Geodesy*,
International Association of Geodesy Symposia 152, https://doi.org/10.1007/1345_2020_104

climate variability as it indirectly observes hydrological and cryospheric mass changes.

Crustal deformation in response to load change has extensively studied using elastic deformation with both forward and inverse modeling techniques (e.g., Davis et al. 2004; van Dam et al. 2007). In particular, studies of hydrological load change found that vertical GPS movements have a capability for estimating changes in terrestrial water content (TWC) changes through inversion of vertical deformation (e.g., Tregoning et al. 2009; Fu et al. 2012). The application of GPS vertical movements for TWC estimation was verified mostly in western North America by comparing the GPS observations to predicted movements derived from hydrological load models and GRACE observations. However, it is not clear if GPS vertical movements can be used for estimating TWC changes in other regions, such as eastern North America.

In this study we hypothesize that vertical crustal movements recorded by long continuous GPS time series (>20 years) can provide an independent measure of climate variability. If so, our study will provide an insight into the sustainability of geodetic reference frames, as climate change progresses. The rational for the hypothesis is the observed crustal response to hydrological load changes, as presented above. Thus, our first attempt is to compare vertical GPS movements with hydrological climate indices, as the Palmer Drought Severity Index (PDSI; Alley 1984). We also compare the observed GPS movements with predicted movements calculated from modeling the crustal response to hydrologic load changes, based on both a hydrological model and GRACE observations. Our results indicate that vertical GPS movements correlate with the predicated hydrological load and PDSI mainly in the western North America. In locations where vertical GPS movements do not correlate well with PDSI, mainly in eastern North America, we explore possible correlations with other climate indices, such as the North Atlantic Oscillation (NAO). However, our results indicate poor correlation with these indices, suggesting that vertical GPS time series have only a limited sensitivity to climate change.

2   Data

Our study relies on multiple data types, as we seek to find casual relations between vertical crustal movements, hydrological load estimates, and climate indices. The vertical crustal movements are determined from long continuous GPS time series. In this study, we use daily solutions provided by the Nevada Geodetic Lab (NGL – http://geodesy. unr.edu/) in the IGS08 reference frame. NGL also provides solutions in the NA12 reference frame; however, in these solutions a significant part of the hydrological signal is removed due to the continental scale spatial filtering (Blewitt et al. 2013).

The hydrological load estimates are calculated based on the Global Land Data Assimilation System (GLDAS; Rodell et al. 2004) and GRACE gravity field observations. We did not include atmospheric and non-tidal load calculations, as these loads are significantly small when compared with the amplitude of the hydrological loading signal. We also used time series of climate indices as provided by the National Oceanic and Atmospheric Administration (NOAA). We used PDSI values of the sub-state divisions, according to the locations of the selected GPS stations, which are available at https://www.ncdc.noaa.gov/cag/divisional/time-series. We also used time series of the North Atlantic Oscillation (NAO), Atlantic Multidecadal Oscillation (AMO), and El Niño/Southern Oscillation (ENSO). All indices are provided by NOAA at https://www.esrl.noaa.gov/psd/data/climateindices/list/.

3   Methodology

The main tool for testing our hypothesis is to conduct a systematic comparison between time series of vertical GPS movements, predicted crustal movements due to hydrologic load changes, and climate indices. Before conducting such a comparison, we need first to select suitable GPS records for the analysis and also compute the predicted movements for the same locations using GLDAS model results and GRACE observations. The GPS site selection was based on the following three criteria: The site (1) must be located in North America [25–55°N; 50–130°W]; (2) have a time series with a time span longer than 20 years; and (3) have at least 6,000 daily solutions. Based on these criteria, we found a total of 177 stations (Fig. 1). However, many of these time series contain steps and transient behavior, which most likely represent artifacts due to equipment/firmware change, co-seismic displacements, and other unexplained behavior. Using timing of offsets provided by NGL, we corrected the time series for the noted offsets. Unreported offsets where detected and corrected using a best-fit step function algorithm. After time series cleaning, we found that only 87% (155 stations) have periods without unusual behavior (strange transients) that can be used in the comparison. Additional pre-comparison processing includes detrending the series, as some, especially those in northern latitudes, contain a significant trend (up to 4 mm/year) reflecting crustal response to Glacial Isostatic Adjustment in this region.

For the selected 155 GPS stations, we predicted crustal movements due to hydrological load changes using GLDAS and GRACE observations. The GLDAS water storage loads were obtain from the GLDAS website (https://ldas.gsfc.nasa.

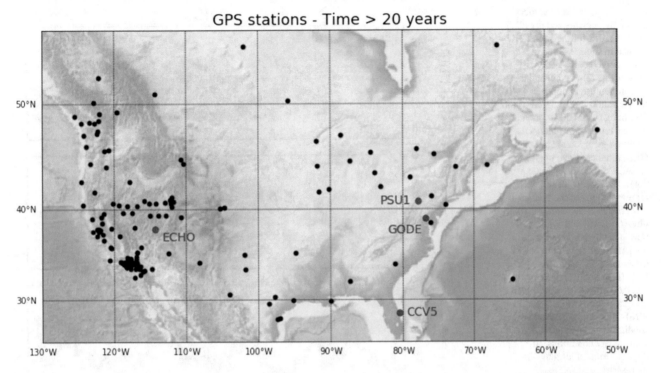

Fig. 1 Location map of GPS station in north America with continuous data acquired over a period of more than 20 years. Red circles and text mark the location of the four GPS time series presented in this study

gov/gldas), which provides water storage values comprised of soil moisture, canopy water, and snow water equivalent with a spatial resolution of one degree grid spacing at monthly intervals. The GRACE-derived load changes were determined by convolving the gravity field coefficients with load Love numbers (in the CF reference frame defined by the GPS). The GRACE data are the CSR R05 products. GRACE C20 is replaced with the C20 determined from SLR. Degree-1 is restored to the spherical harmonic files. The data are then destriped (Swenson and Wahr 2006) and subsequently filtered with a 350 km Gaussian averaging kernel. High frequency atmospheric and oceanic de-aliasing products were added back into the gravity fields. For the GLDAS data, we estimated the crustal response to loading by converting the water equivalent values to surface mass and then convolving them with Farrell's (1972) Green's functions (Gutenberg-Bullen B Earth model) that have been converted into to a center of figure frame. This reference frame is consistent with the frame that is estimated by GPS time series that have been transformed into a reference frame such as IGS08.

The comparisons between time series of vertical GPS movements, predicted crustal movements due to hydrologic load changes, and climate indices are conducted both visually and quantitatively. The visual comparison is conducted by using smooth curves through the time series data point based on a lowpass filter with 0.5-year cutoff, which accounts

for multi-year and seasonal variations, but not daily changes. The daily values have much more variability and are considered as measurement noise. The quantitative comparison is conducted using a correlation analyses of mean monthly values, which are quantified by the Pearson product-moment correlation coefficient (R^2).

4 Results

We analyzed vertical GPS time series of 155 sites with long continuous daily solutions (>20 years) and obtained mixed results. Here we present the results of four representative sites, ECHO, PSU1, GODE, and CCV5 located in both western and eastern parts of the US (Fig. 1). We chose these four sites, as they are located in different environments and climatic conditions. ECHO is located in the Basin and Range Province in eastern Nevada at an elevation of 1,684 m and subjected to semi-arid climate. PSU1 is located in an open area within the campus of Pennsylvania State University at an elevation of 311 m and is subjected to temperate climatic conditions. GODE is also located in an open area at elevation of 14.5 m in the state of Maryland, outside Washington DC, and is also subjected to temperate climate. CCV5 is located within a NASA facility in Cape Canaveral at elevation of 2 m and is subjected to hot and humid subtropical climate.

Fig. 2 Time series of vertical crustal movements and the PDSI climate index for the site ECHO, located in southeastern Nevada. (**a**) Observed daily GPS movements (blue dots) and their smooth representation based on lowpass filter with 0.5-year cutoff (red line). (**b**) Predicted monthly vertical crustal movements due to a hydrological load based on the GLDAS model (blue dots) and GRACE observations (orange dots). Continuous smooth representations of the predicted movements is shown in the solid blue (GLDAS) and cyan (GRACE) based on the same lowpass filter. (**c**) Calculated monthly PDSI values for western Utah (blue dots) [source: NOAA] and their smooth representation using a lowpass filter. Negative PDSI values indicate drought conditions and positive values indicate wet conditions. (**d**) Superposition of all four time-series used for visual comparison between the time series. The PDSI series is plotted inversely to demonstrate the inverse correlation between the climate index and the vertical crustal movements. (**e**) Correlation between observed monthly averaged GPS movements and predicated GLDAS movements. (**f**) Correlation between observed monthly averaged GPS movements and calculated PDSI values

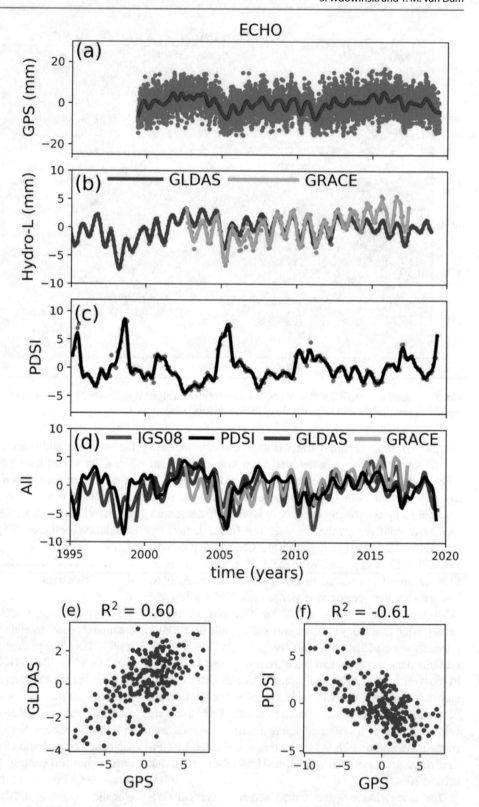

The analysis of ECHO compares time series of vertical GPS movements, predicted movements due to hydrological load changes (GLDAS and GRACE), and the PDSI for south central Nevada (Fig. 2). The GPS time series extends over a period of 20.04 years with 6,443 daily solutions, reflecting 86% temporal coverage of the measurement period. The time series deviates within a range of 20 mm and contains both multi-year and seasonal periodic signals, as emphasized by the lowpass filter smooth curve (red line in Fig. 2a). The predicted movements due to hydrologic load changes by both

Table 1 Correlation coefficient (R^2) and RMS (mm) results of correlation analyses among GPS, GLDAS, GRACE, and PDSI time series at three selected GPS station locations

Site		GPS-GLDAS	GPS-GRACE	GPS-PDSI	GLDAS-GRACE	GLDAS-PDSI	GRACE-PDSI
ECHO	R^2	0.60	0.58	−0.61	0.57	−0.56	−0.36
	RMS	2.59	2.77		1.97		
PSU1	R^2	0.58	0.54	−0.28	0.80	−0.27	−0.11
	RMS	3.01	3.02		1.97		
GODE	R^2	0.31	0.20	−0.20	0.81	0.29	−0.17
	RMS	3.76	4.20		1.41		
CCV5	R^2	−0.22	0.38	−0.20	−0.08	−0.04	−0.28
	RMS	4.16	3.81		1.63		

GLDAS and GRACE are dominated by seasonal variations and some changes from one year to another (Fig. 2b). The PDSI time series is also characterized by seasonal changes and multi-year signals of long troughs (droughts – negative PDSI values) and short duration peaks (wet conditions – positive PDSI). According to the PDSI time series, droughts occurred during 1996–1998, 2002–2004, 2006–2011, and 2012–2017, whereas wet conditions occurred in 1995, 1998, 2005 and 2019 (Fig. 2c). A visual comparison of all four time-series is conducted using the smooth curves of all series (Fig. 2d). The comparison shows an overall very good fit among the four series, as in the trough of 2005, but also some deviations, as the trough of 2011 where the GPS curve (red) is significantly lower than the other curves. Quantitative comparison among the time series using the Pearson product-moment correlation coefficient (R^2) reveal strong positive correlation (0.60) between the GPS and GLDAS series (Fig. 2e) and strong negative correlation (−0.61) between GPS and PDSI series (Fig. 2f). Correlation analysis among all four time-series reveal strong positive or negative correlations (0.55–0.6) except between GRACE and PDSI (−0.36) (Table 1).

The analysis of the PSU1 site is presented in Fig. 3 and Table 1. The GPS time series extends over a period of 21.58 years with 7,168 daily solutions, reflecting 91% temporal coverage of the measurement period. The analysis yields strong positive correlations among the GPS, GLDAS, and GRACE time series (0.54–0.8), but poor negative correlations with the PDSI series (0.11–0.28). The misfit between PDSI and the other series is apparent in Fig. 3d.

The analysis of the GODE data is presented in Fig. 4 and Table 1. The GPS time series extends over a period of 25.24 years with 8,302 daily solutions, reflecting 90% temporal coverage of the measurement period. The analysis yields poor correlations among the GPS, GLDAS, GRACE, and PDSI time series (0.17–0.31), except for a strong positive correlation between GLDAS and GRACE (0.8). The misfit among all series can be seen in Fig. 4d.

The analysis of the CCV5 site is presented in Fig. 5 and Table 1. The GPS time series extends over a period of 21.08 years with 6,716 daily solutions, reflecting 87%

temporal coverage of the measurement period. The analysis yields poor correlations among the GPS, GLDAS, GRACE, and PDSI time series (0.4–0.28).

In sites where we found a poor fit between GPS and PDSI time series, as PSU1, GODE, and CCV5, we also conducted correlation analyses with other climate indices, including NAO, AMO, and ESNO. The rational for such a comparison is that climate patterns can affect non-hydrologic loads, such as atmospheric or non-tidal oceanic loads. However, our results yielded poor correlations between GPS time series and these other three climate indices.

The analysis of all 155 stations with long time series (>20 years) revealed variable results in terms of correlations between the GPS, GLDAS, GRACE, and PDSI time series. The highest correlation levels were found among the GPS-GRACE pairs, in which 88 pairs revealed moderate to high correlation level ($R^2 > 0.4$). Most of the stations with higher correlation levels are located in western North America in inland areas of high elevation (Fig. 6). The correlation level of the GPS-GLDAS and GLDAS-GRACE also showed a fairly good correlation level (60 pairs with $R^2 > 0.4$). However, the correlation level of the PDSI climate index with GPS, GLDAS, and GRACE showed an overall poor fit level.

5 Discussion and Conclusions

We conducted a systematic comparison analysis among observed vertical GPS movements, predicted crustal movements due to hydrologic load changes, and climate indices, in order to test the hypothesis that vertical crustal movements recorded by long continuous GPS time series can provide an independent measure of climate change. Our analysis yielded mixed results, in which we observed good correlations between vertical GPS and PDSI time series at some sites, mainly in western north America. However, our results for eastern north America, yielded poor correlation between vertical GPS and PDSI time series. Our results for this region also yielded poor correlations between GPS and three other climate indices, NAO, AMO, and ESNO. These

Fig. 3 Time series of vertical crustal movements and the PDSI climate index for the site PSU1, located in central Pennsylvania. Explanations for (**a–f**) as in Fig. 2

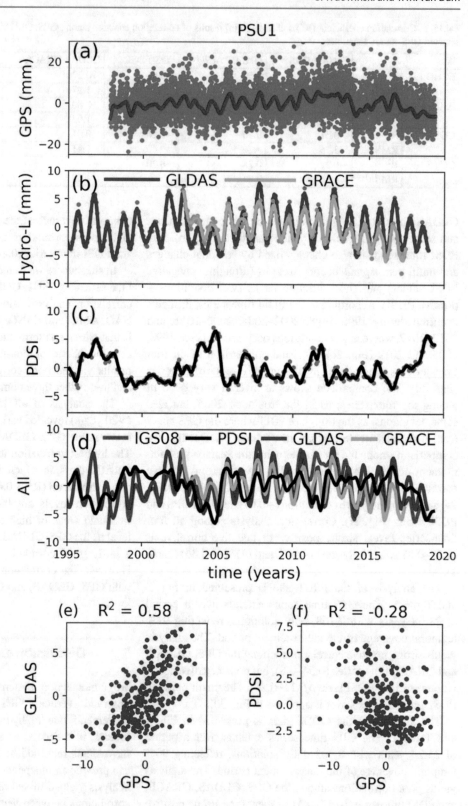

results suggested that vertical GPS time series can represent the climate signal only in some sections of north America, mainly in the mountainous western section of the continent.

We also conducted a systematic analysis between observed (GPS) and predicted (GLDAS and GRACE)

vertical crustal movements due to hydrologic load changes. Our analysis yielded, again, mixed results. We found in some locations, including ECHO and PSU1, moderate correlations between the two time-series, suggesting that vertical GPS time series can serve as a good independent

Fig. 4 Time series of vertical crustal movements and PDSI climate index for the site GODE, located in central Maryland. Explanations for (**a–f**) as in Fig. 2

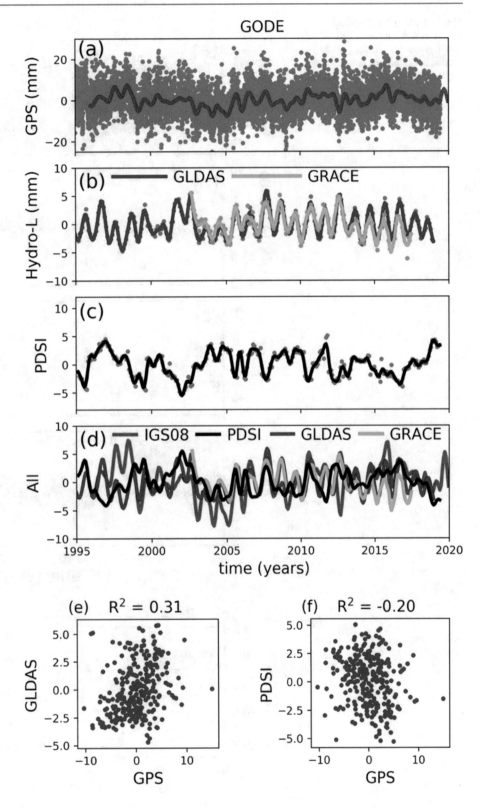

indicator of the hydrological load, which is used often for estimating Continental Water Storage (CWS). However, in some locations, as in GODE, the correlation between the observed GPS movements and predicted movements by hydrological loading changes yield poor correlations. The observed GPS movements, in such cases of poor correlation, represent most likely displacements in response to other physical processes, such as seasonal soil compaction or instability of the monument, which in many cases are located on buildings. These results suggest that vertical

Fig. 5 Time series of vertical crustal movements and PDSI climate index for the site CCV5, located in central Florida. Explanations for (**a–f**) as in Fig. 2

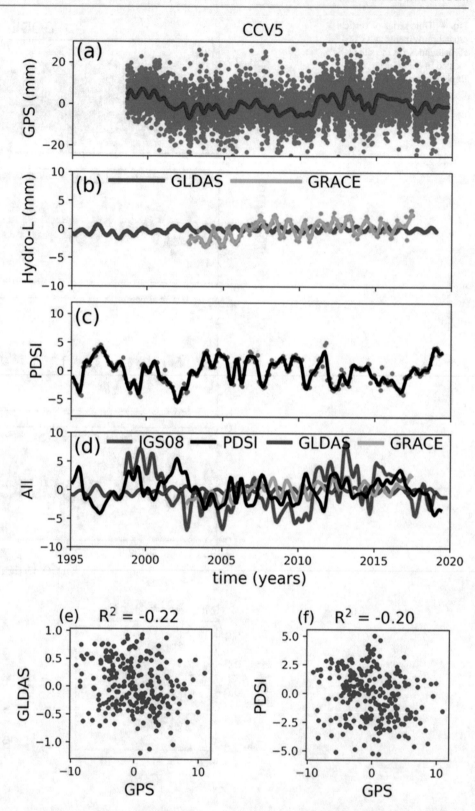

GPS time series cannot always serve as indicator for CWS. Furthermore, the results suggest that a correlation analysis between vertical GPS movements and predicted GLDAS- or GRACE-based movements can serve as a useful tool for considering a GPS site for CWS estimates.

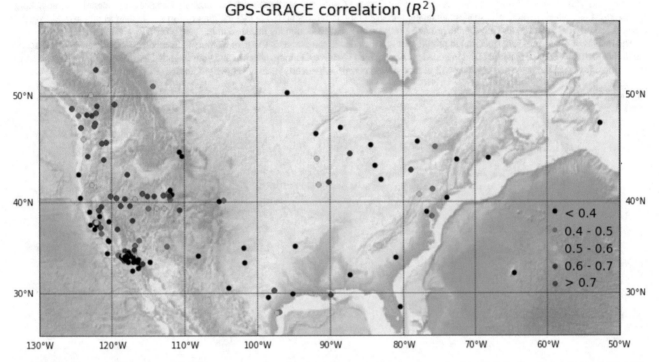

Fig. 6 Location map of GPS stations with long time series (>20 years). The color circles mark the correlation level between GPS and GRACE-predicted vertical movements

Acknowledgements The authors thank Paulo de Tarso Setti Júnior and Qiang Chen for data processing. This study was supported by NASA (Grant Number: 80NSSC17K0098).

References

Alley WM (1984) The Palmer drought severity index: limitations and assumptions. J Clim Appl Meteorol 23(7):1100–1109

Amos CB, Audet P, Hammond WC, Burgmann R, Johanson IA, Blewitt G (2014) Uplift and seismicity driven groundwater depletion in Central California. Nature 509(7501):483–486. https://doi.org/10.1038/nature13275

Bevis M, Wahr J, Khan SA, Madsen FB, Brown A, Willis M, Kendrick E, Knudsen P, Box JE, van Dam T (2012) Bedrock displacements in Greenland manifest ice mass variations, climate cycles and climate change. Proc Natl Acad Sci 109:11944–11948

Blewitt G, Kreemer C, Hammond WC, Goldfarb JM (2013) Terrestrial reference frame NA12 for crustal deformation studies in North America. J Geodyn 72:11–24

Brosa AA, Agnew DC, Cayan DR (2014) Ongoing drought-induced uplift in the western United State. Science 345(6204):1587–1590. https://doi.org/10.1126/science.1260279

Davis JL, Elósegui P, Mitrovica JX, Tamisiea ME (2004) Climate-driven deformation of the solid earth from GRACE and GPS. Geophys Res Lett 31(24):1–4

Farrell WE (1972) Deformation of the earth by surface loads. Rev Geophys 10(3):761–797. https://doi.org/10.1029/RG010i003p00761

Fu Y, Freymueller JT, Jensen T (2012) Seasonal hydrological loading in southern Alaska observed by GPS and GRACE. Geophys Res Lett 39(15):15310

Hammond WC, Blewitt G, Kreemer C (2016) GPS imaging of vertical land motion in California and Nevada: implications for Sierra Nevada uplift. J Geophys Res Solid Earth 121(10):7681–7703

Jiang Y, Dixon TH, Wdowinski S (2010) Accelerating uplift in the North Atlantic region as an indicator of ice loss. Nat Geosci 3:404–407

Mörner NA (1979) The Fennoscandian uplift and late Cenozoic geodynamics: geological evidence. GeoJournal 3(3):287–318

Rodell M et al (2004) The global land data assimilation system. Bull Am Meteorol Soc 85:381–394. https://doi.org/10.1175/BAMS-85-3-381

Swenson S, Wahr J (2006) Post-processing removal of correlated errors in GRACE data. Geophys Res Lett 33:L08402. https://doi.org/10.1029/2005GL025285

Tregoning P, Watson C, Ramillien G, McQueen H, Zhang J (2009) Detecting hydrologic deformation using GRACE and GPS. Geophys Res Lett 36(15):L15401

van Dam T, Wahr J, Lavallée D (2007) A comparison of annual vertical crustal displacements from GPS and Gravity Recovery and Climate Experiment (GRACE) over Europe. J Geophys Res Solid Earth 112(B3):B03404

Wahr J, Khan SA, van Dam T, Liu L, van Angelen JH, van den Broeke MR, Meertens CM (2013) The use of GPS horizontals for loading studies, with applications to northern California and Southeast Greenland. J Geophys Res Solid Earth 118:1795–1806. https://doi.org/10.1002/jgrb.50104

Wdowinski S, Eriksson S (2009) Geodesy in the 21st century. Eos Trans AGU 90:153–155

Tracking Hurricanes Using GPS Atmospheric Precipitable Water Vapor Field

Yohannes Getachew Ejigu, Felix Norman Teferle, Anna Klos, Janusz Bogusz, and Addisu Hunegnaw

Abstract

Tropical cyclones are one of the most powerful severe weather events that produce devastating socioeconomic and environmental impacts in the areas they strike. Therefore, monitoring and tracking of the arrival times and path of the tropical cyclones are extremely valuable in providing early warning to the public and governments. Hurricane Florence struck the East cost of USA in 2018 and offers an outstanding case study. We employed Global Positioning System (GPS) derived precipitable water vapor (PWV) data to track and investigate the characteristics of storm occurrences in their spatial and temporal distribution using a dense ground network of permanent GPS stations. Our findings indicate that a rise in GPS-derived PWV occurred several hours before Florence's manifestation. Also, we compared the temporal distribution of the GPS-derived PWV content with the precipitation value for days when the storm appeared in the area under influence. The study will contribute to quantitative assessment of the complementary GPS tropospheric products in hurricane monitoring and tracking using GPS-derived water vapor evolution from a dense network of permanent GPS stations.

Keywords

GPS · Precipitable water vapor · Tropical cyclones · Hurricane Florence

Electronic Supplementary Material The online version of this chapter (https://doi.org/10.1007/1345_2020_100) contains supplementary material, which is available to authorized users.

Y. G. Ejigu (✉)
Department of Space Science and Application Research Development, Ethiopian Space Science and Technology Institute (ESSTI), Addis Ababa, Ethiopia

Department of Physics, Wolkite University, Wolkite, Ethiopia
e-mail: yohannes.getachew@wku.edu.et

F. N. Teferle · A. Hunegnaw
Geodesy and Geospatial Engineering, Institute of Civil and Environmental Engineering, University of Luxembourg, Luxembourg, Luxembourg

A. Klos · J. Bogusz
Faculty of Civil Engineering and Geodesy, Military University of Technology, Warsaw, Poland

1 Introduction

Historical Tropical cyclones (TC) records from the National Hurricane Center (NHC) indicate that 916 hurricanes were formed in the Atlantic basin between 1851 and 2018, of which 330 are major hurricanes with Saffir-Simpson scale 3, 4 or 5.[1] Over the years, their dissipation of power has increased as the strengths of hurricanes have intensified (Klotzbach 2006; Elsner et al. 2008; Bhatia et al. 2019). Hurricane Florence was one of the category four major hurricanes during the 2018 Atlantic hurricane season that produced the largest freshwater flooding in the Carolinas, USA. The consequence was that produced catastrophe and life-threatening flooding. The fundamental physics of TC is a complex process that is fueled by a combination of warm ocean water, moist air and winds (Emanuel et al. 2006). The

---

[1] https://www.aoml.noaa.gov/hrd/tcfaq/E11.html.

© The Author(s) 2020
J. T. Freymueller, L. Sanchez (eds.), *Beyond 100: The Next Century in Geodesy*,
International Association of Geodesy Symposia 152, https://doi.org/10.1007/1345_2020_100

band of air and wind that surrounds the eye of TC acts as a conduit for vertical wind that drags water vapor up from the warm ocean surface (Emanuel 1999; Kepert 2010). The transfer of this energy occurs mainly through the evaporation of water into the atmosphere consequently boosting the radial circulation within the storm by conserving the rotational momentum of rotating air (Smith 2000; Emanuel 2003). The convection systems rely on the atmosphere of the water vapor and vertical temperature profiles in the convection region. Thus, the role of atmospheric water vapor information is very valuable in the study, monitoring and prediction of the TC.

Water vapor is a key component in the atmosphere's thermodynamics. It transports latent heat, contributes to absorption and emission in a number of bands and condenses into clouds reflecting and adsorbing solar radiation. Therefore, it directly affects the energy balance. Severe meteorological events such as a hurricane emphasize the need to monitor and track using additional observational tools to mitigate the risks. Traditionally, the water vapor data has been acquired primarily through the use of water vapor radiometers (WVR) and radiosondes for many years. These techniques of acquiring water vapor data enabled major progress in understanding the procedures happening in the atmosphere, while suffered drawbacks like their low spatial-temporal resolution and high operating costs (Gaffen et al. 1992).

Nowadays, developments in satellite remote sensing technologies offer a wide range in global observations of various structural features of meteorological parameters. The GPS measurement of water vapor is considered one of the key observational tools because GPS works in all weather conditions, unlike other passive remote sensing techniques (Rocken et al. 1995; Zhang et al. 2015). Therefore, GPS has become a defacto standard technique for measuring water vapor with a notable benefit over other traditional meteorological sensors. The propagation delay that affects GPS signals depends on the amount of water vapor in the atmosphere (Bevis et al. 1992). This water vapor sensitivity of GPS delay measurements has been used to derive the precipitable water vapor (PWV) in the atmosphere. Rocken et al. (1995) conducted the first GPS/STORM experiment using GPS meteorology techniques under extreme weather conditions, in a tornado high risk area in the Midwest of the USA. Zhang et al. (2015) have shown the spatio-temporal distributions of GPS-PWV to monitor and predict the typical aspects of typhoons, albeit using a limited number of GPS stations and capable of capturing the signature of severe weather. Recently, Graffigna et al. (2019) showed a significant change in ZTD gradients before a tropical storm arrived. In regions around dense clouds and heavy precipitation, GPS provides independent measurements of atmospheric processes where visible, infrared and microwave-based satellite measurements are largely contaminated (Vergados et al. 2013).

GPS-derived tropospheric products are currently used to investigate meso-scale weather systems and near-real-time applications (Falvey and Beavan 2002; Nakamura et al. 2004; Kawabata et al. 2013; Bordi et al. 2015). With these troposphere products being assimilated into forecast numerical weather prediction (NWP) models (Wilgan et al. 2015; Zus et al. 2015) and thus demonstrate a positive impact on short-range moisture field forecasts and ultimately improve precipitation predictions for severe rainfall events (Bennitt and Jupp 2012). Developments in real-time GPS and of nowcasting NWP models make this application possible by providing PWV that complementing to some level the meteorological predictions (Li et al. 2015). Since PWV relates to the mass of clouds and the dynamics of the temperature near the surface, it is one of the main drivers of the meteorological processes that determine our weather and, in particular, precipitation. The GPS tropospheric products can be used to determine the spacial and temporal change of water vapor, thus help to understand the complex properties of a storm (Tahami et al. 2017; Ejigu et al. 2019a,b). Furthermore, in areas where a dense network of GPS stations is available, it is possible to produce high resolution GPS-PWV maps for the investigation of spatio-temporal water vapor changes. Thus, the use of GPS tropospheric products for regional severe storm prediction is a current research topic and an application that the geodetic community wishes to promote within the meteorological groups. Besides, recent studies have investigated the accuracy of GPS tropospheric products by identifying factors that affect their quality when GPS data are processed (Kačmařík et al. 2017; Klos et al. 2018; Ejigu et al. 2018).

One of the main motivations of this study is to monitor and track early stages of a hurricane (e.g. Hurricane Florence) in 2018 from GPS-derived PWV. We have derived the PWV from the zenith total delay (ZTD) estimation using mean weighted temperature and surface pressure meteorological information. The accuracy of the temporal distribution of the GPS-derived PWV datasets has been validated by comparing with the precipitation value for days when the storm appeared in the area under influence. Also, the accuracy of the GPS-derived PWV datasets has been validated by comparison with the PWV from the radiosonde. Radiosondes are widely used for understanding meteorological parameters in the vertical direction. We emphasis on capturing the signature of major Hurricane Florence events using atmospheric water vapor field information obtained from GPS measurements in the USA. We show how PWV can be used to track and monitor the future hurricanes. The capability of a dense GPS network (i.e. those with antennas only 10s of kilometers apart) to make observations at the required high spatial resolution is being investigated. In this regard, the study objective is to assess the connection between the temporal evolution of GPS-derived PWV and its influence on TC structure. This GPS-derived PWV product may serve as an additional input parameter for hurricane numerical weather prediction (NWP) models.

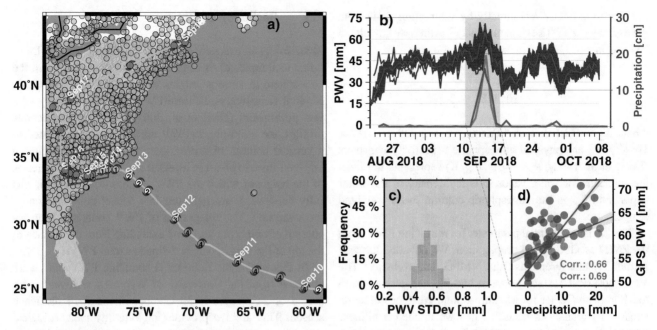

Fig. 1 (**a**) Distribution of GPS stations shown as green circles. The cyan line is the actual path of Florence. The numbers on this line indicates the categories of the hurricane. (**b**) Six-hours resolution stacked time series of GPS-PWV of individual GPS stations (black-lines) and averaged (red-line); stations within the red rectangle shown in panel map (**a**) are employed for a simple stacking. Superimposed are the daily accumulated GPM/IMERG (blue-line) and TRMM (magenta-line) precipitation time series. The light-green shaded region illustrates the periods where Florence shows a maximum change in GPS-PWV and precipitation. (**c**) Histogram of the standard deviation (STDev)

for GPS-PWV. The histogram was generated from the 6-h resolution stacked time series for the set of around 839 distributed stations for a period of 2 months in 2018. (**d**) Scatter plot represents the regression between satellite-derived precipitation (i.e., GPM/IMERG is in blue-circles and TRMM is in magenta-circles) and GPS-PWV. The red line is the estimated linear regression and the green shadow is the 95% confidence interval. The regression is carried out with a 3-h resolution over the time periods of maximum change in GPS-PWV (i.e. 13–18 August 2018, i.e. the light-green shaded regions)

2 Data and Methods

2.1 Retrieval of PWV from GPS ZTD

We selected a dense network of permanent GPS stations (Fig. 1a). The network was in the area affected by Hurricane Florence and includes more than 839 GPS stations for a period of over 2 months (before, during and after Florence), from 24 August to 10 October 2018, distributed of nearly 20-km inter-stations distance across the Carolinas.

We have used the zenith total delay (ZTD) of Tropo-SINEX formatted (Gendt 1997) obtained from the Nevada Geodetic Laboratory (NGL) which employs the GIPSY/OASIS-II software in a precise point positioning (PPP) strategy (Zumberge et al. 1997; Bertiger et al. 2010; Teunissen and Khodabandeh 2014). The solutions are freely available at the NGL Tropo-SINEX ftp data server.[2] The NGL Tropo-SINEX files have a 5-min temporal resolution. The observation cut-off angle was 7°. The satellite clocks, receiver clocks, station coordinates, integer ambiguity and total zenith delays were estimated. The estimated ZTD

delay and the horizontal tropospheric gradients (East-West and North-South) are allowed to vary within random walk $5.0e^{-8}$ km/$\sqrt{\text{sec}}$ and $5.0e^{-9}$km/$\sqrt{\text{sec}}$, respectively. Global Mapping Function (Böhm et al. 2006a) was applied to map slant path delays along GPS signals to ZTD. Details on the GPS observation processing strategy summary are available at the relevant Analysis Center code (ACN) file from NGL website.[3] The ZTD can be further split into zenith hydrostatic delay (ZHD) and zenith wet delay (ZWD). The ZHD can be determined precisely by measuring the surface pressure (P_s) at the station position (Saastamoinen 1972),

$$ZHD = \frac{(2.2768 \pm 0.0015)P_s}{1 - 2.66 \times 10^{-3} \times \cos(2\phi) - 2.8 \times 10^{-7}h} \quad (1)$$

where h is the station height in meters above the ellipsoid. The formal error of Eq. (1) (± 0.0015 mm/hPa) was calculated by considering that all uncertainties of the P_s and h parameters are uncorrelated and this can be found in Davis et al. (1985). We calculated ZHDs using data available from the Vienna Mapping Functions (VMF) gridded files (Böhm et al. 2006b). In GPS meteorology, it is common

---

[2]ftp://gneiss.nbmg.unr.edu/trop/.

[3]http://geodesy.unr.edu/gps/ngl.acn.

practice to extract ZWD directly by subtracting ZHD from the estimated ZTD (Bevis et al. 1992). Further, we calculated PWV using

$$PWV = \frac{ZWD}{10^{-6}(K_2' + \frac{K_3}{T_m})R_v\rho} \quad (2)$$

where $K_2' = 22.1 \pm 2.2 K/\text{hPa}$ and $K_3 = 3.739 \pm 0.012 \times 10^5 K^2/\text{hPa}$ are physical atmospheric reflectivity constants (Bevis et al. 1992), Rv = 461 (Jkg/K) represents the ideal gas constant for water vapor, ρ is the density of the water vapor, and T_m is the atmospheric column mean weighted temperature.

The T_m is an essential parameter for retrieving PWV from the ZWD of GPS signal propagation. We obtained the grid format T_m measurements from VMFG data website.[4] The VMF observed surface pressure and T_m are derived from the European Center for Medium-Range Weather Forecasts (ECMWF) reanalysis dataset. Then we applied a bilinear interpolation technique to extract T_m at the station height. Furthermore, we have carried out the uncertainty of the GPS-derived PWV before conducting the analysis based on an approach similar to the one used by Ning et al. (2013).

2.2   Precipitation Datasets

We obtained the precipitation data from the recent Integrated Multi-Satellite Retrievals for Global Precipitation Measurement (GPM/IMERG) satellite mission. The GPM/IMERG datasets is produced by NASA Goddard Earth Sciences (GES) team, which provides a combined microwave and infrared (IR) satellite gridded precipitation estimates. The GPM/IMERG has 0.1° and half-hour spatio-temporal resolutions for microwave and Infra-red (IR), respectively. The complete algorithm and description are accessible from Huffman et al. (2017). We also compared the precipitation measurements from the Tropical Rainfall Measuring Mission (TRMM)'s 3-h combined Level-3 microwave-IR estimates (3B42 product, version-7). The purpose of the 3B42 product is to produce tropical rainfall measurements and rain gauge-adjusted precipitation rates merged with other satellite mission measurements. This datasets has a 3-h temporal resolution and 0.25° × 0.25° spatial resolution (Huffman and Bolvin 2015). We cover the period of 24th August to 10th October the 2018 hurricane season.

2.3   Radiosonde Data

Meteorologists use radiosondes as one of their main traditional techniques to measure the water vapor content in the atmosphere at various heights with the balloon ascending. Modern radiosondes measure PWV with an accuracy of a few millimeters (Niell et al. 2001). By using radiosonde profiles, the atmospheric PWV content can be estimated in a vertical column of a cross-sectional unit area extending between any two pressure levels. It is usually defined in terms of the height at which the PWV content would lie if it was fully condensed and gathered in a vessel cross-section of the same unit. The integration of PWV contained in every column of unit cross-sectional extending from the surface to the top of the atmosphere is defined as total PWV.

In this paper, the radiosonde sounding PWV data were obtained from the University of Wyoming archives[5] for a period of 24 August to 30 October in the 2018 hurricane season. This archive provides high-quality meteorological parameters such as pressure, temperature and relative humidity at various altitudes (Durre et al. 2016). The radiosonde data has a low temporal rate (mostly operating twice per day, often related to running costs) and that limits their applications in short-term weather forecasting. Furthermore, radiosondes drift laterally than ascending, hence their measurements are not strictly vertical above the radiosonde location.

3   Results

3.1   GPS-Derived PWV and Hurricane Florence

Using a dense ground network of permanent GPS stations, it is possible to pierce through the atmosphere to quantify the distribution of water vapor present and obtain PWV measurement between the GPS station and the GPS satellite in the local zenith direction. Figure 1b presents the stacked GPS-PWV time series for Hurricane Florence, together with the satellite rainfall product from GPM/IMERG and TRMM time series associated with this event. Stations that were within the perimeter of the red rectangle shown in Fig. 1a are employed for stacking to understand the PWV characteristics as the storm approaches (i.e., we take a simple sort of stacking for the PWV 6-h resolution time series of each station and produce a simple average as a representative time series for hurricane Florence landfalls and large distraction area). For these, the influence of Florence is regarded as maxima.

---

[4]http://vmf.geo.tuwien.ac.at.

[5]http://weather.uwyo.edu/upperair/sounding.html.

In the 2018 hurricane season, Florence largely affected the Southeastern United States, and we stacked all available GPS stations' PWV time series around the South Carolina area over a period of 2-months, from 24 August to 10 October. The GPS-PWV time series demonstrated maximum values during the storm front's passage over the respective area. It displays that the time series of PWV at all stations used for stacking are consistent. The GPS-PWV of 74-mm and more was reached on 15/16-September 2018 when Florence crossed from the North Carolina-South Carolina border eastward across Southeastern North Carolina.

We also show the daily cumulative GPM/IMERG and TRMM precipitation attributable to storm from 24-August to 10-October for comparison (Fig. 1b, blue line). Most of the precipitation occurred from 14–17 September. The daily cumulative GPM/IMERG precipitation exceed 20-cm over Southeastern North Carolina as Florence passed over the city. Figure 1c shows the standard deviation (STDev) for the GPS-PWV for all stations stacked. During GPS data processing, the quality of GPS-PWV can be evaluated from the ZTD law of error propagation. The estimation error is extracted from the residual of the constrained least square solution. The maximum STDev for PWV reaches 0.8 mm. However, 95% of the STDev are below 0.6 mm and on average at 0.5 mm. We show scatter plots of GPS-PWV versus precipitation from TRMM (Fig. 1d). We have made a 3-h interval resolution for regression analysis based on the maximum value of GPS-PWV and precipitation events for 1 week (between 14 to 18 September). These parameters exhibit a good strong coupling and follow a similar footprint associated with the storm, with a Pearson correlation of up to 69% with GPM/IMERG and 66% with TRMM.

3.2 Accuracy of GPS-Derived PWV

Radiosondes data are often used as a source for independent validation to demonstrate the quality of the GPS-derived PWV field. However, to correctly assess the GPS PWV precision during severe weather periods, it is essential to use more reliable and more precise devices, such as a water vapor radiometer (WVR). Liou et al. (2001) shows the consistency of radiosonde (RS) PWV estimates depends on the degree of atmospheric in-homogeneity.

To determine the quality of PWV derived from GPS, the radiosonde measurements of PWV from two nearby stations at Charleston (CHS) located at [27.76°N, 97.50°W] and Greensboro (GSO) located at [30.11°N, 93.21°W] were used for the comparison of GPS-derived PWV in addition to assessment of formal error propagation. The RS stations CRP and LCH are approximately located at 13 and 15 km from the GPS stations SCFJ and HITP, respectively. It is required that the horizontal and vertical separation between RS and GPS

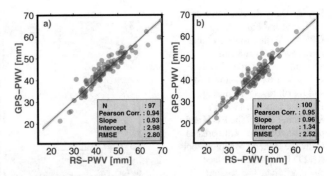

Fig. 2 Regression of PWV from GPS and radiosonde measurements at two stations; (**a**) The RS station at Charleston and GPS station SCFJ, and (**b**) the RS station at Greensboro and GPS station HIPT. The period covers from 24 August to 10 October 2018 at an interval of 12-h temporal resolutions. The red line is the estimated linear regression and the green shadow that underlying in the red line is the 95% confidence interval. The number of data samples (N), Pearson correlation, slope, intercept and root mean square error (RMSE) are given in the legends

stations should be less than 50 km and 100 m, respectively (Wang and Zhang 2008).

Generally, the distribution of RS stations is very sparse and nonuniform. As a result, it is rare to find a RS station very close to a GPS stations. Figure 2 demonstrates the comparison of PWV obtained from the GPS and RS observations for the period between 20 August to 10 October 2018 (covering the time period of pre-, during- and post-Florence). From Fig. 2, we found a strong and consistent degree of correlation with values between 94 and 95%. The difference between RS and GPS derived PWV shows an RMS of 2.8 and 2.5 mm, which is consistent with values from previous comparisons (Deblonde et al. 2005; Ejigu et al. 2018; Li et al. 2003). This provides us an external validation of our GPS-derived PWV estimates data-sets. Our GPS-derived PWV, however, is at higher temporal and spatial resolutions than these from the RS data.

3.3 Estimating GPS PWV Distribution Maps and Monitoring of Hurricane Florence

To quantify how well a storm was captured by GPS-PWV, we constructed water vapor maps using a well-distributed GPS station network and compared the temporal distribution of the GPS-PWV content with the precipitation value for days when the storm appeared in the area. Figure 3 depicts the daily mean GPS-PWV evolution, and the daily cumulative GPM/IMERG precipitation, for Florence from 13 to 18 September 2018. On 30 August, Florence was accompanied by a broad low pressure system that moved off the west coast of Africa. Florence held a steady west-northwest motion at about 15 knots for the next several days as it passed around the Southern periphery of a broad Bermuda-Azores ridge

Fig. 3 Daily distribution maps for Hurricane Florence. GPS-PWV (first row (**a–c**) and third row (**g–i**)) and daily accumulated precipitation from GPM/IMERG (second row (**d–f**) and fourth row (**j–l**)) during Florence between 13 to 18 September 2018. The PWV and precipitation maps were created using the Generic Mapping Tools (GMT) by spherical surface spline griding. The contour interval of the GPS-PWV is 7 mm, and the GPM/IMERG precipitation is 5 cm. The precipitation values are from GPM/IMERG. Florence's path is plotted as the cyan line and the hurricane symbol as brown, respectively

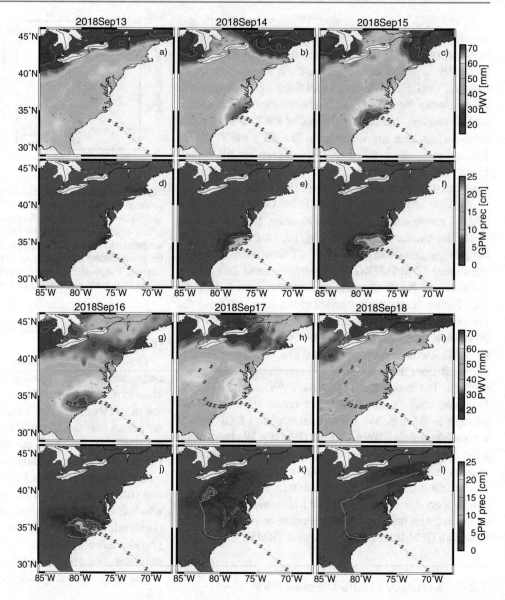

extending from northwestern Africa and Southern Europe west to the east coast of the United States (Stacy and Robbie 2019). It then began to intensify and turn into a hurricane on 04 September. On 13 September, Florence moved closer to the North Carolina coast, and by 14 September it had moved inland. Also, the PWV values extended to rise at a fairly rapid rate. The dramatic increase illustrates the effect of the hurricane. The driver behind this change is mainly owing to the increase of the partial pressure of water vapor (Seco et al. 2009), which is intimately linked to the ZWD. The GPS-PWV increased considerably when Florence reached in the East-Southeast of Wilmington, North Carolina area, around 13–17 September. Significant and intense a daily cumulative precipitation reached up to 20 cm were found over area from Wilmington coast, and by 14 September had moved inland to Elizabethtown, North Carolina. Again it

is clear that the GPS-PWV maps successfully capture the passage of the storm and a significantly elevated PWV (65–71 mm) is observed for 14 September when Florence made landfall at Wrightsville Beach, North Carolina as a Category 1 hurricane. In addition, we see that the GPS-PWV strengthened further and centered on the storm location, and that maximum PWV moved with the storm center. Following this landfall, and then after the passage of the storm, the PWV magnitude decreased, and the daily cumulative precipitation in Carolinas also fell dramatically, to less than 1 mm. Also, the 6-hourly estimated PWV spatio-temporal evolution for the Florence are depicted in the Figures S1, S2 as part of the supplementary material. Likewise, a 6-hourly animated video of the PWV field, which are very well synchronized with Florence for the 1 week period is included as dynamic contents in the supporting materials.

4   Outlook and Conclusions

In this study, we examined a detailed analysis of the impact of the present-day GPS product PWV during hurricane Florence, which affected the Southeastern United States as category four from 13 to 18 September 2018. The time series PWV data for before, during and after Florence storm were examined. The variation in the time series of the PWV content can be correlated with variations in the precipitation value (from GPM and TRMM) coinciding with the passing of a hurricane storm front. According to Sapucci et al. (2016), a sharp increase in the GPS-PWV occurs before rainfall and most of the maximum rainfall occurs near the PWV peak. As it is evident in the comparison, stations around eastern North Carolina (the area where Florence had caused a significant storm surge flooding) showed a sharp peak in PWV during the time Florence was passing. This shows that GPS-derived PWV can be used to monitor Hurricane Florence over the Carolinas. Hence, when the water vapor condenses into clouds or rain, it releases latent heat and thus GPS measurements of water vapor can contribute to track the tropical cyclone's. In particular, for 14–18 September in the Carolinas the PWV surged, to reach up to 74 mm much higher than a typical value of 40 mm. The increase in PWV by an average of 28 mm (with a range of about 46–74 mm) is commonly observed in the general vicinity of the area crossed by a storm for 4–5 days. There is some variability in water vapor due to the local weather processes during pre-storm. As researchers have reported previously (e.g., Bryan and Oort 1984; Cadet and Nnoli 1987), in summer the water vapor intensity is elevated compared to that in other seasons in the Atlantic Ocean. We also observed a sudden drop in the PWV time series values after the storms had passed.

The main advantage of GPS-PWV from a dense continuous GPS network is that it is possible to produce PWV distribution maps in close proximity to real-time or near real time. We constructed the map of maximum possible GPS-derived PWV distribution. We observed a large proportion of PWV in the distribution maps of PWV and were able to determine the feature of the storm. The findings verified that the temporal variation in GPS-PWV is tightly related to the route of the hurricane, possibly tracking and monitoring hurricane operations that rises typically at least several hours prior to the storm's arrival.

Further, the accessibility of ground-based GPS in all weather conditions around the globe and inexpensive GPS receiver are cost-effective meteorological sensors. Also, real-time GPS makes PWVs possible at 5-min updates. These developments provide the background for the inclusion of real-time GPS in nowcasting models for severe events. Moreover, assimilated GPS-PWV could significantly improve the moisture fields within the nowcasting NWP model, resulting in a better description of the water budget, and thus will further improve our ability to track and monitoring storms and their impact on e.g. coastal communities.

Acknowledgements The ZTD Tropo SINEX data are provided by the Nevada Geodetic Laboratory (NGL) at the University of Nevada, Reno. We used precipitation data from the TRMM and GPM/IMERG satellite missions, provided by the NASA Goddard Space Flight Center and accessed at https://pmm.nasa.gov/data-access/. Anna Klos is supported by the Foundation for Polish Science (FNP). Janusz Bogusz is funded by the Polish National Science Center grant no. UMO-2016/21/B/ST10/02353. Addisu Hunegnaw is funded by the Luxembourg National Research Fund/Fonds National de la Recherche (FNR) with project code O18/12909050/VAPOUR/.

References

Bennitt GV, Jupp A (2012) Operational assimilation of GPS Zenith total delay observations into the met office numerical weather prediction models. Mon Weather Rev 140(8):2706–2719

Bertiger W, Desai SD, Haines B, Harvey N, Moore AW, Owen S, Weiss JP (2010) Single receiver phase ambiguity resolution with GPS data. J Geod 84(5):327–337

Bevis M, Businger S, Herring TA, Rocken C, Anthes RA, Ware RH (1992) GPS meteorology: remote sensing of atmospheric water vapor using the global positioning system. J Geophys Res Atmos 97(D14):15,787–15,801

Bhatia KT, Vecchi GA, Knutson TR, Murakami H, Kossin J, Dixon KW, Whitlock CE (2019) Recent increases in tropical cyclone intensification rates. Nat Commun 10(1):635

Böhm J, Niell A, Tregoning P, Schuh H (2006a) Global mapping function (GMF): a new empirical mapping function based on numerical weather model data. Geophys Res Lett 33(7):L07304

Böhm J, Werl B, Schuh H (2006b) Troposphere mapping functions for GPS and very long baseline interferometry from European centre for medium-range weather forecasts operational analysis data. J Geophys Res Solid Earth 111(B2). https://doi.org/10.1029/2005JB003629

Bordi I, Raziei T, Pereira LS, Sutera A (2015) Ground-based GPS measurements of precipitable water vapor and their usefulness for hydrological applications. Water Resour Manag 29(2):471–486

Bryan F, Oort A (1984) Seasonal variation of the global water balance based on aerological data. J Geophys Res Atmos 89(D7):11,717–11,730. https://doi.org/10.1029/JD089iD07p11717

Cadet D, Nnoli N (1987) Water vapour transport over Africa and the Atlantic ocean during summer 1979. Q J R Meteorol Soc 113(476):581–602

Davis JL, Herring TA, Shapiro II, Rogers AEE, Elgered G (1985) Geodesy by radio interferometry: effects of atmospheric modeling errors on estimates of baseline length. Radio Sci 20(6):1593–1607. https://doi.org/10.1029/RS020i006p01593

Deblonde G, MacPherson S, Mireault Y, Héroux P (2005) Evaluation of GPS precipitable water over Canada and the IGS network. J Appl Meteorol 44:153–166. https://doi.org/10.1175/JAM-2201.1

Durre I, Vose RS, Yin X, Applequist S, Arnfield J (2016) Integrated global radiosonde archive (IGRA) Version 2. https://doi.org/10.7289/V5X63K0Q

Ejigu YG, Hunegnaw A, Abraha K, Teferle FN (2018) Impact of GPS antenna phase center models on zenith wet delay and tropospheric gradients. GPS Solut 23(4):659–680

Ejigu GY, Teferle FN, Klos A, Bogusz J, Hunegnaw A (2019a) Improved monitoring and tracking hurricanes using GPS atmospheric water vapor. Geophys Res Abstr 21:EGUGA p 17823

Ejigu YG, Teferle FN, Hunegnaw A, Klos A, Bogusz J (2019b) Tracking hurricanes Harvey and Irma using GPS tropospheric products. AGU Fall Meeting, Dec 2019

Elsner JB, Kossin JP, Jagger TH (2008) The increasing intensity of the strongest tropical cyclones. Nature 455:92–95

Emanuel K (1999) Thermodynamic control of hurricane intensity. Nature 401:665–669

Emanuel K (2003) Tropical cyclones. Ann Rev Earth Planet Sci 31(1):75–104

Emanuel K, Ravela S, Vivant E, Risi C (2006) A statistical deterministic approach to hurricane risk assessment. Bull Am Meteorol Soc 87(3):299–314

Falvey M, Beavan J (2002) The impact of GPS precipitable water assimilation on mesoscale model retrievals of orographic rainfall during SALPEX'96. Mon Weather Rev 130:2874–2888

Gaffen DJ, Elliott WP, Robock A (1992) Relationships between tropospheric water vapor and surface temperature as observed by radiosondes. Geophys Res Lett 19(18):1839–1842. https://doi.org/10.1029/92GL02001

Gendt G (1997) SINEX TRO–solution (Software/technique) independent exchange format for combination of TROpospheric estimates Version 0.01, 1 March 1997. https://igscb.jpl.nasa.gov/igscb/data/format/sinex_tropo.txt

Graffigna V, Hernández-Pajares M, Gende M, Azpilicueta F, Antico P (2019) Interpretation of the tropospheric gradients estimated with GPS during hurricane Harvey. Earth Space Sci 6(8):1348–1365

Huffman GJ, Bolvin D (2015) TRMM and other data precipitation data set documentation. NASA pp 1–44

Huffman GJ, Bolvin D, Eric N J (2017) Integrated multisatellite retrievals for GPM (IMERG) technical documentation. NASA pp 1–46

Kačmařík M, Douša J, Dick G, Zus F, Brenot H, Möller G, Pottiaux E, Kapłon J, Hordyniec P, Václavovic P, Morel L (2017) Inter-technique validation of tropospheric slant total delays. Atmos Meas Tech 10(6):2183–2208

Kawabata T, Shoji Y, Seko H, Saito K (2013) A numerical study on a mesoscale convective system over a subtropical island with 4D-Var assimilation of GPS slant total delays. J Meteorol Soc Jpn 91:705–721

Kepert JD (2010) Global perspectives on tropical cyclones. World Scientific, New Jersey

Klos A, Hunegnaw A, Teferle FN, Abraha KE, Ahmed F, Bogusz J (2018) Statistical significance of trends in zenith wet delay from re-processed GPS solutions. GPS Solutions 22(2):51

Klotzbach PJ (2006) Trends in global tropical cyclone activity over the past twenty years (1986–2005). Geophys Res Lett 33(10):L10805

Li Z, Muller JP, Cross P (2003) Comparison of precipitable water vapor derived from radiosonde, GPS, and moderate-resolution imaging spectroradiometer measurements. J Geophys Res Atmos 108(D20). https://doi.org/10.1029/2003JD003372

Li X, Dick G, Lu C, Ge M, Nilsson T, Ning T, Wickert J, Schuh H (2015) Multi-GNSS meteorology: real-time retrieving of atmospheric water vapor from BeiDou, Galileo, GLONASS, and GPS observations. IEEE Trans Geosci Remote Sens 53(12):6385–6393

Liou YA, Teng YT, Van Hove T, Liljegren JC (2001) Comparison of precipitable water observations in the near tropics by GPS, microwave radiometer, and radiosondes. J Appl Meteorol 40(1):5–15

Nakamura H, Koizumi K, Mannoji N, Seko H (2004) Data assimilation of GPS precipitable water vapor to the JMA mesoscale numerical weather prediction model and its impact on rainfall forecasts. J Meteorol Soc Jpn 82. https://doi.org/10.2151/jmsj.2004.441

Niell AE, Coster AJ, Solheim FS, Mendes VB, Toor PC, Langley RB, Upham CA (2001) Comparison of measurements of atmospheric wet delay by radiosonde, water vapor radiometer, GPS, and VLBI. J Atmos Ocean Technol 18(6):830–850

Ning T, Elgered G, Willén U, Johansson JM (2013) Evaluation of the atmospheric water vapor content in a regional climate model using ground-based GPS measurements. J Geophysl Res Atmos 118(2):329–339. https://doi.org/10.1029/2012JD018053

Rocken C, Hove TV, Johnson J, Solheim F, Ware R, Bevis M, Chiswell S, Businger S (1995) GPS/STORM—GPS sensing of atmospheric water vapor for meteorology. J Atmos Ocean Technol 12(3):468–478

Saastamoinen J (1972) Atmospheric correction for the troposphere and stratosphere in radio ranging satellites. The use of artificial satellites for geodesy. Geophys Monogr Ser 15:247–251

Sapucci LF, Machado LAT, Menezes de Souza E, Campos TB (2016) GPS-PWV jumps before intense rain events. Atmos Meas Tech Discuss 2016:1–27

Seco A, González P, Ramírez F, García R, Prieto E, Yagüe C, Fernández J (2009) GPS monitoring of the tropical storm delta along the canary islands track, November 28–29, 2005. Pure Appl Geophys 166(8–9):1519–1531

Smith R (2000) The role of Cumulus convection in hurricanes and its representation in hurricane models. Rev Geophys 38:465–489

Stacy R, Robbie B (2019) Tropical Cyclone Report: Hurricane Florence (AL062018). Tech. rep., South Carolina, USA

Tahami H, Park J, Choi Y (2017) The preliminary study on the prediction of a hurricane path by GNSS derived PWV analysis. In: Proceedings of the ION 2017 Pacific PNT Meeting, Honolulu, Hawaii

Teunissen P, Khodabandeh A (2014) Review and principles of PPP-RTK methods. J Geod 1–24. https://doi.org/10.1007/s00190-014-0771-3

Vergados P, Mannucci AJ, Su H (2013) A validation study for GPS radio occultation data with moist thermodynamic structure of tropical cyclones. J Geophys Res Atmos 118(16):9401–9413

Wang J, Zhang L (2008) Systematic errors in global radiosonde precipitable water data from comparisons with ground-based GPS measurements. J Climate 21(10):2218–2238. https://doi.org/10.1175/2007JCLI1944.1

Wilgan K, Rohm W, Bosy J (2015) Multi-observation meteorological and GNSS data comparison with numerical weather prediction model. Atmos Res 156:29–42

Zhang K, Manning T, Wu S, Rohm W, Silcock D, Choy S (2015) Capturing the signature of severe weather events in Australia using GPS measurements. IEEE J Sel Top Appl Earth Observ Remote Sens 8(4):1839–1847

Zumberge J, Heflin MB, Jefferson D, Watkins M, Webb F (1997) Precise point positioning for the efficient and robust analysis of GPS data from large networks. J Geophys Res 102:5005–5017

Zus F, Dick G, Heise S, Wickert J (2015) A forward operator and its adjoint for GPS slant total delays. Radio Sci 50(5):393–405

Continuous Monitoring with a Superconducting Gravimeter As a Proxy for Water Storage Changes in a Mountain Catchment

Quentin Chaffaut, Jacques Hinderer, Frédéric Masson, Daniel Viville, Jean-Daniel Bernard, Solenn Cotel, Marie-Claire Pierret, Nolwenn Lesparre, and Benjamin Jeannot

Abstract

In mountainous area, spring water constitutes the only drinking water resource and local economy is highly dependent on forest health and productivity. However, climate change is expected to make extreme water shortage episodes more and more frequent. Forest is therefore more and more exposed to water stress. It appears necessary to quantify the drought induced by water deficit to evaluate forest vulnerability and to plan the future of forest management. In this study we quantified the 2018 water deficit experienced by the forest in the Strengbach catchment, located in the French Vosges mountains. Three methods for estimating catchment water storage changes (WSC) have been compared. The first relies on superconducting gravimeter monitoring while the second relies on catchment water balance. The third one relies on global hydrological model MERRA2. We show that WSC estimated from measured gravity changes correlate well with WSC estimated from catchment water balance while WSC inferred from MERRA2 significantly differs. The Strengbach catchment water cycle is mostly annual but exhibits significant interannual variability associated with the 2018 drought episode: August 2018 has a water deficit of 37 mm (as inferred from catchment water balance) or 76 mm (as seen with superconducting gravimetry) compared to August 2017. We illustrate here the use of superconducting gravimeter monitoring as an independent proxy for WSC in a mountainous catchment while most of hydro-gravimetric studies have been conducted on relatively flat areas. We therefore contribute to expand the area of use of high precision gravity monitoring for the hydrological characterization of the critical zone in mountainous context. This innovative method may help to assess forest vulnerability to drought in the context of climate change.

Keywords

Hydro-gravimetry · Hydrological modeling · Mountain catchment · Superconducting gravimetry

Q. Chaffaut (✉) · J. Hinderer · F. Masson · J.-D. Bernard
IPGS-EOST, CNRS-UMR 7516, University of Strasbourg, Strasbourg, France
e-mail: qchaffaut@unistra.fr

D. Viville · S. Cotel · M.-C. Pierret · N. Lesparre
LHyGeS-EOST, CNRS-UMR 7517, University of Strasbourg, Strasbourg, France

B. Jeannot
LHyGeS-ENGEES, CNRS-UMR 7517, University of Strasbourg, Strasbourg, France

1   Introduction

Spring water constitutes the only drinking water resource for villages located in the French Vosges mountains. Furthermore, local economy (tourism, hunt, logging, wood transformation) is highly dependent on forest health and productivity. However, the forest welfare is highly sensitive to water shortage associated with severe drought e.g. the 2003 western European drought episode or the 2012–2015 California drought (Bréda et al. 2006; Asner et al. 2016). Unfortu-

J. T. Freymueller, L. Sanchez (eds.), *Beyond 100: The Next Century in Geodesy*,
International Association of Geodesy Symposia 152, https://doi.org/10.1007/1345_2020_105

nately, climate change is expected to increase temperature variability and to enhance the frequency and the severity of such drought events, especially in the northern hemisphere (Seneviratne et al. 2012). This may compromise long-term tree survival in some part of the Vosges mountains, especially because of combined effect of water stress, decrease of fertility and parasite attack. It is thus necessary to assess water storage changes (WSC) at the catchment scale to compute a posteriori water deficit experienced by the forest.

Catchment WSC result from the water fluxes acting in the landscape, balancing precipitation, evapotranspiration and runoff. Classically, there are two ways for estimating WSC at the catchment scale in mountainous areas: one may rely on local prediction from global hydrological models or on catchment water balance derived from local hydro-meteorological measurements. On the first hand, global hydrological models provide soil WSC but with a sparse spatial resolution. On the other hand, catchment water balance is representative of catchment WSC but is still particularly difficult to assess from local hydro-meteorological measurements in a mountainous context because topography or land-cover variations make rainfall and evapotranspiration fluxes highly heterogeneous and consequently difficult to monitor (Shamir et al. 2016).

A third option to estimate WSC is the use of in situ time-variable gravimetry that is a non-invasive method, in contrast to traditional point scale measurements used for directly measuring soil WSC e.g. with neutron probes (Hector et al. 2013). Time-variable gravimetry is also directly sensitive to integrated WSC and so bridges the gap between point scale measurements and large-scale estimates of WSC (Fores et al. 2017). It therefore appears as a well-suited method to assess WSC at the catchment scale independently from hydro-meteorological measurements.

With a nominal precision of 0.1 nm.s^{-2}, the superconducting gravimeter (SG) is the most sensitive relative gravimeter available (see Hinderer et al. 2015 for a review). The SG gravity signal contains several geophysical, atmospheric and hydrological contributions listed hereafter. The strongest contribution is the tidal signal from oceans and solid Earth due to the attraction of Moon and Sun. The tidal signal produces gravity variations up to $2,800$ nm.s^{-2}. The polar motion is another signal of external origin resulting from the motion of the Earth rotational axis, it produces gravity variations up to 100 nm.s^{-2}. Hydrological and atmospheric signals seen by the SG strongly depend on climatic conditions and on the local geomorphological context. Both hydrological and atmospheric signal contains a local contribution as well as a non-local contribution resulting from large-scale atmosphere and hydrology. Large-scale contributions are estimated thanks to global atmospheric and hydrological models (Llubes et al. 2004; Boy et al. 2002). The total (local + non-local) hydrological contribution produces grav-

ity variations usually up to 150 nm.s^{-2} and the total atmospheric contribution produces gravity variations of the same order of magnitude. Nowadays, the hydro-gravimetric signal i.e. the gravity signal corrected for every other well modelled instrumental and geophysical contributions becomes relatively easy to extract. As a result, hydro-gravimetric studies using SG dedicated to local WSC monitoring become more and more common (e.g. Hector et al. 2014; Fores et al. 2017).

However, most of hydro-gravimetric studies relying on SG focus on the hydrology of relatively flat areas like plains or plateaus. Here we apply in situ superconducting gravity monitoring to the hydrological characterization of a mountainous catchment. Here hydrological contributions are due to fast changes in soil water storage associated with precipitations (hourly timescale or less) or slower changes associated with underground flow which produces gravity variations up to 150 nm.s^{-2}. Due to its strong topography, our study site exhibits a very reactive hydrological behavior as well as an a priori complex spatial distribution of water. This last point certainly constitutes our main challenge because gravity does not depend only on the water amount but on the spatial distribution of water too.

Here we assume that SG hydrological residual signal acts as a daily proxy of WSC at the catchment scale in a forested mountainous area and brings new independent constrains to test catchment WSC derived from catchment water balance on one hand or predictions of global hydrological model MERRA2 (Reichle et al. 2017) on the other hand.

2   Study Site: The Strengbach Catchment

The Strengbach catchment is a small (0.8 km^2) forested catchment located on the western side of the Vosges mountains in France (Fig. 1). It corresponds to the site of OHGE i.e. Hydro-Geochemical Observatory of the Environment (http://ohge.unistra.fr/) which is part of the OZCAR network for the study of the critical zone (http://ozcar-ri.prod.lamp.cnrs.fr). A part of water from four springs located on the site is taken as drinking water for the village. The Strengbach catchment is a granitic catchment with altitudes ranging between 883 m at the outlet and 1,146 m at the summit. The bedrock mainly consists in Brezouard granite and is covered by a granitic saprolite whose thickness varies between 1 and 9 m (Pierret et al. 2018). This thin superficial layer is expected to host the active aquifer i.e. the main contributor of the stream draining of the Strengbach catchment (Weill et al. 2017). Catchment runoff is measured at the outlet station while meteorological measurements are provided by the summit weather station and by a network of pluviometers distributed across the catchment (Fig. 1). Evapotran-

Fig. 1 Strengbach catchment topography and localization of hydro-meteorological and superconducting gravity measurements. Vertical distance between iso-level lines is 10 m

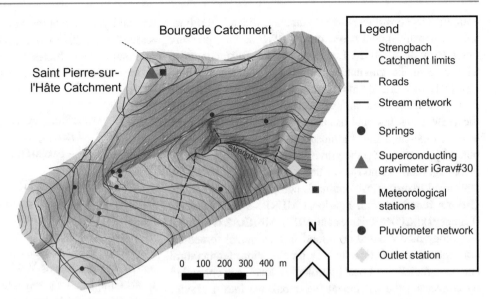

3 Data and Methodology

3.1 Extraction of Hydro-Gravimetric Signal from Superconducting Gravimeter Data

The raw output of SG is a voltage that needs to be converted into gravity units by means of calibration. The calibration process consists in adjusting the scale factor of the SG with side by side observations during several days using an absolute gravimeter FG5#206 from Micro-g Lacoste Inc. (Rosat et al. 2018). At least one other absolute measurement

spiration is modeled from meteorological measurements. In this way, OHGE observatory provides the catchment water balance computed from the modeled and monitored hydro-meteorological fluxes.

In the framework of the CRITEX project (https://www.critex.fr) a new superconducting gravimeter iGrav#30 (SG) from GWR Instruments Inc. has been installed in June 2017 at the summit of the Strengbach catchment in the vicinity from the meteorological station (Fig. 1). SG is installed on the edge of a 8.4 m × 4.4 m shelter with concrete foundations but no gravimetric pillar. In this way WSC occur only at a smaller altitude than the SG and every area located in the footprint of the gravimeter contribute positively to the gravity signal measured by the SG i.e. a water storage increase induces a gravity increase. This specific location maximizes the hydro-gravimetric signal and enable the use of SG hydro-gravimetric signal as a proxy of local WSC. Note that the SG shelter acts as a mask which prevent water to infiltrate beneath the gravimeter (Creutzfeldt et al. 2010; Deville et al. 2013; Reich et al. 2019). The resulting mask effect is quantified in a next section.

is necessary to constrain the long-term instrumental drift. The calibration factor of SG is -919 ± 3 nm.s^{-2}.V^{-1} and the instrumental drift is 70 nm.s^{-2}.year^{-1}. Since we only have two absolute gravity measurements at our disposal, we couldn't assess the linearity of the long-term drift. However, this last one is well approximated by linear polynomial in a long-term view as indeed observed on other SGs (Fores et al. 2017; Rosat et al. 2018).

Being part of the IGETS network (previously GGP see Crossley and Hinderer 2010), level 1 SG raw data are available at https://isdc.gfz-potsdam.de/igets-data-base/. First pre-processing step consists in decimating second samples into minute samples using a standard low pass filter. Then spikes resulting from visits in the SG shelter or earthquakes are removed. After this preprocessing step, data are corrected for the long-term instrumental drift and for polar motion the last one being provided by the International Earth Rotation Service (ftp://hpiers.obspm.fr/iers/eop/eopc04/).

We separate tidal, hydrological and atmospheric contributions that are present in the SG gravity signal by introducing a priori atmospheric and hydrological corrections prior to the tidal model adjustment. We compute the input data of tidal analysis by correcting iGrav#30 gravity from instrumental drift, polar motion, theoretical annual and semi-annual tides (delta factor 1.16 and lag 0°), atmospheric and hydrological loading. Then shorter period tides (monthly to half-daily tides) are adjusted by the tidal analysis version ET34-X-V71 of ETERNA (Schueller 2015). Both atmospheric and hydrological loading processes are decomposed into a local newtonian component (distance < 11 km from the gravimeter) and a non-local newtonian+elastic component which account for large-scale (non-local) atmosphere and hydrology contributions. Non-local atmospheric loading is computed by convolving Green's function with a 2.5D atmospheric density model based on ECMWF surface pressure fields (Boy et al. 2002). Local atmosphere (distance up to 11 km from the gravimeter and 30 km above the topography) is discretized into prisms. The air density profile is derived from local surface pressure measurement using the perfect gas law and assuming hydrostatic equilibrium at each time-step. Local atmospheric loading is then computed by summing up the gravity effect of all prisms using the integration method described in (Leirião et al. 2009). Non-local hydrological loading is computed by convolving Green's functions with non-local MERRA2 cell elements (Llubes et al. 2004; Reichle et al. 2017). MERRA2 is a global hydrology model based on a land surface model forced by atmospheric parameters such as precipitation, temperature and solar radiations. Local hydrological loading is computed by converting the catchment water balance into a gravity signal assuming a homogeneous water coverage over the topography (see next section for a detailed explanation).The last processing step for extracting the local hydrological contribution (the so-called hydrological residual) from the SG signal consists in correcting the tidal analysis input data by removing the adjusted tidal model and adding back the modeled local hydrological loading (removed prior to the tidal analysis).

3.2 Converting the Local Hydro-Gravimetric Signal into Water Storage Changes at the Catchment Scale

The Strengbach catchment is defined topographically so that the watershed limit corresponds to the crest line. In such hydrological catchment, borders act as no-flow conditions and runoff is collected fully at the outlet. Therefore, catchment water storage depends only on input flux i.e. rainfall and output flux i.e. runoff and evapotranspiration. Runoff is measured at the outlet of the catchment, mean annual runoff is 697 mm for the period 2014–2018. Evapotranspiration is modeled using the BILJOU model (Granier et al. 1999) by considering the forest cover and soil type and using solar radiation, temperature, humidity and wind speed measurements from the summit weather station; its mean annual value is 418 mm for the same 2014–2018 period. Rainfall is measured every 10 min by an automatic rain gauge located at the summit weather station. In parallel, repeated measurements (2-week sampling rate) of a network of rain gauges distributed across the catchment allow to measure spatial heterogeneity of rainfall. The combination of both datasets allows to upscale the catchment mean hydrological rainfall (the rainfall amount that effectively reaches the surface) from the automatic rain gauge measurement. The mean

annual hydrological rainfall is 1,194 mm including 20% of snow for the 2014–2018 period. Based on measured input and output water fluxes, we compute the catchment WSC at a daily time step according to the water-balance equation:

$$
\begin{aligned}
\mathrm{WaterStorageChanges}\,(t) \\
= \mathrm{Rain_{cumulated}}\,(t) - \mathrm{Runoff_{cumulated}}\,(t) \\
- \mathrm{EvapoTranspiration_{cumulated}}\,(t)
\end{aligned}
\tag{1}
$$

For comparison with catchment water balance and MERRA2 local, we need to convert SG hydro-gravimetric signal expressed in $\mathrm{nm.s^{-2}}$ into WSC expressed in mm of water. The measured gravity response associated with catchment WSC depends on the amount of WSC as well as on the location of WSC. The hydro-gravimetric signal is also impacted by the mask effect: the SG shelter acts as an impermeable layer which reduce WSC in the close surrounding of the gravimeter (Creutzfeldt et al. 2010; Deville et al. 2013; Reich et al. 2019). Here we assume that the ground WSC occurs mainly in the thin granitic saprolite soil layer which hosts the active aquifer of the Strengbach catchment (Weill et al. 2017). Based on this consideration, we model catchment WSC as a spatially homogeneous water layer of time-variable thickness placed at a given depth below the ground surface. A water layer of nominal thickness (0.1 m) is discretized into prisms whose horizontal extension is 0.5 m $\times$ 0.5 m. Then we compute the water admittance at the SG location i.e. the gravity response of a water layer of nominal thickness evaluated at SG location (Fig. 2). We do this by summing up the gravity effects of prisms using the integration method described in (Leirião et al. 2009): for a normalized distance (i.e. the ratio between prism size and prism distance from SG) below 25 [−] we use the prism formula, for a normalized distance between 25 and 36 [−] we use the Macmillan formula (an approximation of the prism formula) and for a normalized distance bigger than 36 [−] we use the point-mass formula.

As a first step, we compute the unmasked water admittance i.e. we neglect the mask effect by summing up the gravity effect of all prisms for integration radius ranging from 0 to 30 km away from SG. We compute it for a water layer depth of 0.1 m or 10 m (Fig. 2). It converges to (respectively) 0.71 $\mathrm{nm.s^{-2}.mm\ Water^{-1}}$ or 0.73 $\mathrm{nm.s^{-2}.mm\ Water^{-1}}$, it means that the unmasked admittance asymptotic value is almost not sensitive to the depth of the water layer. However, it should be noted that the vertical distribution of water has a significant effect on the unmasked admittance up to 100 m of distance from the SG (note the difference between solid lines for an integration radius ranging from 0 to 100 m on Fig. 2). In a next step we compute the masked admittance by excluding the prisms from below the shelter:

Fig. 2 Solid lines: unmasked admittance for a water layer depth of 0.1 m (light grey) or 10 m (dark grey). Dotted lines: masked admittance for a water layer depth of 0.1 m (light grey) or 10 m (dark grey)

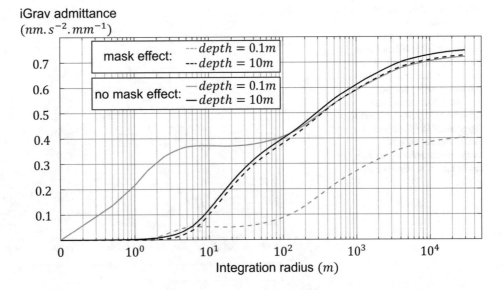

i.e. we assume there is no WSC occurring below the shelter. For a depth of 0.1 m masked admittance reaches only 0.43 nm.s^{-2}.mm Water^{-1} while it converges to 0.71 nm.s^{-2}.mm Water^{-1} for a depth of 10 m, which almost correspond to the unmasked admittance value. Therefore, the magnitude of the mask effect and then the masked admittance asymptotic value depends heavily on the vertical distribution of water: the more the water layer is shallow, the more the mask effect is significant. In the absence of any other observational constraints on the vertical distribution of water around the SG, it results that masked admittance ranges between 0.43 and 0.71 nm.s^{-2}.mm Water^{-1}.

Both masked and unmasked admittance reach 90% of their asymptotic value for an integration radius of 5 km, which gives an estimate of the SG footprint (Fig. 2) It means that most of the local hydrological signal comes from a circle of 5 km radius centered at SG location. SG is therefore sensitive to WSC occurring in the three contiguous catchments (see Fig. 1).

4 Results and Discussion

SG hydro-gravimetric signal and catchment water balance are compared in terms of WSC expressed in mm of water (Fig. 3). The admittance value used to convert SG gravimetric signal (in nm.s^{-2}) into WSC (in mm of water) is adjusted by scaling the SG signal on the catchment water balance. The adjusted admittance we found is 0.60 ± 0.02 nm.s^{-2}.mm Water^{-1} which lies within the range of computed masked admittance values (Fig. 2).

The root mean square difference between zero-averaged SG and catchment balance WSC is 36 mm of water. Gravity and hydro-meteorological estimates of WSC are conse-

quently in good agreement which is remarkable considering the numerous corrections applied on gravity data as well as the simplistic hypothesis we made to convert water storage into gravity. Note that a part of the remaining discrepancy between gravity WSC and catchment WSC may be due to a residual instrumental drift in the SG signal. Instrumental drift will be better constrained thanks to new absolute measurements planned in the future. SG is located at the junction point between three different catchments: The Strengbach catchment on the South East, the Bourgade catchment on the North East and the Saint Pierre Sur L'Hâte catchment on the South West (Fig. 1). It is therefore sensitive to WSC occuring in these tree catchments. As we success to reproduce the measured hydro-gravimetric signal by extrapolating the Strengbach catchment WSC within the footprint area of the gravimeter, it suggests that: (1). SG hydro-gravimetric signal may be considered as a new independent proxy of water storage in the Strengbach catchment. (2) Despite their different orientations, geology, forest cover, etc. these three catchments may have a similar hydrological behavior.

Both SG and catchment water balance exhibit fast (daily to weekly) WSC but the water cycle is dominated by a seasonal component with significant interannual variability. Minimum and maximum water storage occurs merely at the same time every year. For both 2017 and 2018 the minimum water storage occurs at the end of August and the maximum water storage occurs at the end of January in 2017 and at the end of February in 2018. However, summer 2018 is significantly drier than summer 2017. For the month of August, catchment water balance indicates a decrease of 37 mm between 2017 and 2018 and SG indicates a decrease of 76 mm equivalent water thickness. This strong water deficit observed by both catchment water balance and gravity measurements may result from the rainfall deficit and slightly higher evapotranspiration associated with the 2018

Fig. 3 Top: Comparison between daily SG hydro-gravimetric signal, catchment water balance and MERRA2 local hydrology using an SG admittance scaled on catchment water balance. Bottom: Misfit histograms between SG hydro-gravimetric signal and MERRA2 local (in black) and between SG hydro-gravimetric signal and catchment water balance (in green)

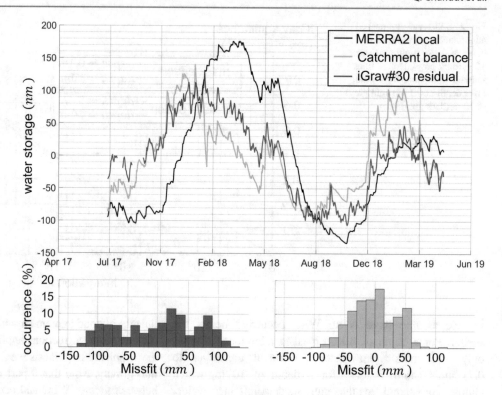

drought. For the period 2014–2018 the mean annual rainfall is 1,194 mm and mean annual evapotranspiration is 418 mm, while in 2018 the mean annual rainfall was 1,079 mm and evapotranspiration was 424 mm.

We compared the SG hydro-gravimetric signal to the local component of the hydrological model MERRA2 (Reichle et al. 2017) computed at EOST (http://loading.u-strasbg.fr/). MERRA2 local WSC exhibits less fast WSC than observed by the SG and in addition there is a significant phase shift between both signals. The adjusted admittance we found by scaling the SG signal on MERRA2 local is 0.349 nm.s^{-2}.mm Water^{-1} which is outside the range of computed masked admittance values (Fig. 2). This merely point out that MERRA2 local water storage changes are too strong compared to catchment water balance WSC and SG WSC. The mean squared difference between SG (using the admittance scaled on catchment water balance: 0.597 nm.s^{-2}.mm Water^{-1}) and MERRA2 is 66 mm of water. This relatively poor fit may result from the low spatial resolution (70 km in latitude and longitude) of MERRA2 model as well as the use of satellite data to force it. The need for local measurements of hydrometeorological parameters to conduct hydro-gravimetric studies dedicated to local hydrology was indeed highlighted previously by other studies (e.g. Fores et al. 2017). Our study therefore clearly demonstrates the benefits, especially in a mountainous catchment, of in situ gravity observations with a superconducting gravimeter compared to global hydrological model for the characterization of catchment hydrology.

5   Conclusion

Considering the numerous corrections applied onto the measured gravity signal as well as the hypothesis made to convert the SG hydro-gravimetric signal into WSC (including the simplistic approach used to take into account the building mask effect), it is remarkable to have such a good agreement between WSC derived from SG monitoring and from catchment water balance (mean squared difference of 36 mm of water). It shows that the SG hydro-gravimetric signal is a valuable proxy of WSC in the Strengbach catchment and that nearby catchments may have a similar hydrological behavior. It also demonstrates the benefits of in situ gravity changes observations compared to MERRA2 global hydrological model, especially in mountainous areas with strong topography.

The water cycle is dominated by an annual component but exhibits a strong water deficit due to drought. August 2018 has a water deficit of 37 mm (as inferred from catchment water balance) or 76 mm (as seen by SG) compared to August 2017. Although strong interannual fluctuations of rainfall and catchment WSC have been documented since the beginning of the hydro-meteorological monitoring (1986), such very long and intense drought episodes are unprecedented.

In this study we demonstrated the benefit of superconducting gravimeter observations as a proxy of WSC in a very hydrologically reactive mountainous catchment while most of the existing hydro-gravimetric studies are focused on the hydrological characterization of relatively flat areas. It therefore expands the area of use of superconducting gravimeters for the hydrological characterization of the critical zone in mountainous contexts. This study is also a step necessary to assess Strengbach forest vulnerability to drought in the context of climate change, and hence to allow the future management of the local forestry.

Acknowledgements We thank J.-P. Boy for providing surface loading computation of MERRA2 hydrological loading and ECMWF atmospheric loading through the EOST loading service (http://loading.u-strasbg.fr). iGrav#30 superconducting gravimeter has been funded by EQUIPEX CRITEX (https://www.critex.fr). We thank OHGE (http://ohge.unistra.fr/) for providing all hydro-meteorological data used in this study. We also thank the doctoral school ED413 from the Strasbourg university as well as ANR HYDROCRIZSTO ANR-15-CE01-0010-02 which provided the funding for this study. We are grateful for the work done by anonymous reviewers which considerably improved this manuscript.

References

Asner GP et al (2016) Progressive forest canopy water loss during the 2012–2015 California drought. Proc Natl Acad Sci 113(2):E249–E255

Boy JP et al (2002) Reduction of surface gravity data from global atmospheric pressure loading. Geophys J Int 149:534–545. https://doi.org/10.1046/j.1365-246X.2002.01667.x

Bréda N et al (2006) Temperate forest trees and stands under severe drought: a review of ecophysiological responses, adaptation processes and long-term consequences. Ann For Sci 63:625–644. https://doi.org/10.1051/forest:2006042

Creutzfeldt B et al (2010) Reducing local hydrology from high precision gravity measurements: a lysimeter-based approach. Geophys J Int 183:178–187. https://doi.org/10.1111/j.1365-246X.2010.04742.x

Crossley D, Hinderer J (2010) GGP (Global Geodynamics Project): an international network of superconducting gravimeters to study time-variable gravity. In: Gravity, geoid and earth observation. Springer, Berlin, pp 627–635

Deville S et al (2013) On the impact of topography and building mask on time varying gravity due to local hydrology. Geophys J Int 192(1):82–93. https://doi.org/10.1093/gji/ggs007

Fores B et al (2017) Assessing the precision of the iGrav superconducting gravimeter for hydrological models and karstic hydrological

process identification. Geophys J Int 208:ggw396. https://doi.org/10.1093/gji/ggw396

Granier A et al (1999) A lumped water balance model to evaluate duration and intensity of drought constraints in forest stands. Ecol Model 116:269–283. https://doi.org/10.1016/S0304-3800(98)00205-1

Hector B et al (2013) Gravity effect of water storage changes in a weathered hard-rock aquifer in West Africa: results from joint absolute gravity, hydrological monitoring and geophysical prospection. Geophys J Int 194:737–750. https://doi.org/10.1093/gji/ggt146

Hector B et al (2014) Hydro-gravimetry in West-Africa: first results from the Djougou (Benin) superconducting gravimeter. J Geodyn 80:34–49. https://doi.org/10.1016/j.jog.2014.04.003

Hinderer J et al (2015) Superconducting gravimetry. In: Schubert G (ed) Treatise on geophysics, vol 3, 2nd edn. Elsevier, Oxford, pp 59–115

Leirião S et al (2009) Calculation of the temporal gravity variation from spatially variable water storage change in soils and aquifers. J Hydrol 365:302–309. https://doi.org/10.1016/j.jhydrol.2008.11.040

Llubes M et al (2004) Local hydrology, the Global Geodynamics Project and CHAMP/GRACE perspective: some case studies. J Geodyn 38:355–374. https://doi.org/10.1016/j.jog.2004.07.015

Pierret MC et al (2018) The Strengbach catchment: a multidisciplinary environmental sentry for 30 years. Vadose Zone J 17(1):180090. https://doi.org/10.2136/vzj2018.04.0090

Reich M et al (2019) Reducing gravity gravity data for the influence of water storage variations beneath observatory buildings. Geophysics 84(1):EN15–EN31. https://doi.org/10.1190/GEO2018-0301.1

Reichle et al (2017) Assesment of MERRA-2 land surface hydrology estimates. J Clim 30(8):2937–2960. https://doi.org/10.1175/JCLI-D-16-0720.1

Rosat S et al (2018) A two-year analysis of the iOSG-24 superconducting gravimeter at the low noise underground laboratory (LSBB URL) of Rustrel, France: environmental noise estimate. J Geodyn 119:1–8. https://doi.org/10.1016/j.jog.2018.05.009

Schueller K (2015) Theoretical basis for earth tide analysis with the new ETERNA34-ANA-V4.0 program. Bull Inf Marées Terrestres 149:024–012

Seneviratne SI et al (2012) Changes in climate extremes and their impacts on the natural physical environment. In: Managing the risks of extreme events and disasters to advance climate change adaptation, a special report of working groups I and II of the Intergovernmental Panel on Climate Change (IPCC). Cambridge University Press, Cambridge, pp 109–230

Shamir E et al (2016) The use of an orographic precipitation model to assess the precipitation spatial distribution in Lake Kinneret watershed. Water 8(12):591. https://doi.org/10.3390/w8120591

Weill S et al (2017) A low-dimensional subsurface model for saturated and unsaturated flow processes: ability to address heterogeneity. Comput Geosci 21:301–314. https://doi.org/10.1007/s10596-017-9613-8

Least-Squares Spectral and Coherency Analysis of the Zenith Total Delay Time Series at SuomiNet Station SA56 (UNB2)

Anthony O. Mayaki, Marcelo Santos, and Thalia Nikolaidou

Abstract

Zenith Total Delay (ZTD) from ground-based Global Navigational Satellite Systems (GNSS) observations plays an important role in meteorology. It contains information about the troposphere due to the interactions that GNSS signals have with the atmosphere while traveling from satellites to ground receivers. Since almost all weather is formed in the troposphere, the analysis of a collection of ZTD time series would provide insight about the periodic characteristics of the weather of a place. It would also provide insight about the influences that meteorological parameters such as pressure, temperature and relative humidity have on the weather's periodic nature. In this study, the least-squares spectral analysis approach is employed to determine the periodic oscillations in a 7-year time series of ZTD obtained from collocated GNSS and meteorological stations at the University of New Brunswick, Fredericton. Least-Squares Coherency Analysis of the time series spectra of the ZTD and its component hydrostatic and wet delays, and pressure, temperature and relative humidity is also performed. This is done to evaluate the level of contributions those parameters have in the periodicities inherent in the ZTD time series. Except for the zenith hydrostatic delay and pressure which show no annual periodic oscillation, the spectra of all the other time series show strong annual and semi-annual oscillations. Being the most dominant oscillation in the ZTD time series, the annual oscillation is largely driven by temperature, and this is maybe due to the high temperature variation characteristic of the climatic zone Fredericton falls under.

Keywords

Global Navigation Satellite Systems · Least-Squares Coherency Analysis · Least-Squares Spectral Analysis · Zenith Total Delay

1   Introduction

The Zenith Total Delay (ZTD) is an essential parameter that can be used to describe the various temporal and spatial characteristics of the weather and climatic processes of a place through the analysis of a time series of observations. The ZTD is estimated during the analysis of Global Navigation Satellite Systems (GNSS) observations for accurate positioning application. In meteorology however, these estimates are very useful for improving the short-term forecasting accuracies of the various numerical weather (prediction) models that they are assimilated into. The ZTD is made up of the zenith hydrostatic and zenith wet delays, ZHD and ZWD respectively. While the ZHD varies predictably and can be modeled sufficiently from observed meteorological parameters, the ZWD varies unpredictably and is difficult to model. Although the ZWD can be mod-

A. O. Mayaki (✉) · M. Santos · T. Nikolaidou
Geodesy and Geomatics Engineering, University of New Brunswick, Fredericton, NB, Canada
e-mail: omayaki@unb.ca; msantos@unb.ca; thalia.nikolaidou@unb.ca

© The Author(s) 2020
J. T. Freymueller, L. Sanchez (eds.), *Beyond 100: The Next Century in Geodesy*,
International Association of Geodesy Symposia 152, https://doi.org/10.1007/1345_2020_110

eled using water vapor pressure, temperature and relative humidity (Mendes and Langley 1998; Younes 2016), due to its unpredictability, it is estimated when processing GNSS observations.

Various studies have been conducted to determine the periodic oscillations of the ZTDs from GNSS observations collected at stations in various parts of the world (Bałdysz et al. 2015; Isioye et al. 2017; Jin et al. 2007; Klos et al. 2016). These studies have shown the presence of dominant annual (first harmonic) periodic components in the ZTD time series, with varying amplitudes and phases based on the station's location in the world. ZTD is a function of meteorological parameters such as pressure, temperature and relative humidity. These parameters are subject to short- and long-term oscillations/variations typically caused by disturbances within the atmosphere. These disturbances are influenced directly or indirectly by solar radiation, resulting in the periodic oscillations of the parameters; oscillations that could be diurnal or seasonal in nature. Time series analyses of a collection of observations of these parameters allow for the determination of the inherent oscillations, the knowledge of which is vital for weather forecasting and climatology (Kipp and Zonnen n.d.).

The focus of this study is to evaluate the contributions of the meteorological parameters to the periodicities in a ZTD time series. In this study, Least-Squares Spectral Analysis -LSSA- (Vaníček 1969, 1971; Wells et al. 1985; Pagiatakis 1998) of observations from collocated GNSS and meteorological stations respectively in Fredericton, New Brunswick (NB), Canada is performed. These observations are the ZTD with its ZHD and ZWD components from UNAVCO's SuomiNet SA56 GNSS station (also known as UNB2), and pressure, temperature and relative humidity observations from the collocated meteorological station. Least-Squares Coherency Analysis -LSCA- (Pagiatakis et al. 2007; Mtamakaya 2012) was also performed to determine the contributions of the meteorological parameters to the periodic oscillations in the ZTDs. Fredericton is located inland in the province of NB and because of this, its climate resorts under the humid (warm summer) continental climate class, "dfb", as defined by the Köppen climate classification. Due to Fredericton's inland location, its climate has warmer summers and colder winter nights than other surrounding coastal areas. On average, the warmest month is July while the coldest month is January.

The paper is structured as follows. The data used, and the methodology employed are discussed in Sect. 2. In Sect. 3, we present the results with discussions on the periodicities in the ZHD, ZWD and meteorological parameters time series and their effects on the periodicities in the ZTD time series. Conclusions finalize the paper.

2   Data and Methodology

Daily SA56 GNSS RINEX and meteorological observation files, spanning the years 2009–2015, with data logging intervals of 30 s and 1 min respectively were obtained from the UNAVCO ftp server. The RINEX files were processed using the GNSS Analysis and Positioning Software -GAPS- (Leandro et al. 2007) to obtain GPS-only ZTD estimates. GAPS employs the precise point positioning - PPP- technique (Zumberge et al. 1997) for the processing of GNSS observations. The adopted processing options follow the ones used in Mayaki et al. (2018). The plots of the time series of the ZTD, ZHD and ZWD, with outliers removed and pressure, temperature and relative humidity are given in Figs. 1 and 2 respectively. From visual inspection, there were no discernible offsets in the plots of the time series and no record of instrumentation change from the station's site log documentation. Therefore, data homogenization was not done on the time series before processing. However, outliers in the ZTD and ZWD time series were removed by applying the three-sigma rule using the median.

To compute the least-squares spectra of the time series, the LSSA version 5.02 program was used and it can be obtained from the website of the department of Geodesy and Geomatics Engineering, University of New Brunswick.[1] LSSA is based on the developments by Vaníček (1969, 1971) with improvements and implementation done by Wells et al. (1985) and Pagiatakis (1998). Notable advantages provided by LSSA are: (1) the analysis of time series with data gaps and unequally spaced values without pre-processing, (2) no limitations for the length of the time series, (3) time series with an associated covariance matrix can be analyzed, (4) the systematic noise can be rigorously suppressed without causing any shift in the existing spectral peaks, and, (5) statistical tests on the significance of spectral peaks can be performed. The choice of LSSA was additionally supported by its use in other studies for the proven integrity of its results (Mayaki 2019; Mtamakaya 2012; Hui and Pagiatakis 2004). In LSSA, the observed time series f is considered as a function of time t_i, $i = 1, 2, \ldots n$. Here, the time series may or may not have equally spaced values. The main objective of LSSA is to determine and clarify the periodic signals in f, especially when f includes both random and systematic noise. Comprehensive details about the LSSA are given in Vaníček (1969, 1971), Wells et al. (1985) and Pagiatakis (1998).

The computation was done at every 3 h, and due to the length of the time series, three bands of 2,000 spectral values each were used to represent the spectra of the

---

[1] http://www2.unb.ca/gge/Research/GRL/LSSA/leastSquares.html.

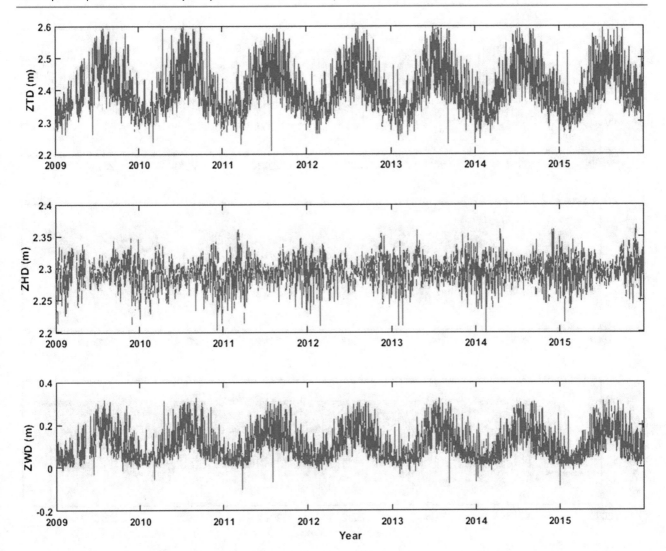

Fig. 1 SA56 ZTD, ZHD and ZWD time series

time series. These bands were chosen to portray the yearly, monthly and daily periods respectively. The first band captures the periods between 4,740 h (a little more than half of a year or 197.5 days) to the extent of the time series (61,344 h). The second band captures the periods between 576 h (24 days) to 4,740 h, and the third band is between 3 h and 576 h. The critical level for detecting significant peaks and the level of significance for statistical testing is defined on a 99% confidence level, which represents the most stringent option. Several executions of the LSSA were carried out, with the strongest period (that is, the period with the largest significant percentage variance) suppressed in succeeding executions. The percentage variance is the least-squares spectrum. The suppression of strong periods in subsequent executions give rise to new significant periods which may have been weak or invisible in previous execu-

tions. Also, from the suppression, the amplitudes and the phases of the strong periods from the preceding executions are given. Only detected significant periods up to half of the length of the time series under analysis are considered.

For the LSCA, the products of the LS spectra of the ZTD with those of the ZHD, ZWD, pressure, temperature and relative humidity are computed. This was done to speculate which of pressure, temperature or relative humidity is the main contributor to the periodic oscillations/periodicities found in the ZTD by way of using the ZHD and the ZWD. The product spectra are determined by the summation of the natural logarithms of the percentage variances obtained from the LS spectra of the ZTD with those of the ZHD, ZWD, pressure, temperature and relative humidity (Mtamakaya 2012; Elsobeiey 2017).

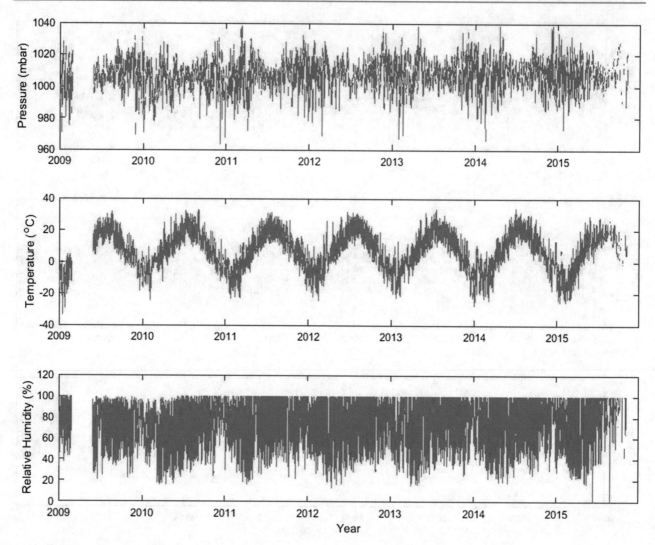

Fig. 2 SA56 pressure, temperature and relative humidity time series

Meteorological observations from the Fredericton CDA (Canadian Department of Agriculture) CS (Campbell Scientific) meteorological station, less than 6 km from the SA56 station, were also obtained from Environment Canada and processed for comparison with the results obtained from the SA56 station. These observations include the dew point and dry bulb temperatures, the corrected mean sea level and uncorrected station pressure observations, and the relative humidity.

3 Results and Discussions

3.1 LSSA of Time Series

Figures 3, 4, 5, 6, 7 and 8 show the least-squares (LS) spectral plots of the first execution (with their corresponding three bands) of the ZTD, ZHD, ZWD, pressure, temperature

and relative humidity time series respectively. Considering the percentage variances, the most dominant peak across the three bands as seen in the Figs. 3(i), 5(i), 7(i) and 8(i) is centered about the annual period (first harmonic). Also visible is the peak around the semi-annual (second harmonic, 6 months) period as seen in Figs. 3(ii), 5(ii), 7(ii) and 8(ii). In Figs. 4(ii) and 6(ii) however, the semi-annual period has the highest peaks in terms of percentage variances. Tables 1 and 2 contain the periods and phases of the strongest periodic components from the LSSA for the SA56 station and the Fredericton CDA CS meteorological station respectively. The standard deviation of the phases as estimated by the LSSA are also provided. The results show similar periods and phases (times of occurrences) of the annual periodic component in the ZTD, ZWD and temperature time series, at around 366 days and between 198° and 203° (corresponding to days in the month of July). The semi-annual periodic component seen in the ZHD and

Fig. 3 Spectral plot of SA56
ZTD least-squares spectrum

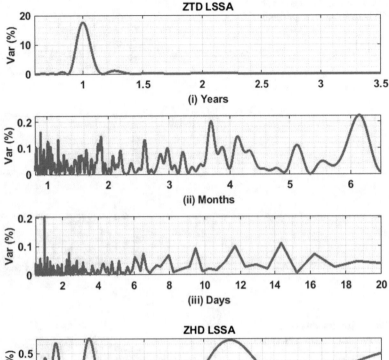

Fig. 4 Spectral plot of SA56
ZHD least-squares spectrum

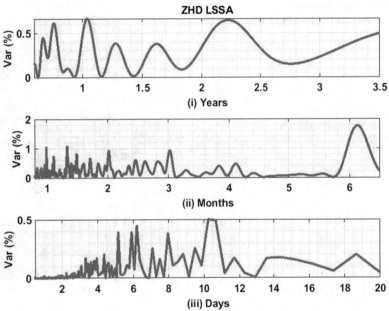

pressure time series occur approximately every 186 days between 180° and 185°. The similarities in results for the ZHD and pressure are expected to a certain degree since the ZHD from PPP is primarily modeled as a function of pressure. A study by Pikridas (2014) compared ZWD estimated from PPP to ZWD modeled through the application of Saastamoinen (1972), which uses temperature and the partial water vapor pressure. The results showed good agreement between the estimated and modeled ZWDs and so, it is conceivable that the ZWD estimated from GAPS PPP would show similar spectral results to a modelled ZWD.

3.2 LSCA of Time Series

Figures 9, 10, 11, 12 and 13 show the plots of the product of the time series LS spectra of the ZTD with those of the ZHD, ZWD, pressure, temperature and relative humidity. The contributions from the meteorological parameters to the periodicities in the ZTD can be observed since these parameters, through the ZHD and ZWD, can be used to model the ZTD. According to Jin et al. (2007), although the ZWD makes up 10% of the ZTD, the variations seen in the ZTD are caused by the ZWD. Figure 10 shows that the ZWD contributes more than the ZHD (Fig. 9) to the annual

Fig. 5 Spectral plot of SA56
ZWD least-squares spectrum

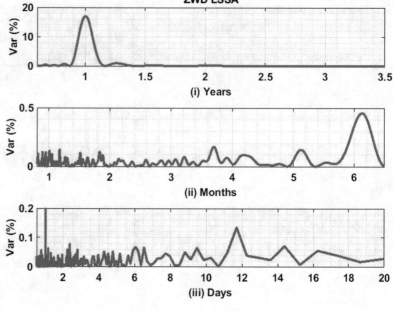

Fig. 6 Spectral plot of SA56
pressure least-squares spectrum

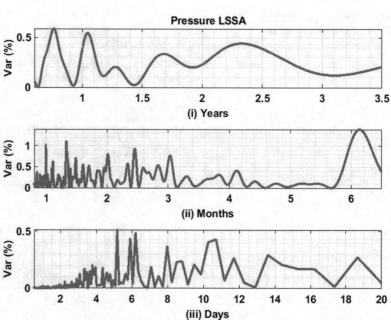

periodicity in the ZTD. Also, since the ZWD can be modeled using relative humidity and temperature, their contributions to the annual periodicity in the ZTD are seen to be higher in Figs. 12 and 13 compared to the pressure contributions in Fig. 11.

4 Conclusion

The Zenith Total Delay (ZTD) is an important parameter that reflects the state of the weather and climatic processes of a place. The time series analysis of a collection of observations of the ZTD facilitates the understanding of the periodic nature of the weather. In this work, the 3-h temporal resolution time series of the ZTD, the zenith hydrostatic and wet delays (ZHD and ZWD respectively), and meteorological parameters (pressure, temperature and relative humidity) are analyzed. These data were from the SA56 UNAVCO station in Fredericton, New Brunswick, Canada and span 2009–2015. Least-Squares Spectral Analysis was performed on the time series to determine their inherent periodicities. Least-Squares Coherency Analysis was also performed to evaluate the contributions of the meteorological parameters to the periodicities in the ZTD time series.

Fig. 7 Spectral plot of SA56 temperature least-squares spectrum

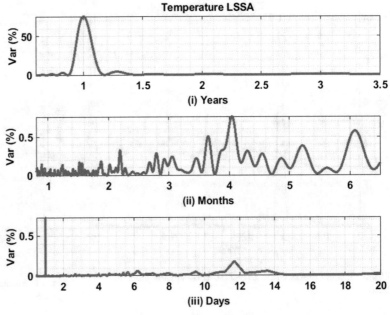

Fig. 8 Spectral plot of SA56 relative humidity least-squares spectrum

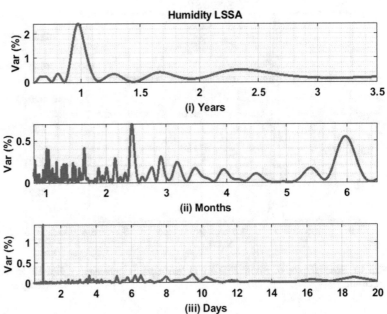

Table 1 Periods and phases of the dominant peaks for SA56

Time series	Period (Days)	Phase with std dev (Degrees)
ZTD	366.51	201.67 ± 0.07
ZHD	186.70	180.47 ± 0.01
ZWD	366.20	201.25 ± 0.07
Pressure	186.70	185.39 ± 5.93
Temperature	366.20	199.51 ± 3.68
Relative humidity	357.03	299.18 ± 13.17

Table 2 Periods and phases of the dominant peaks for Fredericton CDA CS

Time series	Period (Days)	Phase with std dev (Degrees)
Sea level pressure	186.70	180.85 ± 5.31
Station pressure	186.70	181.39 ± 5.27
Dew point temperature	366.51	203.60 ± 3.29
Dry bulb temperature	366.51	198.53 ± 3.19
Relative humidity	366.51	279.29 ± 10.27

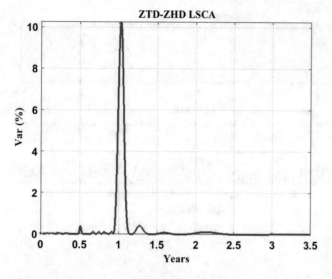

Fig. 9 Spectral plot of SA56 ZTD-ZHD least-squares coherency spectrum

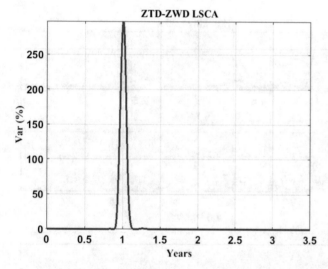

Fig. 10 Spectral plot of SA56 ZTD-ZWD least-squares coherency spectrum

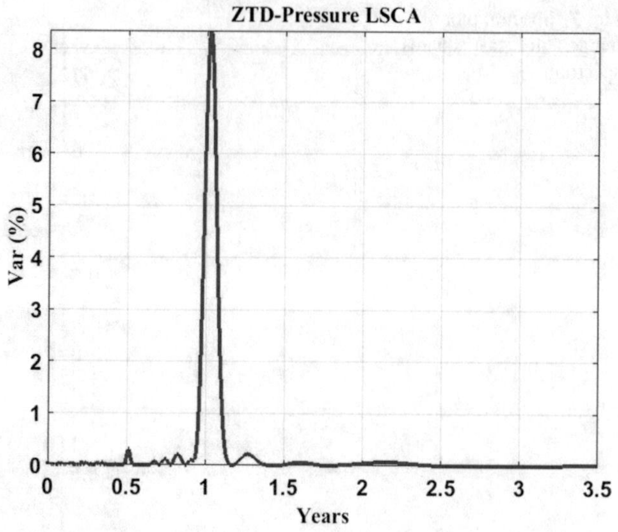

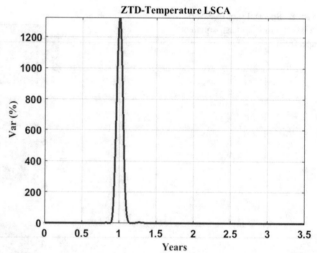

Fig. 11 Spectral plot of SA56 ZTD-Pressure least-squares coherency spectrum

Fig. 12 Spectral plot of SA56 ZTD-Temperature least-squares coherency spectrum

Annual periodicities of approximately 1 year (366 days) are detected in the ZTD, ZWD and temperature time series, with their phases between 198° and 203°. Semi-annual periodicities are detected in all the time series but are strongest in those of the ZHD and pressure. The annual periodic oscillation detected in the ZTD time series is primarily due to the temperature. The phases of these annual variations also coincide with days in the month of July. The results from this study agree with those from previous studies.

The continuation of this study would include similar analyses for the ZTD time series of other stations with co-located GNSS and meteorological instrumentation, enabling the characterization of the climate based on GNSS observations.

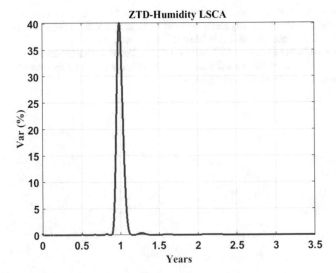

Fig. 13 Spectral plot of SA56 ZTD-Relative humidity least-squares coherency spectrum

Acknowledgements The authors acknowledge the GAGE Facility operated by UNAVCO Inc., with support from the National Science Foundation (NSF) and the National Aeronautics and Space Administration under NSF Cooperative Agreement EAR-1724794 for open access to the GNSS and surface meteorological RINEX data. The authors also acknowledge Environment Canada for providing the meteorological data from the Fredericton CDA CS station, the Natural Sciences and Engineering Research Council of Canada for the funding of this research and Dr. F.G. Nievinski for useful discussion and support.

References

Bałdysz Z, Nykiel G, Figurski M, Szafranek K, KroszczyńSki K (2015) Investigation of the 16-year and 18-year ZTD time series derived from GPS data processing. Acta Geophys 63(4):1103–1125

Elsobeiey M (2017) Advanced spectral analysis of sea water level changes. Model Earth Sys Environ 3(3):1005–1010. https://doi.org/10.1007/S40808-017-0348-2

Hui Y, Pagiatakis S (2004) Least squares spectral analysis and its application to superconducting gravimeter data analysis. Geo-Spatial Inf Sci 7(4):279–283. https://doi.org/10.1007/BF02828552

Isioye O, Combrinck L, Botai O (2017) Evaluation of spatial and temporal characteristics of GNSS-derived ZTD estimates in Nigeria. Theor Appl Climatol 132(3–4):1099–1116. https://doi.org/10.1007/s00704-017-2124-7

Jin S, Park JU, Cho JH, Park PH (2007) Seasonal variability of GPS-derived zenith tropospheric delay (1994–2006) and climate implications. J Geophys Res 112:D09110. https://doi.org/10.1029/2006JD007772

Kipp and Zonnen (n.d.) Parameters of meteorology. Retrieved August 13, 2019, from https://www.kippzonen.com/Knowledge-Center/Theoretical-info/Solar-Radiation/Parameters-of-Meteorology

Klos A, Hunegnaw A, Teferle FN, Abraha KE, Ahmed F, Bogusz J (2016) Noise Characteristics in Zenith Total Delay from homogeneously reprocessed GPS time series. Atmospheric measurement techniques (Discussions)

Leandro R, Santos M, Langley R (2007) GAPS: the GPS analysis and positioning software – a brief overview. In: Proceedings of the 20th international technical meeting of the science for a sustainable planet, International Association of Geodesy Symposia, p 139

Mayaki AO (2019) Evaluation of precise point positioning derived Zenith Total Delays from the Nigerian GNSS reference network. M.Sc.E. thesis, Department of Geodesy and Geomatics Engineering Technical Report No. 319, University of New Brunswick, Fredericton, New Brunswick, Canada, 144 pp

Mayaki AO, Nikolaidou T, Santos M, Okolie CJ (2018) Comparing the Nigerian GNSS reference Network's Zenith Total Delays from precise point positioning to a numerical weather model. In: International association of geodesy symposia. Springer, Berlin. https://doi.org/10.1007/1345_2018_43

Mendes VB, Langley RB (1998) Tropospheric Zenith Delay Prediction accuracy for airborne GPS high-precision positioning. In: Proceedings of the institute of navigation 54th annual meeting, Denver, CO, U.S.A., 1–3 June, pp 337–347

Mtamakaya JD (2012) Assessment of atmospheric pressure loading on the international GNSS REPRO1 solutions periodic signatures. Ph.D. dissertation, Department of Geodesy and Geomatics Engineering, Technical Report No. 282, University of New Brunswick, Fredericton, New Brunswick, Canada, 208 pp

Pagiatakis SD (1998) Stochastic significance of peaks in the Least-Squares Spectrum. J Geod 73:67–78

Pagiatakis SD, Yin H, El-Gelil MA (2007) Least-squares self-coherency analysis of superconducting gravimeter records in search for the Slichter triplet. Phys Earth Planet Inter 160(2):108–123. https://doi.org/10.1016/j.pepi.2006.10.002

Pikridas C, Katsougiannopoulos S, Zinas N (2014) A comparative study of zenith tropospheric delay and precipitable water vapor estimates using scientific GPS processing software and web based automated PPP service. Acta Geodaet Geophys 49(2):177–188. https://doi.org/10.1007/s40328-014-0047-7

Saastamoinen J (1972) Atmospheric correction for the troposphere and stratosphere in radio ranging of satellites. In the Use of Artificial Satellites for Geodesy, Geophysics Monograph Series 15:247–251

Vaníček P (1969) Approximate spectral analysis by least squares fit. Astrophys Space Sci 4:387–391

Vaníček P (1971) Further development and properties of the spectral analysis by least-squares. Astrophys Space Sci 12:10–33

Wells DE, Vaníček P, Pagiatakis SD (1985) Least squares spectral analysis revisited; Technical report 84. Department of Surveying Engineering, University of New Brunswick, Fredericton

Younes SA (2016) Modeling investigation of wet tropospheric delay error and precipitable water vapor content in Egypt. Egypt J Remote Sens Space Sci 19(2):333–342. https://doi.org/10.1016/j.ejrs.2016.05.002

Zumberge JF, Heflin MB, Jefferson DC, Watkins MM, Webb FH (1997) Precise point positioning for the efficient and robust analysis of GPS data from large networks. J Geophys Res 102:5005–5017

Associate Editors

© The Author(s) 2022
J. T. Freymueller, L. Sanchez (eds.), *Beyond 100: The Next Century in Geodesy*,
International Association of Geodesy Symposia 152, https://doi.org/10.1007/1345

List of Reviewers

Ågren, Jonas
Amos, Matthew
Ampatzidis, Dimitrios
Araszkiewicz, Andrzej
Assumpçao, Marcelo
Barzaghi, Riccardo
Bogusz, Janusz
Böhm, Johannes
Borries, Claudia
Boy, Jean-Paul
Brockmann, Jan
Brunini, Claudio
Brzezinski, Aleksander
Caporali, Alessandro
Chao, Ben
Combrinck, Ludwig
Dabove, Paolo
Ditmar, Pavel
Fernández, Laura
Fu, Yuning
Graffigna, Victoria
Gruber, Thomas
Hadas, Tomasz
Heinkelmann, Robert
Heki, Kosuke
Hinderer, Jacques
Kaplon, Jan
Karegar, Makan
Khodabandeh, Amir
Klos, Anna
Kreemer, Corné
Krzan, Grzegorz
Kvas, Andreas
Lambert, Sebastien
Li, Fei
Lidberg, Martin
MacMillan, Daniel

Melbourne, Timothy I.
Melgar, Diego
Métivier, Laurent
Meurers, Bruno
Moore, Michael
Motagh, Mahdi
Nastula, Jolanta
Nordman, Maaria
Pálinkáš, Vojtech
Pavlis, Erricos C.
Pavlis, Nikolaos Konstantinos
Paziewski, Jacek
Perosanz, Felix
Próchniewicz, Dominik
Rebischung, Paul
Sabadini, Roberto
Sánchez, Laura
Schmidt, Michael
Schuh, Wolf-Dieter
Shimabukuro, Milton
Spada, Giorgio
Stanaway, Richard
Steffen, Holger
Teunissen, Peter J. G.
Tomaszewski, Dariusz
Torres-Sospedra, Joaquín
Trojanowicz, Marek
Van Malderen, Roeland
Vedel, Henrik
Vergos, Georgios
Walpersdorf, Andrea
Wziontek, Hartmut
Zaminpardaz, Safoora
Zhou, Hao
Zingerle, Philipp
Zus, Florian

© The Author(s) 2022
J. T. Freymueller, L. Sanchez (eds.), *Beyond 100: The Next Century in Geodesy*,
International Association of Geodesy Symposia 152, https://doi.org/10.1007/1345

Author Index

283

Subject Index

A
Absolute gravity, 15, 154–156, 158, 163, 164, 166, 263
Africa, 72, 77–85, 110, 114, 119, 199, 255, 256
Argentinean stations to the IHRF reference network, 19

B
Baltic Sea, 161–179

C
Caribbean, 169, 199–201, 204, 205, 210, 212, 214, 228
Chronometric levelling, 5
Climate indices, 242–248
Clock networks, 3–9
Co-location, 29–35
Combined solutions, 58, 61–64
Cooperative localization (CL), 125, 127
Crustal deformation, 144, 181, 187, 197, 200, 210, 242

D
Doppler observation, 129–133

E
Early warning systems, 209–211, 215
Earth orientation parameters (EOP), 22, 39, 93, 99–103
Earth rotation models, 39
Earth rotation theory, 99–103
Earth's transfer function, 113–119
Earth surface kinematics, 198–200, 204, 205

G
Gaia, 21–27
Galactic aberration (GA), 21–26
Geodetic reference systems, 13, 154, 197, 242
Geodetic time series, 48
Geoid determination, 80, 83–84
GEOIDE-Ar 16, 13, 15, 19
GEONET, 181–187, 210, 211
Global Geodetic Observing System (GGOS), 29, 99–103, 215, 220
Global Navigation Satellite System (GNSS), 12, 29, 37, 45, 83, 113, 124, 129, 135, 143, 153, 162, 169, 181, 198, 209, 219, 227, 269
Global positioning system (GPS), 15, 37, 58, 110, 131, 136, 144, 156, 170, 198, 210, 252, 270
GNSS-denied environments, 124, 127
GNSS position time series, 45–52

G (cont.)
GRACE, 57–62, 64, 67–74, 154, 155, 157, 221, 223, 242–246, 248, 249
GRACE Follow-On (GRACE-FO), 58, 64, 67, 68, 71, 74
Gravity field, 3–6, 17, 39, 42, 58–60, 62, 64, 67–74, 77, 83, 107, 154, 242, 243
Gravity field recovery, 68–70
Gravity interpolation, 78
Green's function, 115, 190, 264
Ground-based GNSS, 210
Groundwater, 155–157, 181–187, 241
Guadeloupe, 170, 171, 173, 175–179

H
Height system unification, 4, 7–9
Hurricane Florence, 251, 252, 254–256
Hydro-gravimetry, 262–267
Hydrological loading, 114, 118, 119, 181, 242, 244, 247, 263, 264
Hydrological modeling, 114–118, 182, 242, 262, 266

I
IAG and IGFS service, 58, 64, 107
ICRF, 22, 103
IGFS Product Center, 58
Indoor positioning, 124, 127
Inertial Measurement unit (IMU), 143–149
Inertial Systems (INS), 135–140
Integrated water vapour (IWV), 227, 228, 232, 234, 235, 237, 238
Interferometric synthetic aperture radar (InSAR), 153–157, 185, 186, 241
International Combination Service for Time-Variable Gravity Fields (COST-G), 57–64
International Height Reference System and Frame (IHRS/IHRF), 3–9, 11–19
Inverse Laplace transform, 190–193

K
Kinematic platform, 129, 130, 133

L
Land subsidence, 153–157, 165, 185, 187
Land uplift, 161–166
Latin America, 197–205, 227–239
Least-squares coherency analysis, 269–277
Least-squares spectral analysis, 269–277
Lesser Antilles, 169–178
Local ties, 29–32, 35, 38
Low-cost navigation, 143–149

J. T. Freymueller, L. Sanchez (eds.), *Beyond 100: The Next Century in Geodesy*,
International Association of Geodesy Symposia 152, https://doi.org/10.1007/1345

Printed in the United States
by Baker & Taylor Publisher Services